Y0-EFY-149

DATE DUE

OCT 1 8 1978
NOV 0 8 1978
APR 1 8 1979
MAY 9 1979
OCT 1 7 1979
NOV 1 8 1981
MAR 0 2 1983
JUN 0 8 1987
NOV 2 3 1987
MAY 1 7 1988
FEB 2 0 1990
MAR 1 2 1990
NOV 1 1 1990
MAR 1 6 1992
MAR 1 5 1993

GAYLORD PRINTED IN U.S.A.

Problems and Controversies in Television and Radio

Problems and Controversies in Television and Radio

Basic Readings selected and edited by

Harry J. Skornia and Jack William Kitson

Pacific Books, Publishers Palo Alto, California

Copyright © 1968 by Pacific Books, Publishers. All rights reserved. No part of this book may be used or reproduced in any manner whatsoever without written permission from the publisher, except in the case of brief quotations embodied in critical articles and reviews.

Library of Congress catalog card No. 67–20825.

Manufactured in the United States of America.

Pacific Books, Publishers
P. O. Box 558
Palo Alto, California 94302

384.54
Sk5p

Preface

FOR SOME YEARS, since the birth of courses in radio and television, professors teaching graduate seminars, or courses in criticism, problems, and issues, have had difficulty in assembling adequate materials. Many of the best "position statements" have been in the form of speeches or testimony which were never formally published. Until Newton Minow's *Equal Time* appeared, bringing his "vast wasteland" and other speeches together in book form, the only thing that could be done with his clear statement and others like it was to use them in mimeographed form. Other timeless and historical materials, like the early piece by H. V. Kaltenborn, included here, or the Friedrich and Sternberg study, are now out of print. Many excellent periodical articles repose in old files, unknown to students of today. There are, in fact, few libraries in which many of the most basic documents needed for the study of American broadcasting can even be found. *Problems and Controversies in Television and Radio* seeks to begin to meet this need.

The essays and speeches assembled here attempt to provide the basis for intelligent reading and discussion. Though they aim at as fair a balance as possible, they illustrate the fact that there are not necessarily two sides to every question: there are often four or five or more. Network spokesmen often do not agree. Advertising agency executives sometimes are more critical of certain advertising or broadcast practices than are critics or professors. Some senators are consistently pro- (or anti-) industry. And, some take quite different positions, depending upon the time and the occasion and the stakes.

Timelessness and general unavailability, rather than recency, have been important considerations in the choice of the items included, though as many valid and up-to-date pieces as possible have been included. Some of the positions assumed are perhaps too critical. Others are too defensive. But if they touch clearly on basic issues, they qualify for inclusion here.

Some readings which we would have liked to include are not to be found in this collection because we could not secure permission to reprint

Speech
5-9-78
138170

them. Other editors might well have chosen differently, or sought different kinds of balance. The present editors have felt that self-congratulation is less needed than efforts to improve broadcasting in America. Certainly intelligent discussion is always useful.

We do not feel that we, as editors, should take issue in this volume with positions taken in some of the pieces—though as teachers in the classroom we might well do so. We leave it to the readers and teachers who may use this book to provide the balance needed—supplementing what is here provided with other materials which they feel fairness may require. This book is intended to provide only part of the materials—the least generally accessible ones—needed for intelligent discussion of American radio and television issues, problems, and policies.

Harry J. Skornia
Jack William Kitson

Urbana, Illinois

Contents

I. Introduction: Frames of Reference

THE CONDITIONS of access, regulation, and operation are quite different for broadcasting than they are for the other journalistic media or the entertainment industries. It is essential to know what the major differences are, and what some of the unique characteristics of broadcasting are.

This section includes a diversity of viewpoints, as well as an insight into several solid economic and historical facts about broadcasting.

With the exception of the articles by Norman Cousins, editor of the *Saturday Review*, and Marshall McLuhan, director of the Centre for Culture and Technology, University of Toronto, the essays here are either unpublished, out of print, or in limited circulation, brochure, or official document form. Each deserves wider circulation and discussion than has been possible heretofore because of the limited availability of the materials.

1. The Public Still Owns the Air

by Norman Cousins

Mr. Cousins is editor of The Saturday Review. *This article appeared in that publication on May 9, 1959, and is reprinted here by permission of the author.*

A question that has always existed with respect to television in the United States has now become critical and insistent. The question: To what extent does a television station have to fulfill a public service role? Also: Exactly how is public service to be defined?

First, some background. Radio and television broadcasting, like newspaper and magazine publishing, is in the field of mass communications. Both deal in information and entertainment, in varying degrees. And both depend on advertising for revenue and profit, in whole or part.

There is, however, one significant difference. There is no limit on the number of newspapers that may be published in a city, other than that imposed by the natural laws of competition. No prospective newspaper or magazine publisher has to apply to the United States government for permission to go into business. This is not the case with a radio or television station. The most valuable ingredient is owned by the American people and is furnished without charge to the broadcaster. This is the air channel. Since the air can carry only a limited amount of broadcasting traffic, the government has to choose, from among the many applicants, the ones who are to be given the franchise.

Yet the theory that the U.S. owns the air lanes does not hold when a broadcaster decides to sell.

Consider the case of the short-term operator. He will acquire a television franchise from the Federal Communications Commission, often against a wide field of applicants. He will install telecasting equipment worth, let us say, three-quarters of a million dollars. Three years later he will sell not only the equipment but the franchise for several million dollars. In short, when a television operator decides to dispose of a station, the air rights do not revert to the American people. The FCC, to be sure, must approve the sale of a station. But it is up to the seller to determine who gets the franchise, and, also, what he wants to be paid for it.

Even in the allocation of channels, the public service factor has not always been observed. In New York City, for example, not a single channel in the range of regular television frequencies has been reserved for educational TV. The FCC, in its defense, claims it has set aside some bands in the ultra-high frequency range. The trouble with this argument, however, is that television sets in New York are not equipped to receive ultra-high frequencies. Realistically, therefore, the only chance that the nation's largest city will have educational telecasting is if it acquires one of the present regular channels. Two years ago, it appeared that this might be the case. The owners of a commercial channel decided to sell. New York University tried to save the day for educational TV by putting up a very substantial bid for the station. It had the prospective backing of one of the large foundations which quite properly felt that educational television might not become a major force nationally unless a flagship station were established in New York. Unfortunately, New York University was outbid. The owner couldn't be blamed for getting as much as he could. He hadn't been very successful as a telecaster. His equipment was worth not much more than a half million dollars. But he sold for almost four million.

Just as the public service factor seems to be diminishing in the allocation of stations, so it seems to be diminishing in the operation of the stations themselves. In television, sponsored entertainment more than dominates; it overwhelms. Most of the quality information or educational programs are now herded together in what has not inappropriately been termed a Sunday TV ghetto. Good programs on other days are being scrapped. Informational television has never known anything finer than *See It Now*, produced by Edward R. Murrow and Fred W. Friendly. *See It Now* cannot now be seen; it has been backed off the air by commercial entertainment, much of it in the spurs-and-bang-bang variety.

It is not precisely true that the only way a high-quality program can be seen is if a sponsor is willing to pay for it. The stations are now pursuing a policy designed to eliminate programs of high quality EVEN IF SPONSORS ARE WILLING TO PAY FOR THEM. Take the case of the *Firestone Hour*, for years a regular feature on network television. The *Firestone Hour* was neither highbrow nor lowbrow; it represented a comfortable level of quality and good taste. But the *Firestone Hour* is now being shunted off the air, despite the eagerness of the sponsor to continue. The reason is that the station managers fear that the surrounding programs may suffer. Any "weak" spot in the evening programming—measuring "weakness" by ratings—must not be allowed to affect the big winners.

It is possible that the Federal Communications Commission believes that such developments are not its concern. But one thing is certain. The leaders of articulate public opinion in other areas are not going to remain inarticulate in this area for very long. The station owners are bound to discover that opposition to their policies does not have to achieve the mass numbers of a *Gunsmoke* audience in order to be effective. The entire history of effective reform in America is related to this power of sharp, swift action by leaders of public opinion. The present permissive policies of the present FCC cannot last. Congress will not remain disinterested. The public still owns the air.

A basic situation exists that requires correction. The correction will come. How basic and sweeping this correction will be depends on the speed with which the present drift can be reversed by commercial television itself.

2. Congress and the Control of Radio-Broadcasting *

by Carl J. Friedrich and Evelyn Sternberg

The following essay was the first of two which resulted from an extensive radio research project carried on at the Graduate School of Public Administration at Harvard University, under a grant from the Rockefeller Foundation. It appeared in the October, 1943, issue of the American Political Science Review. The second part, "Congress and Radio in Wartime," though also excellent in describing what are described as "confused efforts to 'regulate' a very young industry," was published in the December, 1943, issue. It is not included here because of space limitations. As will be noted on page 64, at the time this book went to press, Dr. Friedrich was Eaton Professor of the Science of Government at Harvard. We are grateful to the authors for permission to publish this essay here.

I. THE PROBLEM

Ever since the first regularly scheduled public radio-broadcast in 1920, Congress has played a unique and central role in the control of radio-broadcasting. As an agency for legislation, it has created the regulatory mechanisms under which the radio industry functions, and it has written the laws which govern this important area of communications. Congress, in fact, has set the pattern within which the various groups and interests operate, subject, of course, to the working rules of the capitalist order. In doing so, Congressmen have been at the beck and call of millions of constituents interested in radio as listeners or broadcasters, as educators or clergymen, as big or little business men. In caring for all of these varying interests, Congress has concerned itself with a few broad problems: what is heard on the radio, who shall control what is heard, who is able to hear what goes over the air, and who profits from radio. But Senators and Representatives are not merely the puppets of various pressures; they have a distinct political interest in programming, profits, and control. They have in radio a potent molder of public opinion—a powerful instrument which can help them to victory or defeat in the next election—and they have used it and will continue to use it to serve their personal fortunes, their parties, and their platforms. Thus in their own interest as well as in the interest of their constituents, Congressmen themselves form a pressure group, or rather a number of small but intensive pressure groups, influencing, cajoling, threatening, or entreating the regulatory Commission which they have created.

Congress makes itself felt not only as an agency for legislation, but also as an agency for criticism. There is a long history of Congressional attempts at investigation of the Federal Communications Commission or the industry. Also, the Senate, in confirming members of the Commission, and the House, in considering annual appropriation bills, have taken opportunity to criticize and question the Commission.

But even a cursory study of Congressional debates or hearings reveals the inadequacies of Congress in handling matters of technical complexity. Actually, Congress has always been a step behind technical progress in the radio field, following new developments with legislation only when these have grown big and important enough to demand the attention of Congress. These difficulties may be inherent in the nature of radio and in the nature of our Congressional system as well.[1] How can a Congress of laymen, working under complicated machinery, intelligently and with dispatch handle a technical phenomenon which has important social implications? How can Congress best make use of the "experts," and who should these experts be? What can be the job of Congress in regulating radiobroadcasting, and how can it best be done?

---

* The Radio-broadcasting Research Project has been carried on in conjunction with the Communications Seminar at the Graduate School of Public Administration, aided by a grant from the Rockefeller Foundation. But neither the Foundation nor the School is in any sense responsible for the views expressed by the authors. The article except for minor changes was completed in January, 1943.

II. THE RADIO ACT OF 1927

Congress did not enact regulatory legislation until seven chaotic years after radio-broadcasting had made its debut. Meanwhile, the growing industry functioned under the Radio Act of 1912, which authorized the Secretary of Commerce to issue station licenses. Though this act had been drafted with reference to radio-telegraphy and radio-telephony, it was used by the Secretary in an attempt to allocate frequencies in the broadcasting band, regulate hours of operation, and fix power. But in 1926, when the Secretary of Commerce brought suit against a radio station for jumping its wave length and operating at hours not designated by its license, a Federal court decided the case in favor of the station.[2] This ruling was corroborated in an opinion of the Attorney-General stating that the Secretary could not designate frequencies, hours, or power, and could not refuse a license. Thus the act ceased to have any effect for radio-broadcasting and the need for new legislation became urgent. The idea of radio legislation was certainly not an entirely new one in Congress. Between 1921 and 1927, more than fifteen bills had been introduced in both houses to "regulate radio communications" and several more to amend the 1912 act to meet the new situation; but these died in committees, most often without hearings.

As Representative White put it, the new act was introduced at the request of listeners, industry, the fourth National Radio Conference, and two Chief Executives.[3] Its history is a fine illustration of legislative confusion and wrangling. This was owing in part to the number of slightly differing bills presented in both houses,[4] which lengthened the legislative process, while the necessity for immediate enactment of some kind of law before the end of the short session was extremely pressing. In part, too, it was owing to the difficulties of legislating for a rapidly changing medium and to the ignorance of all but a few Congressmen concerning the technicalities involved.

The debates on the 1927 act were marked by lengthy arguments on proper parliamentary procedure (an hour and ten minutes having once been consumed on the question of whether the bill could properly be called up), confusion over radio terms and technicalities, criticism of the conduct of the debates, and constant irrelevancies.[5] There is no question that the legislation was largely determined before it reached the floor of the House, and the long debates were often involved and unnecessary, leaving the majority of the members as much in the dark as ever about the intricacies of radio and radio legislation. Representative Davis's warning, "You are dealing with what is going to be the most powerful political instrument of the future," did not make much impression on the final bill.[6]

In January, 1927, the conference committee submitted a compromise bill.[7] The House wanted the Secretary of Commerce to retain authority to issue licenses, subject to review by a commission, while the Senate was interested in the establishment of a permanent radio commission. By the compromise, the Federal Radio Commission was established on a temporary experimental basis for a year, after which powers and authority of the Commission except as to revocation of licenses would revert to the Secretary of Commerce and the Commission would continue as an appellate body only. The compromise was not enthusiastically received. Debating acceptance of the conference report, the House agreed that the legislation was not perfect, but that since something had to be done immediately, it was better than nothing. To Representative White, the new radio legislation established "that the right of the public to service was superior to the right of any individual to use the ether." [8]

The Commission was given power to classify radio stations, assign frequencies and wave-lengths, and regulate interference. Congress passed the Radio Act only a few days before it adjourned, and failed to make any appropriation for the Commission, which therefore was left to function without funds.

III. THE COMMUNICATIONS ACT OF 1934

From 1927 to 1930, the Federal Radio Commission existed as the licensing authority on a year-to-year basis, being renewed annually until made permanent. Its uncertainty of tenure, its complete dependence on the Congress for existence, made the FRC a timid agency, sensitive to Congressional criticism and appeals. It continued to function until the Federal Communications Act in 1934. For some years, various government officials had expressed an interest in the establishment of an over-all commission concerned with communications. The existing set-up was unsatisfactory, since authority was divided between the ICC and FRC and the Department of Commerce. In 1932–1933, a legislative attempt to combine the radio division of the Department of Commerce with the Commission was stymied by President Hoover's pocket veto after the bill had been passed by both houses.[9]

Early in 1934, President Roosevelt set up an interdepartmental communications committee including Secretary of Commerce Roper, Senator Dill, Representative Rayburn, Dr. Irwin Stewart, Dr. W. M. W. Splawn, and Major-General Charles Saltzman to urge legislation in Congress for the establishment of a federal communications commission with authority over both wire and radio companies. The understanding was that if the attempt at legislation failed, Senator Dill would ask the return of the radio commission to the Department of Commerce.[10] The committee submitted a report to the President which he in turn transmitted to the interested Congressional committees. Briefly, the Committee recommended "the transfer of existing diversified regulation of communications to a new or single regulatory body, to which would be committed any further control of two-way communications and broadcasting."[11] The *New York Times* saw in the report "new evidence of a trend toward government regulation of public necessities," and Mr. David Sarnoff's lecture before the Army Industrial College was widely interpreted to mean that the radio industry was now reconciled to government control.[12]

With the approval of the President, Senator Dill and Representative Rayburn started drafting bills[13] after agreement that controversial subjects should be omitted. In other words, the bill was to be minimum legislation, leaving the way open for the new commission to study the problem with a view to further legislation. Despite word that Roosevelt wanted the Dill bill made law, it immediately became a center of controversy, because it called for repeal of the Radio Act of 1927. Under this act, the industry had jockeyed itself into a position of legal security which would be lost if the act were repealed.

A revised bill,[14] product of the hearings before the Senate subcommittee, was introduced by Dill on April 5. Meanwhile the House Committee on Interstate and Foreign Commerce was holding a hearing on the Rayburn bill, which enjoyed the support of the industry because it did not repeal the Radio Act of 1927 but simply abolished the FRC.

Senate leaders met with President Roosevelt to determine his wishes concerning the legislation. "It was apparent," said the *New York Times*, "that his desires would control to a large extent."[15] The greatest controversy was over the so-called Wagner-Hatfield amendment, which would require that twenty-five per cent of facilities be allotted to religious, cultural, agricultural, cooperative, labor, and similar nonprofit organizations. *Variety* reported that "the NAB were in a panic checking off names of Senators and trying to pull wires and get votes." The NAB wrote to all Senators asking them "not to destroy the whole structure of American broadcasting." Senator Dill, though in favor of educational and religious broadcasting, disliked the method suggested in the amendment and argued that the so-called noncommercial stations would have to sell almost seventy-five per cent of their time to be self-supporting. Dill preferred to see the commercial stations required to give a certain part of their time to nonprofit organizations.[17] The amendment was rejected, but *Variety* noted that "the Dill bill went whooping through the Senate without even the formality of a record vote after less than four hours' debate."[18]

The Rayburn bill, meanwhile, was favorably reported from the House committee and passed after some blistering remarks by Representative McFadden against the CBS-NBC monopoly and censorship. "The strong hand of influence," said McFadden, "is drying up the independent broadcasting stations in the United States and the whole thing is tending toward centralization of control in these two big companies. . . ."[19]

A new bill emerged from the conference committee as a substitute for both the Senate and House bills.[20] It established the Federal Communications Commission, with seven commissioners to serve seven-year terms, and, with a provision for the Commission to fix its own divisions, not to exceed three. The Federal Radio Commission was abolished and the Radio Act of 1927 was repealed, though the radio provisions enacted in Title III of the new bill were similar to the former act. The Commission was given power to issue licenses, classify stations, assign frequencies, determine locations, and inspect apparatus. The system of equal allocation of broadcasting facilities within five zones was continued. However, additional licenses outside the quota for stations not over 100 watts were authorized in an attempt to cure the inadequacies of the quota system. The act further provided that the Commission should have authority to regulate chain broadcasting. Censorship was prohibited, lotteries banned, and equal facilities for candidates for public office provided. There were detailed regulations for issuance of licenses to aliens and to corporations in which aliens were interested. The FCC was asked to study "new uses for radio, provide for experimental use of frequencies, and generally encourage the larger and more effective use of radio in the public interest." The Commission was further asked to study the possibility of allocating a percentage of radio-broadcasting facilities to nonprofit activities. After debate in the House, where certain Congressmen were unwilling to "rubber stamp a Senate bill," [21] the conference report was adopted. President Roosevelt signed the bill, and on July 1 the act went into effect.

IV. RELATIONS BETWEEN CONGRESS AND THE COMMISSION

That Congress was seldom satisfied with the administration of the Federal Communications Commission is evidenced by the plethora of attempts at investigation and the harsh criticism in debates and hearings. Even as the bill abolishing the Federal Radio Commission was being discussed, there were three Senate resolutions to investigate it,[22] but none of these was acted upon. In 1936, a House resolution was introduced asking for investigation of the Commission, especially with regard to abuses in granting of licenses, broadcasts of obscene programs, and monopolistic practices.[23] The next year, Representative Connery again introduced the resolution,[24] but the Rules Committee refused to report it out. Connery's zeal for investigation was occasioned by the clergy's wrath at certain supposedly obscene broadcasts, and by the refusal of licenses and time to religious organizations. Meanwhile, Senator White, in a speech condemning monopoly, newspaper ownership of stations, trafficking in licenses, and chain broadcasting, urged investigation of the act and the Commission.[25] The Senator's chief concern was whether the FCC was carrying out the congressional purpose. Said he: "I do not view with complacence administrative disregard of legislative purpose." [26] It was the function of Congress and not of the administrative agency to determine questions of government policy. The FCC, by its rulings, was nullifying the Congressional intent and in fact creating legislation of its own. The White resolution [27] authorizing the ICC to investigate the industry was unanimously reported from the Interstate Commerce Committee and barely missed passage. The next year the Senate Audit and Control Committee reported out Senator White's resolution, and he and Senator Wheeler were very anxious to have it come to a vote. Friction in the Commission had given new point to the resolution. As Representative O'Connor said: "There is a division in this Communications Commission as to whether there should be an investigation. It is an internal family row. . . ." [28] Commissioners Payne and Craven had declared they welcomed a Congressional probe.[29] In February, 1938, Chairman McNinch, appointed the summer before in an effort to straighten out Commission affairs, aired charges in a network address against some of his colleagues who had disagreed with him on matters of policy and practices. But it was pointed out that the FCC itself had undertaken a study of the alleged monopoly and was, as its opponents claimed, in the ridiculous position of investigating itself.

But MacFarlane and Connery in the House had not given up hope of an investigation. At long last, the Rules Committee reported out the Connery Resolution,[30] one of several pending before it, and there followed a long and stormy session on the floor of the House.[31] Representative O'Connor of New York, chairman of the Rules Committee, declared that he had never seen such lobbying against a resolution, and although he cared little about the fate of the resolution, he hated to see the House ruled by a lobby. Throughout the debate there were references to the lobbyists. Said Representative O'Connor: "You will find difficulty in getting through the lobby because of the crowd of radio lobbyists," and Representative Connery quoted the *Washington Merry-Go-Round*, to wit: "Apparently the RCA is worried about a congressional investigation. [It has sent a] high powered publicity agent scurrying around the Halls of Congress to mold public opinion." At the last minute, some of its staunchest supporters, Representative MacFarlane, for example, voted against the resolution for one of two reasons. Either they decided to rely on the Temporary National Economic Committee's investigation of monopolies, or they saw in the investigation a chance for Republicans to smear Democrats. Representative Fish declared: "I am willing to bet dollars to doughnuts on a new campaign hat that the Democratic majority does not dare adopt this resolution." But when Representative Celler asked O'Connor: "Does the gentleman feel . . . that this resolution will give great comfort to the . . . Republican party?," O'Connor replied: "I do not think there is anything to that at all. My concern is to preserve the Democratic party against political scandals which exist in the FCC. . . . My misguided Democrats, submit to this pernicious lobby if you will, but I feel you are making a grave mistake." At any rate, the resolution was defeated by a good majority, and two days later the White resolution was passed in the Senate.[32]

Less than a year later, Connery again asked for an investigation [33] of the radio monopoly, and in 1940 Senator Tobey submitted a resolution to investigate, among other things, the administration of the Communications Act of 1934.[34] Representative Wigglesworth presented a resolution in 1941 to investigate the FCC. [35] Early in 1942, Representative Cox, criticizing Chairman Fly, announced his intention of offering a resolution for investigation of the Commission, and the following month he introduced the resolution to probe the organization, personnel, and activities of the FCC.[36] Fly was stoutly defended by Representative Rankin and the resolution blocked. On the first day of the new session in 1943, Cox reintroduced the resolution for a select five-man committee inquiry, charging the FCC with "terroristic control" of radio, and denouncing Chairman Fly in no uncertain terms.[37] The House voted almost unanimously for the Cox resolution.[38]

By keeping tabs on personnel, Congress has had another means of determining direction of the Commission. After the passage of the 1934 act, there was a good deal of political activity to block the appointment of the Federal Radio Commissioners to the new agency, and *Variety* reported leading Congressmen "working tooth and nail to grab off patronage." [39] The hearing on renomination of Colonel Thad Brown in 1940 [40] is a notable example of Congressional power over personnel. Senator Tobey launched a one-man crusade against a favorable report and used the renomination hearing to open up the monopoly charges against both major networks and to condemn the Commission's handling of the problem. Since Brown had been chairman of the FCC committee investigating monopoly, Senator Tobey attempted to make Brown's renomination depend on the adequacy of the committee's report. Though all appointments are determined in part by politics, the political tie-up in Brown's case was particularly evident. The background of this situation is typical, for Senator Tobey's concern for the public interest came in response to specific urgings. The story, as pieced together from "inside" information and from the trade press, was roughly as follows: A priest of the Roman Catholic Church, hailing from Portsmouth, N.H., who had done some broadcasting and wanted to do more, had been given the cold shoulder by CBS. Thereupon, he insisted that radio was a vicious monopoly and that something should be done about it. He demanded that Senator Tobey find out what had become of the Commission's long-heralded monopoly investigation. Colonel Brown, it would appear, had not been especially active as chairman of the FCC committee charged with that particular investigation. To make it worse, Brown was generally credited with being a "spokesman" for the industry on the Commission, though it became quite clear that other members of the Commission as well were at times unduly responsive to the industry's viewpoint. The local political angle lent further unsavory odor to the situation when it developed that Senator Tobey's political opponent, Senator Moses of New Hampshire, had acted as an intermediary in securing special favors for Sarnoff, president of RCA and of the National Broadcasting Company. In any case, Commissioner Brown's renomination was rejected by the Senate, and he died shortly thereafter.

Congress can be effective not only in confirming personnel, but in granting or withholding appropriations. This power over appropriations has been important from the very first year of the Federal Radio Commission, when Congress failed to appropriate any funds at all, to last year, when a House committee adopted an amendment denying funds for salary to an FCC foreign broadcasting agent whose appointment had been criticized.[41]

But these recognized Congressional activities do not begin to suggest the extent and intimacy of the relations between Congress and the Commission. In 1940, the Attorney-General's Committee on Administrative Procedure noted that "it is a widely and firmly held belief that the FCC had been subjected to constant external pressure, particularly by members of Congress"; [42] but an adequate job of documenting these pressures has never been done and, indeed, would be almost impossible to do. While it is true that all communications submitted to the FCC in connection with any case must be "on the record," there is no way, for example, of checking informal telephone conversations and social meetings. Being only human, it would seem impossible for Commission members, in the words of Senator Tobey, to avoid subconscious influence, and, like Caesar's wife, be above suspicion. If the FCC is to be above suspicion, a duty devolves upon Congressmen as well as upon Commissioners. Congress condemns the FCC for being oversensitive to political pressures for which it is itself responsible. Congress might do its part to preserve the independence and integrity of the Commission it created. As the Acheson committee monograph points out, Congressional response to constituents in the matter of licenses and frequencies is heightened because of the political value of radio and radio-broadcasters to the Congressman in his home town or state. "Attempts by Congressmen to utilize their official positions as an excuse for special pleading (under the guise of explaining 'peculiarities of local situations') are made with some degree of frequency from the time an application is filed until the Commission has rendered its final order." [43] Commenting on the report, *Variety* notes that the "errand boy" Congressman has become increasingly active in radio matters, and that this is one of the most vicious aspects of the back-door radio lobby in Washington.[44]

The situation calls for reconsidering and redefining the relations between Congress and the Commission.

V. CONGRESSIONAL INTERESTS IN RADIO-BROADCASTING

(1) *Broadcasting by Congressmen.* Having recognized the political value of radio, Congressmen have made good use of it. For example, CBS reports that from 1929 to 1940, Senators addressed radio audiences over Columbia sustaining programs more than 700 times and Representatives more than 500 times. Often Congressmen resent the fact that radio stations outside their immediate constituencies refuse to carry their speeches when these are offered by the networks. But in general Congressmen have little to say to the people of the country and direct their remarks toward their own "local corner of the republic." There has been little occasion for complaints about the use of radio by Congressmen. During election campaigns, however, when political candidates pay for time, there have sometimes been charges of discrimination by stations or slander by speakers. Such charges were brought by Senator Stiles Bridges of New Hampshire against Station WMUR of Manchester as a result of its activities in the elections of November, 1942. Senator Bridges asked the FCC to suspend WMUR's license to compel it to cease engaging in false and malicious propaganda. Francis Murphy, who unsuccessfully opposed Bridges for election, is chief stockholder and director of Radio Voice of New Hampshire, Inc., which operates the station. Bridges charged that the station broadcast as news political statements promoting Murphy and slandering Bridges.[45]

There has been from time to time a movement in Congress to broadcast proceedings of the Senate and House directly from the floor, and at least four bills to this effect have been introduced, but they have not been acted upon.[46] It is interesting to speculate on the effect a radio audience would have on Congressional debates and procedures.

(2) *Concentration of Control.* In the main, the interests of Congressmen in their dealings with the regulatory commission and in their attempts at legislation have been marked out for them by a number of well-organized pressure groups which represent the Congressman's constituents more effectively than they could ever represent themselves. In general, these interests fall into three categories: control, program content, and adequacy of service; but in fact this can be only an artificial grouping, since the three types of problems are interrelated. Probably the hottest issue politically has been that of control, and from the point of view of time consumed in hearings and debates, Congress has been more concerned with the problems of monopoly than with any other aspect of the radio industry. Competition has always been considered desirable in the American economy, but particularly in radio has Congress been vigilant to preserve

competition because of the nature of radio as a molder of public opinion and an instrument of political power.

The burden of safeguarding the growing industry from the dangers of monopoly was given to the Commission at the onset, when, in the 1927 act, Congress directed the Commission to refuse licenses to those found guilty of unlawful monopoly. But in 1931, RCA, found guilty of violating the anti-trust laws, did not lose its licenses. Congress was dissatisfied with the job the Commission was doing in preventing monopoly, and indeed it had reason to be. The debates on the Communications Act of 1934 were marked by blasts against the so-called radio trust. "Broadcasting in the United States is rapidly becoming a monopoly in the hands of these two systems" [NBC and CBS], said Representative McFadden.[47] Congress was influenced by a flood of articles like that of Eddie Dowling, who asserted that the radio industry was a "private monopoly of immense power . . . playing both ends against the middle and subject to no authority or control except a purely technical supervision of wave-length assignments." [48] The 1934 act provided that a court may revoke a license if the licensee is found guilty of violating the anti-trust laws. In 1935, Representative Monaghan presented a bill [49] for government control in which radio monopoly was cited as the chief abuse of the present system and as proof of the necessity of government control. In the course of the debate, one Congressman said: "The president of NBC publicly admitted that the primary purpose for which his company was organized was not to serve the public interest but to serve the radio manufacturing industry and the Bell Telephone Company." [50]

Concern over concentration of control reached a new high in 1937, when there were pending at the same time no less than four resolutions [51] for investigation of monopolistic practices. Stating that a "colossal fraud was being perpetrated on the American people," MacFarlane claimed that the monopolies controlled all forty clear channels, all stations of over 1,000 watts operating at night, and ninety-three per cent of transmitting power.[52] In his resolution, Senator White noted that "there has come about a monopolistic concentration of ownership or control of stations in the chain companies of the United States." [53] The monopolies were accused of stock racketeering, of faulty financial practices, of trafficking in licenses. Later, when the problem of introducing television came to the fore, the radio monopoly was condemned for handling patents and production in a manner against the public interest.

None of the resolutions for investigation of monopoly was ever passed, but in 1938, under Congressional pressure, the FCC appointed a committee of three Commissioners particularly to determine regulations for chain broadcasting and the growing networks which had occasioned the monopoly charges in Congress. The report was not issued until May, 1941,[54] and meanwhile Congress had grown impatient and started hearings of its own. In the summer of 1940, in considering the renomination of Thad Brown to the Commission, the Senate Committee on Interstate Commerce held lengthy hearings at which the financial manipulations of the major networks and of the RCA were exposed.[55]

Although they made a loud noise, the actual number of Representatives and Senators speaking against the radio monopoly and demanding investigations could be counted upon the fingers of one hand. Those Congressmen on the other side of the fence were mostly inarticulate because the popular trend was to crusade against monopoly and they saw no need to make themselves unnecessarily unpopular. But there is little doubt they did a good deal of off-the-record work with the Commission for those who had important financial interests in the industry.

(3) *Adequacy of Service—Technical.* Congressional concern with concentration of control was occasioned not only by interest in financial manipulations and business practices of the licensees, but also by fear that monopolies would not provide adequate service to listeners. Whether service is considered adequate will depend in part on one's idea of the function of radio, and in part on the group whose interests are being considered primarily. It is obvious that the listening public ought to be the chief concern of the broadcaster. There is not one public, however, but several publics to be served. The needs of audiences in various parts of the country, in rural and urban areas, must be weighed against each other and some kind of balance achieved, since restricted facilities cause problems of allocation to persist. Although the actual job of allocating frequencies has fallen to the Commission, Congress has aired its views at length on the maximum and optimum use of the limited broadcasting band and has incorporated its ideas in general legislation to form the basis of an allocations policy for the Commission. That it has not always been successful has been due to its own inadequacies as a body of nontechnical persons legislating in an engineering sphere, to its faulty use of experts' knowledge, and to the continually advancing state of radio.

In the Radio Act of 1927, Congress's vague dictum to the Commission was to distribute facilities "among the different states and communities so as to give fair, efficient, and equitable radio service to each of same." This proved unsatisfactory, since it left too much discretion to the Commission, and in general Congressmen were not pleased with the job. The next year, therefore, the so-called Davis Amendment was passed, providing for equal allocation of station facilities to each of five zones, and allocation to states within each zone according to population. But the theory behind the Davis Amendment proved fallacious. The five zones were unequal in population distribution and in area, so that while the number of stations was equal, the amount and type of service were not. Early in 1932, Senator Shipstead introduced a bill [56] to correct the inequalities created by the Davis Amendment by distributing radio facilities by state rather than by zone, but this was killed in the Interstate Commerce Committee. At that time Congress was concerned with a bill to revise the 1927 act,[57] and the debates offered an opportunity for representatives from Western states to stump for a more favorable allocations structure. "Every district in the United States is entitled to at least a small station, and if they are country people they are entitled to rights just the same as city people," said Representative Blanton.[58] In the 1934 act, the inequality of the zone system was partially corrected when a section was provided to allow additional licenses outside the quota for stations not over 100 watts. The zone system was finally abolished by Senator Wheeler's bill, which became law in 1936.[59]

But it was not only the problems of allocation of frequencies that worried Congress, but also problems of power. On this subject, there are two schools of thought in Congress, as there are in the industry. There are those who would go all out for a small number of superpower and clear channel stations, and on the other hand, those who are interested in large numbers of small-power, local stations. Of course, this is stating the case in extremes, since there are any number of Congressmen who think it would be nice to have a few superpower stations and some low-powered stations, some clear channels and some locals, etc. This is not the place to discuss the varied ramifications of the complicated issues of superpower and clear channels. Since in certain respects the subject is a sectional one, Congressmen have been busy presenting the views of their respective constituents to the Commission, which has held a number of hearings on the problem. In 1938, the Senate adopted a resolution introduced by Senator Wheeler of Montana,[60] who has been a leader in the fight against superpower, to prohibit stations with power above fifty kilowatts, claiming that such stations were against the public interest because of their adverse effect on small stations and their tendency to aid concentration of control. When the FCC revoked WLW's 500-kilowatt license in 1939, Representative Sweeney of Ohio objected strenuously, demanding "radio parity for rural listeners." [61] Most recently, the hearings on the Sanders bill [62] before the House Committee on Interstate and Foreign Commerce have been the occasion for Congressional reconsideration of superpower and clear channel.[63] The Sanders bill called on the Commission to report to Congress on its administration of Section 307 (b) of the Communications Act of 1934. By this section, the FCC was directed so to distribute licenses and power as to provide "fair, efficient, and equitable" radio service to the several states and communities. Clear Channel Associates, through Victor Sholis, demanded more superpower stations.[64] Against them, Paul Spearman, representing a group called Network Affiliates at the hearings, recommended, among other things, that Congress limit power to fifty kilowatts.[65] The testimony of the Clear Channel Associates was in some degree counteracted by that of E. K. Jett,[66] chief engineer of the Commission, who claimed that the station serving the widest rural area is not a clear-channel station at all. The whole subject of allocations promises to be one of the most pressing issues before Congress in postwar radio legislation.

(4) *Adequacy of Service—Programming.* The problem of adequacy of service involves not only the technical engineering aspects of allocation, but the social aspects of program content as well. For a long time, one of the arguments against clear-channel stations has been that the culture of local areas will suffer without local stations. Similarly, those who object to widespread network broadcasting are afraid that programs emanating exclusively from the big cities and talent centers of the country will make New Yorkers and Hollywoodites out of all Americans, and local American music, dancing, and humor will be forgotten.

It has been suggested that there is a basic conflict between the method of allocating facilities and the manner of using them. Though the Commission allocates facilities scientifically, it has no comparable criteria for evaluating the utilization of them. Indeed, Congress did not intend the FCC to have an interest in programming outside of the very general interest implied in the phrase "public convenience, interest, or necessity." At various times, Congress has sought to remedy this vagueness by providing for more definite programming control. The most notable effort in this direction was the amendment to the act of 1934, introduced by Senators Wagner and Hatfield, for twenty-five per cent allocation of facilities to nonprofit organizations—religious, educational, labor, agricultural, etc.[67] Wagner and Hatfield made it clear that the nonprofit groups were demanding legislation. The chief proponents of the amendment were the Paulist Fathers, whose station WLWL in New York had suffered a cut in broadcasting hours. Testifying before the Senate Committee on Interstate Commerce in March, 1934, Father Harney suggested that twenty-five per cent of facilities be assigned to "human welfare" organizations, leaving to the Commission the power to divide the twenty-five per cent between the various groups, and allowing the noncommercial stations to sell time to the extent of supporting themselves. He cited the "beggarly and outrageous" allocation of facilities to educational stations.[68] Father Harney quoted from a letter which Sykes, chairman of FRC, had written to Representative Merritt, a member of the House Committee, in which Sykes said: "Many special interests are able to appeal to Congress, or to particular members of a Congress, and time does not permit a complete hearing on the question at issue. It seems most desirable, therefore, that all cases be heard by the administrative body." [69] But Harney was determined that the issue should come directly to the Congress, and indeed it was brought to the House by

Representative Rudd as an amendment to its Federal Communications Act and to the floor of the Senate by Senators Wagner and Hatfield as an amendment to the 1934 Federal Communications Bill, and debated at length before it was defeated.[70] At that time, it was pointed out that eighty per cent of the time given to education by commercial stations was sustaining time, when the stations would be presenting programs at their own expense anyway. The amendment was defeated, not so much because Congressmen objected to educational and religious broadcasting, but chiefly because of the faulty drafting of the amendment and the administrative difficulties envisaged in carrying out the particular plan proposed.

There was also growing dissatisfaction among certain Congressmen with the use of radio facilities for commercial advertising. In a speech before the meeting of the American Association of Advertising Agencies in 1932, Representative Edwin Davis declared that "radio is not maintained to sell goods," [71] and that the only justification for advertising is to maintain radio financially. He suggested legislation by Congress on the length and quality of advertising. This suggestion is particularly interesting in view of the recent campaign by listeners and broadcasters against objectionable advertising. Liquor advertising, especially, was in the orbit of Congressional concern, and from 1934 on, no less than ten bills were introduced to prohibit liquor advertising on the air.[72] "There comes over the radio nightly," said Representative Culkin, "a glorification of booze." [73] It was insisted that liquor was as offensive as lotteries, which had been banned from the radio by the act of 1934. Although the bills never came out of committee, the forces which had been working on Congressmen to ban liquor advertising had done an effective job on the NAB, which decided to prohibit such advertising on the air.

(5) *Censorship and Free Radio.* In general, Congress has from the outset been concerned with the type and quality of programs on the air, and there was some recognition of this concern in the Federal Communications Act when it prohibited lotteries and "obscene, indecent, or profane language." At the same time, however, Congress showed its concern for free speech and freedom from censorship when it wrote into the act that nothing "shall be understood or construed to give the Commission the power of censorship over the radio communication . . ." and prohibited the Commission from fixing rules "which shall interfere with the right of free speech." [74] Congress said also "the LICENSEE shall have no power of censorship over the material broadcast." [75] These two concepts of freedom of speech and program quality, while certainly not paradoxical, have frequently involved the Commission, the industry, and Congress in conflict situations. On not a few occasions, Congressmen have been agitated at allegedly "obscene and indecent programs" on which the Commission had taken no action. But always in their remarks in House or Senate they were

treading on precarious ground, since the line where censorship infringed on free speech was ill-defined at best. Though it was agreed that programs should be fit for all to hear, the idea of censorship was anathema to a large number of Congressmen. Yet Congressmen particularly deplored the broadcasts by "Doc" Brinkley,[76] who continued to broadcast into the country from Mexico when the FCC refused him a license, a Mae West broadcast over a major network,[77] and a Mexican program with an obscene song.[78] Congress called upon the broadcasting industry to "clean its stables," [79] and action was demanded of the Commission; but beyond this Congress did not go.

Religious and political broadcasting were most affected by the problems of censorship vs. free speech, and these were the two fields with which Representative McFadden was concerned in a bill introduced in 1934 to abolish radio censorship and provide equal treatment for political candidates.[80] His bill was directed particularly to the protection of persons running for public office, and of religious propaganda over the air. The hearing was mostly taken up with testimony by Jehovah's Witnesses, who claimed to have been shut off the air, as indeed they had been, though, as Bellows of CBS put it, not through the exercise of censorship, but through the exercise of selection.[81] While it was true that licensees had no power to censor, he said, they did have the right to select and were under the constant necessity of doing so. Thus the broadcasters got around the issue.

After passage of the Communications Act, three bills were introduced to clarify the confusion over censorship.[82] These would have required stations to devote certain hours to unrestricted discussion of public issues without payment, and would have protected licensees from libel suits on the broadcasting of public questions. The whole issue was highlighted when Father Coughlin was forced off the air, and again in 1941 when the discussion of isolationism vs. interventionism deeply stirred the country. At that time, Senator Wheeler was accused of having used his influence to prevent broadcasts of Walter Winchell's programs over a Montana network. The head of this network, the Z-Bar, declared that stations had deleted the commentator's programs "as a protest to a very unfair system of presentation that is not conducive to free speech." "Under the present system," he said, "the networks have no way of even calling to the attention of listeners . . . the time a speaker of differing views will be heard." [83]

The most recent legislative attempt to clarify the censorship issue was the bill introduced by Representative Ditter of Pennsylvania at the beginning of 1940.[84] *Variety* saw in the bill "an avowed move to end backdoor censorship and political jockeying." [85] Recalling the Congressional intent in the act of 1934 to deny the FCC any censorship powers, Ditter declared that the "public convenience, interest, or necessity" clause was being used as an excuse to censor programs.[86] Congress and the courts, not the Commission, should determine program standards in the public interest. Furthermore, Ditter contended that program standards should be set up in advance and objected to "ex post facto censorship" (as Commissioner Craven called it) at hearings on license renewals. Though he conceded that there were limitations on free speech, he said: "We certainly never intended to delegate to this Commission the power to impose its judgment as to what are good programs and what are bad programs." [87] The bill would completely deny the FCC the right to consider program offenses in passing on renewals. To date, Congress has not acted on this problem. However, the confusion under which the industry and the Commission previously labored has been mitigated by the directives provided by the War Information and Censorship agencies.

NOTES

[1] A number of general studies of radio-broadcasting which touch incidentally on the problems of government and control have been published in recent years. Among these should be noted: Francis Chase, Jr., *Sound and Fury* (Harper, 1942); S. H. Dryer, *Radio in Wartime* (Greenberg, 1942); C. J. Friedrich, *Radiobroadcasting and Higher Education* (Studies in the Control of Radio, No. 4); R. J. Landry, *Who, What, Why Is Radio?* (Stewart, 1942); P. F. Lazarsfeld, *Radio and the Printed Page* (Duell, 1940); C. B. Rose, Jr., *National Policy for Radio Broadcasting* (Harper, 1940); Jeanette Sayre, *An Analysis of the Radiobroadcasting Activities of Federal Agencies* (Studies in the Control of Radio, No. 3); C. A. Siepmann, "Radio and Education" (*Studies in Philosophy and Education*, Vol. IX, No. 1, 1941). Very important also is R. E. Cushman, "Independent Regulatory Commissions," in *Report of the President's Committee on Administrative Management* (1937).

[2] U.S. *v*. Zenith Radio Corporation, 12 F (2nd) 614.

[3] *Cong. Rec.*, 69th Cong., 1st Sess., Mar. 12, 1926, debate in House of Representatives on H. R. 9971.

[4] H. R. 5589; H. R. 9108; H. R. 9971; and see H. Rept. No. 464, 69th Cong., 1st Sess., and Sen. Rept. No. 772, 69th Cong., 1st Sess.

[5] *Cong. Rec.*, 69th Cong., 1st Sess., Mar. 12, 1926, debate in House of Representatives on H. R. 9971.

[6] *Ibid.*

[7] See Conference Rept. on H. R. 9971, H. Rept. No. 1836, and Sen. Doc. No. 200, 69th Cong., 2nd Sess.

[8] *Cong. Rec.*, *op. cit.*

[9] H. R. 7716. For the history of the Commission, see Laurence F. Schmeckebier, *The Federal Radio Commission; Its History, Activities, and Organization* (The Brookings Institution, 1932).

[10] *New York Times*, Jan. 11, 1934, p. 29.

[11] *Ibid.*, Jan. 28, 1934, p. 28.

[12] *Ibid.*, Feb. 3, 1934, p. 12.

[13] S. 2910; H. R. 8301; see Hearings before the Commission on Interstate Commerce on S. 2910, U.S. Senate, 73rd Cong., 2nd Sess., Mar. 9–15, 1934.

[14] S. 3285.

[15] *New York Times*, Apr. 14, 1934, p. 7.

[16] *Variety*, May 22, 1934.

[17] *Cong. Rec.*, 73rd Cong., 2nd Sess., June 2, 1934, debate in House of Representatives on S. 3285.

[18] *Variety*, May 22, 1934.

[19] *Cong. Rec.*, 73rd Cong., 2nd Sess., June 2, 1934, debate in House of Representatives on S. 3285.

[20] See Conference Report on S. 3285, H. Rept. 1981, 73rd Cong., 2nd Sess.

[21] *Cong. Rec.*, 73rd Cong., 2nd Sess., June 4, 1934, debate in House of Representatives on Conf. Rept. on S. 3285.

[22] S. Res. 250; S. Res. 260; S. Res. 275.

[23] H. Res. 394, by Representative Connery.

[24] H. Res. 61.

[25] *Cong. Rec.*, 75th Cong., 1st Sess., Mar. 17, 1937, p. 2332.

[26] *Ibid.*

[27] S. Res. 149.

[28] *Cong. Rec.*, 75th Cong., 3rd Sess., p. 9323.

[29] *Ibid.*, p. 5284 *ff.*

[30] H. Res. 92.

[31] See *Cong. Rec.*, 75th Cong., 3rd Sess., June 14, 1938, pp. 9313–9325, for this debate and the quotations below.

[32] *Ibid.*, p. 9578.

[33] H. Res. 462.

[34] S. Res. 300.

[35] H. Res. 51.

[36] H. Res. 426.

[37] H. Res. 21.

[38] *Cong. Rec.*, 78th Cong., 1st Sess., Jan. 19, 1943. See also H. Res. 55 by Rep. Sparkman, Jan. 18, 1943, to broaden investigation to cover the industry.

[39] *Variety*, June 13, 1934.

[40] See Hearings before the Committee on Interstate Commerce, U.S. Senate, 76th Cong., 3rd Sess., on Nomination of Thad H. Brown for Reappointment as Federal Communications Commissioner.

[41] *Cong. Rec.*, 77th Cong., 2nd Sess., Jan. 22, 1942.

[42] *Administrative Procedure in Government Agencies*, Part III, p. 59. Sen. Doc. No. 186, 76th Cong., 3rd Sess.

[43] *Ibid.*

[44] *Variety*, Feb. 21, 1940.

[45] *Ibid.*, Nov. 4, 1942.

[46] S. Res. 28, Dec. 7, 1931; S. Res. 71, Dec. 14, 1932; S. Res. 29, Mar. 15, 1933; S. Res. 93, Mar. 15, 1937.

[47] *Cong. Rec.*, 73rd Cong., 2nd Sess., June 2, 1934, p. 10309.

[48] "Radio Needs a Revolution," in *Forum*, Feb., 1934.

[49] H. R. 8475.

[50] *Cong. Rec.*, 75th Cong., 1st Sess., Aug. 23, 1935, pp. 14310–14316.

[51] H. Res. 61; H. Res. 92; S. Res. 149; H. Res. 321.

[52] *Cong. Rec.*, 75th Cong., 1st Sess., July 19, 1937, p. 7282.

[53] *Cong. Rec.*, 75th Cong., 1st Sess., July 6, 1937, p. 6786.

[54] *Report on Chain Broadcasting*, FCC Commission Order No. 37, Docket No. 5060, May, 1941.

[55] Hearings before Committee on Interstate Commerce, U.S. Senate, 76th Cong., 3rd Sess., on Nomination of Thad H. Brown for Reappointment as Federal Communications Commissioner, July–Aug., 1940.

[56] S. 3649.

[57] H. R. 7716.

[58] *Cong. Rec.*, 72nd Cong., 1st Sess., Feb. 10, 1932, pp. 3681–3705.

[59] S. 2243.

[60] S. Res. 294.

[61] *Cong. Rec.*, 76th Cong., 1st Sess., May 16, 1939, pp. 2020–2022 (Appendix).

[62] H. R. 5497, 77th Cong., 2nd Sess.

[63] Hearings before Committee on Interstate and Foreign Commerce, House of Representatives, 77th Cong., 2nd Sess.

[64] See Hearings on H. R. 5497, Part 2, p. 687 *ff*.

[65] See Hearings on H. R. 5497, Part 1, p. 397 *ff*.

[66] See Hearings on H. R. 5497, Part 3, p. 1032 *ff*.

[67] See *Cong. Rec.*, 73rd Cong., 2nd Sess., May 15, 1934, pp. 8828 *ff*.

[68] Hearings before Senate Committee on Interstate Commerce, 73rd Cong., 2nd Sess., on S. 2910, Mar. 9–15, 1934, pp. 184–192.

[69] Hearings before Committee on Interstate and Foreign Commerce, House of Representatives, 73rd Cong., 2nd Sess., on H. R. 8301, Apr. 10, 1934, p. 151.

[70] *Cong. Rec.*, 73rd Cong., 2nd Sess., May 15, 1934, p. 8842.

[71] See *Cong. Rec.*, 72nd Cong., 1st Sess., Apr. 22, 1932, p. 8699.

[72] S. 3015, Mar. 10, 1934; H. R. 8404, June 7, 1935; H. R. 3140, Jan. 18, 1937; S. 3550, Feb. 25, 1938; H. R. 9624, Feb. 25, 1938; H. R. 251 and H. R. 252, Jan. 3, 1939; S. 517, Jan. 10, 1939; S. 575, Jan. 12, 1939; H. R. 123, Jan. 3, 1941; H. R. 6785, Mar. 13, 1942.

[73] *Cong. Rec.*, 74th Cong., 1st Sess., June 18, 1935, p. 9613.

[74] Federal Communications Act, June 19, 1934, ch. 652, sec. 326, 48 Stat. 1091.

[75] *Op. cit.*, sec. 315, 48 Stat. 1088.

[76] See *Cong. Rec.*, 72nd Cong., 1st Sess., Apr. 11, 1932, p. 7862. See also "The Development of the Control of Advertising on the Air," by Carl J. Friedrich and Jeanette Sayre, *Studies in the Control of Radio*, No. 1 (1940).

[77] *Cong. Rec.*, 75th Cong., 3rd Sess., Jan. 14, 1938, p. 560; Jan. 26, 1938, p. 357 (Appendix).

[78] *Cong. Rec.*, 74th Cong., 2nd Sess., Jan. 15, 1936, pp. 417–422; June 20, 1936, p. 10660.

[79] *Cong. Rec.*, 77th Cong., 2nd Sess., June 9, 1942, p. 5260.

[80] H. R. 7986.

[81] Hearings before Committee on Merchant Marine, Radio, and Fisheries, House of Representatives, 73rd Cong., 2nd Sess., on H. H. 7986, Mar. 15–20, 1939, p. 158.

[82] H. R. 9229, 74th Cong., 1st Sess., Aug. 23, 1935, p. 14399; S. 2755, S. 2756, S. 2757, 75th Cong., 1st Sess., July 8, 1937, p. 6893; S. 3515, 76th Cong., 3rd Sess., Mar. 5, 1940, p. 2337; H. R. 1082, 77th Cong., 1st Sess., Jan. 3, 1941, p. 20.

[83] *Broadcasting*, July 28, 1941; see also July 14, 1941.

[84] H. R. 8509.

[85] *Variety*, Feb. 21, 1940, p. 23.

[86] *Cong. Rec.*, 76th Cong., 3rd Sess., p. 806 *ff*. (Appendix).

[87] *Ibid.*

3. Great Change-overs for You

by Marshall McLuhan

Mr. McLuhan, renowned Director of the Centre for Culture and Technology at the University of Toronto, occupied the Albert Schweitzer Chair in Humanities and Social Sciences at Fordham University in 1967–68. His article is reprinted from Vogue *magazine, July, 1966, by permission of the author.*

My friend, Fr. John Culkin, S.J., has pointed out that although we do not know who discovered water, it was almost certainly not a fish. Anybody's total surround, or environment, creates a condition of nonperception. At the same time it creates a clear image of the preceding environment or situation. It is the preceding environment that is taken to be the present situation. It has been said that "the future of the future is the present." Only the artist, however, has the courage or the sensory training to look directly into the present. Wyndham Lewis said years ago, "The artist is engaged in writing a detailed history of the future because he alone is capable of seeing the present."

At present we are on the verge of a large change-over in our entertainment industry. Like many large changes, much has been hidden from view until the last minute. The American public is about to enter the entertainment industry as participant. While attention remains riveted on the rear-view mirror of audience ratings and packaged programs, the audience, in fact, has moved ever closer to an active role. Vision of this spectacular flip has been obscured by many factors including a misunderstanding of the quiz shows and their fate. A few years ago the American public had a sense of involvement in the quiz shows. Suddenly they learned that the shows had been "rigged" and that they had really been left out of the act all along.

With the aid of punch cards and computer processing, it is now possible for millions of people to participate directly in programs in prime time. That is to say, large audiences can be briefed in prime time in top problems in the sciences. Robert Oppenheimer is fond of saying, "There are kids playing here on the sidewalk who could solve some of my toughest problems in physics because they have modes of sensibility and perception that I lost forty years ago."

The greatest scientific discoveries appear quite simple in retrospect. The greatest difficulties in science would appear equally simple to two or three members of an audience of thirty millions. What eight or nine scientists might puzzle over for decades could be penetrated by a mass audience at once. For centuries the Western world has dealt with the audience as a target and consumer area. Electric technology has transformed the nature of the audience. On the one hand the audience becomes eligible for custom-made servicing instead of uniform packaging. On the other hand the audience itself becomes an actor in the show. For some time the world of business has taken on the character of show business. The idea of the audience as passive is a good bit of rear-view mirror hindsight, a sort of *derrière pensée.*

While writing *The Gutenberg Galaxy* I referred to it as the "Gut Gal," and this often became the "Cute Gal," and it seems to go quite naturally with Batman. The big flip-over from Gut Gal to Batman is a flip from pix to icons, or from pictures to cartoons. To understand the change from the world of pictures to the world of icons is to understand why our present world has moved from gradualness and continuity to a world in which everything is abrupt interface. The world of the icon is a world of abrupt encounters. The world of interface is not organized pictorially by gradation and continuity. It is a world of Happenings. The world of the Happening is an electronic world of all-at-onceness in which things hit into each other but in which there are no connections. Gutenberg technology, on the other hand, gradually eliminated the all-at-onceness of oral culture by creating a means of organizing human energy in strongly visual and classified terms. Separateness and privacy acquired prime importance with typographic culture and technology.

If you have seen *Doctor Zhivago,* you will have noticed something unusual about the sound track. In that film, if a train gets under way, it doesn't gradually reach a crescendo. Instead, there is a sudden Wham! Bang! much like the sounds in *Batman.* There is no gradation in the sound buildup, everything is just sudden interface. The French have been trying to write novels this way: The Robbe-Grillet type of novel is one in which there are no characters in the ordinary sense; there are just the abrupt encounters à la Batman of cartoon icons. The world of the icon in art or narrative is a world without gradations and without chiaroscuro. This kind of world is now developing on many fronts.

When you have these interfaces, you have Happenings; you don't have a story line. We'll come back to that in a moment. The world of the interface is a world of Happenings because the surfaces of events grind against each other and create new forms, much as the action of dialogue creates new insight. The world of the interface is the world of the Happening and the world of the cartoon. TV had been protected from this interface by the movie industry and by the effects of the movie industry on TV production, but now TV is affecting the movie industry. As TV has become the environment, or surround, of the movie, it has transformed the movie into an art form. It will be the turn of TV to become an art form when the new satellite environment goes around it.

There is a book by Owen Barfield called *Saving the Appearances* in which he points to one of the peculiarities of our Western world as deriving from the uses of the alphabet. Unlike all tribal societies, alphabetic cultures can be detached; they can avoid involvement; they have the means of detribalizing themselves. By contrast tribal societies are involved, they experience a *participation mystique.* The experience of mystical participation in the cosmos that is shared by all preliterate societies is one in which people are eager to merge with the cosmic powers. Beginning with the Greeks, however, and the phonetic alphabet, there came a habit of detachment and noninvolvement, that is, a kind of uncooperative gesture toward the universe. This has been the basis of all Western society. From this wonderful power of refusing to be involved in the world we live in, the Western world has derived detachment and objectivity. Owen Barfield's study is ably seconded by Eric Havelock in his *Preface to Plato.* Havelock demonstrates in detail how tribal man moved from the involving auditory culture to the Platonic world of ideas and classified knowledge. Edward T. Hall in his studies *The Silent Language* and *The Hidden Dimension* has revealed the changing forms of human perception as our perceptions encounter different cultural arrangements. For example, in *The Hidden Dimension* he explains the fascinating diversities of spatial awareness in America, and in Iran, and in Japan. On our continent, anyone minding his own business, who stays put and immobile, is inviolate, and anybody who barges into him is behaving boorishly. It is quite otherwise in the Arab world. There, it is the moving person who is inviolate and the stationary person who is fair game. In our literary and visual culture it is the specialist, the man who stays with one subject, who is inviolate. He is an expert. But anybody who crosses boundaries, and who keeps moving across boundaries, will lack respect and prestige. As we move into a new world of electronic information, there is much confusion about our older values. We are moving, as it were, from the American into the Arab orbit with regard to space and specialism.

Under conditions of instant information retrieval, classified knowledge loses its older prestige and significance. Retrieval itself becomes a means of discovery at electronic speeds, changing the whole purpose of storage systems. Under visual conditions of classified knowledge, retrieval merely took the form of reference. There are other areas in which major flips are tending to build up very quickly at present. Quite apart from the entry of the audience as work-force to end all rating systems, there is a similar reversal that is becoming apparent in the educational establishment. After centuries of stress on instruction, we have begun to move into a world where education becomes a form of discovery. Today it is the environment itself that is made of information. The world outside the classroom is so loaded with data that Jacques Ellul has observed that the twentieth-century child works harder than any child who ever lived. Sheer data processing confronts the ordinary child with a situation of information overload for which the instinctive response is mythic pattern-making. That is to say, the ordinary young person in our electronic environment moves naturally into a habit of myth-making as a way of coping with an environment made up of information. It is this habit arising from a deep need that is so strongly at variance with the world of the classroom and the curriculum where knowledge is still arranged in unrelated categories.

The psychic strategies needed to cope with information overload make the curriculum and the classroom seem ludicrous and "square" to young people accustomed to TV and the electronic environment. What would seem to be indicated is that instead of undergoing a process of stencilling and instruction, the student population is ready to turn to the arts of discovery and investigation. Small teams of students can be assigned to look into large varieties of problems that concern the entire community. The techniques of the Peace Corps are quite as relevant for our own world as for distant places. What is so magnificently right about the Peace Corps is that it is a totally involving process. If our classroom and educational procedures were extended to the Peace Corps, it would collapse at once.

Another kind of flip that has begun on a very large scale in our electric world of information belongs to the field of advertising. Just as the painters and poets more than a century ago gave priority to the effect over the product, so with advertising today. As our means of information become more pervasive and inclusive, it becomes natural to make the ad a substitute for the product. Advertisers have been puzzled by the tendency of viewers and readers to pay special attention to the ads for products that they already owned. It is as if people used the ad to strengthen their impression of the product, and to get "cued in" as to the means of relating themselves to it. In a word, advertising has become a service industry, as much as a salesman. It is only natural, therefore, that the old-fashioned salesmen, the Willy Lomans, should be quite derailed by this development of the marketing process.

The world of uniform, fixed prices is undergoing rapid alteration in the age of the discount house. Indeed, the printed book was the first uniform and repeatable "commodity." Prior to printing, the book market had been a second-hand market. Such is the world of the antique and of Old Masters today. With the advent of a uniform and repeatable commodity, a totally new type of selling and marketing began. Uniform prices were as revolutionary a thing in the sixteenth century as they are today in India, or Africa. A considerable degree of literacy is the necessary prelude to a pricing system. Today, the venerable technology of movable types has contracted a "shotgun marriage" with electric circuitry and xerography. One result is that the reader can become publisher and author once more, as in the days of the old scribes. This decentralizing of a process that has long been centralized has created a crisis which so far has been localized in the area of copyright. In point of fact, xerography will alter the relation of the book to the market, to the language, and to other media. One media development has been to open the possibility for the student or reader to request a book made to his specifications and needs when existing libraries become linked by computer to Xerox services. The printed book, having begun as the first commodity package, is now ready to become an information service to individual needs.

Print had created the Public, that is, a large group of separate individuals accessible through a common language and a common national territory. Electric circuitry substitutes the Mass for the Public. In contrast to the Public, the Mass consists of people quite deeply involved in one another by virtue of enormous speedup of information services. Electric speeds of information in effect pull out the times and spaces between people, as can be noted in the makeup of the daily paper. Many people talk as if the Mass represented merely a much larger group of people than the old Public. In point of fact, mass has less to do with size than speed. That is why the most trivial events, when circulated at electric speeds, can acquire enormous potential and influence. To think of the mass audience as merely larger and more vulgar than the old reading public is a good example of the rear-view mirror vision of the world. The wedding of the book and electricity points to a new type of custom-built selectivity that is quite the reverse of the older mechanized technology. The metamorphosis of the book thus serves to highlight a structural change in cultural patterns as between the older world of mechanized products and the new world of electric services. It is the same change earlier noted with regard to the audience as potential work-force instead of audience as passive consumer.

One of the paradoxical effects of the change-over from typography and wheel to the world of the electric circuit is manifested by the disappearance of the "story line" in the arts and in other areas as well. For example, it has disappeared in the world of the joke, and of popular humour to a great extent. The story line has disappeared from the recent forms of movie, whether it is the work of Fellini, or Vanderbeek, or Warhol, or Bergman. Oddly enough, the disappearance of the story line creates a much higher degree of involvement for the viewer or reader. The discovery of this means of involving the audience had been made more than a century ago by symbolist poets. Edgar Allan Poe had used the same technique in his invention of the detective story. By the use of scrambled time sequences, the detective story requires the reader to be co-author. When the telegraph entered journalism, it was quickly discovered that no story line could accommodate the total field of information produced at instant speeds.

The newspaper has only one unifying factor: a dateline. There are no connections between any of the items in a newspaper, save on the editorial page which retains the story line and point of view of the book. In an electric world it is not only the story line that disappears, but also the clothes line, and the stag line, and the party line.

The alternative to a story line, and to the art of connecting events, is the art of the interval. Oriental art doesn't use connections, but intervals, whether in the art of flower arrangement or in the poetry of Zen Buddhism. The Western world first intuited the onset of the electric age and the change-over to the art of the interval in symbolism, on one hand, and the primacy of musical structures, on the other hand. Walter Pater had observed the tendency for all the arts to approach the condition of music, that is to say, the art of timing and of interval.

James Joyce in *Finnegans Wake* took over the art of the interval as a means of retrieving the fantastic wealth of perception and experience that is stored in ordinary human language. As used by Joyce, the dispensing with the story line became the means of instant grasp of complex wholes, whether in phrases like "casting his perils before swains," or "Jung and easily Freudant," or "though he might have been more humble there's no police like Holmes."

Insights are like humour, in being both instant and total. The eureka moment of scientific discovery has a structure very much like the joke and the quip. It is a moment of interface or encounter between seemingly unrelated events.

If the story line has disappeared from much art and humour and social organization, this fact has received much attention and popularity in the art form called the Happening. The Happening is an artistic event of all-at-onceness in which there is no story line. It nearly resembles the newspaper and also the ordinary human environment. To put a tomato can in the Guggenheim Museum or to bring the unintended noises of the ordinary environment into the concert hall is an important way of announcing that in the electric age we must begin to consider the environment itself as an art form. We approach a time when the total human situation must be considered as a work of art.

This is taken for granted by the makers of space capsules. The capsule has to include the planet, as it were, in order to be usable. Buckminster Fuller has observed that the capsule is the first completely designed human environment. The capsule thus brings us close to the state of the Balinese, who are puzzled by our concept of art. They say: "We have no art, we do everything as well as possible."

The world of the Happening announces that our involvement in the conditions of life on this planet is such that we must begin to do, not some things, but everything as well as possible. We are approaching the condition of King Oedipus of Thebes. Thebes was a tribal society, and when the King set about investigating the responsibility for misery and disorder he found out that *he* was the criminal.

Under electric conditions the seamless web of human involvement becomes as obvious to specialized men as to tribal men. An electric world is an all-at-once world. That is to say, the world in which the meaning is a Happening. Back in the 1920's there used to be much concern about the "meaning of meaning." At that time the discovery that meaning was not statement so much as the simultaneous interaction of many things came as an exciting surprise.

When I say that "the medium is the message" I am merely stating the fact that "meaning" is a Happening, the multitudinous interplay of events. I have found sometimes that it helps to say "The medium is the massage," because the medium is a complex set of events that roughly handles and works over entire populations. It changes their postures and their outlook.

The "safety car" is another curious example of change in mood. For decades the world had accepted the car as a great improvement over the horse and the cart. The car developed new power and autonomy. Suddenly, however, a new image of the car has emerged. It has begun to take on the lurid aspect of what Harley Parker, the painter, has called the "carsophagus." What seems to be changing is the general feeling for the car as it makes its impact on the audience, or the environment.

If the car is seen, not as an isolated fact, but as a Happening, as something that has many hideous consequences, then people suddenly feel the need to include the consequences in the object itself. The safety car has to be designed like the safety pin. Instead of just being pointed outwards, it folds back into itself. Even the driver is provided in the safety car with a padded cell to remind him that he is a potential killer.

Like modern art, the safety car is designed to include the public both as audience and as participant. The safety car is a modern art form in that it is not just sent out into the environment hopefully, but it includes the environment and the effect of the car on the environment as part of its design.

A great headache of our time as expressed in the popular philosophy of Existentialism, centres on the question: "Who am I?" In a world of electric all-at-onceness, as everybody begins to include everybody else, many people are inclined to feel that they have lost their private identity altogether. Instead of feeling enriched, they feel deprived. Fifty years ago, a man could say with pride and confidence: "I am a Hungarian, I am a dentist, I have six kids." That would have served as a card of identity in the past, but today such classifications do not seem acceptable as a means of identity. It is quite common to hear undergraduates today explain: "We are not job-oriented. We simply want to know what's going on." They seem to reject the idea of job as a form of identity, or a mark of significance. Instead, they want roles.

The job as a kind of organization of work is a highly specialized and repetitive activity. It had been preceded by the role, and today the role is returning as a replacement for the job. A man, say a top executive, doesn't have a job; he has fifty jobs, sixty jobs—that's a role. A mother doesn't have a job; she has sixty jobs—that's a role.

In the electric age when all forms of activity have become interrelated, job specialism begins to look precarious and vulnerable. "Come into my parlour," said the computer to the specialist. Roles = depth = involvement = commitment, while the old job tended to represent classification and noninvolvement. In effect, instant circuitry abolishes the world of the specialized job in the sphere of work, much as it abolishes the separate subject in the sphere of learning. Indeed, in the information environment, work and learning become the same kind of action.

Bonanza, a TV show seen by 350 million people each week in sixty-two different countries, is the perfect instance of the rear-view mirror image. It is the latest suburban world seeing itself in terms of the previous nineteenth-century environment. The ideal image that the suburb forms of itself relates not to our time, but the preceding time. This is not a freak situation, but a deeply ingrained human habit. Whenever we encounter a new situation, we translate it back into a previous situation as best we can.

The human need to learn by going from the familiar to the unfamiliar, brings us into a trap whereby we are unable to make direct contact with the unfamiliar except by pretending that it is something we have already experienced. The consequences of this form of self-deception were severe enough in simpler ages when events moved at a relatively slow pace. In our instant age the rear-view mirror approach is as impractical and pointless as it would be in a space capsule. One of the effects, therefore, of the great speedup of change in human arrangements is well expressed in the popular saying, "If it works, it's obsolete."

By the same token, backward countries can now leapfrog out of the dimmest human pasts into the twentieth century. This is actually happening and the meaning of it can be quite varied. For example, the West Coast never had a nineteenth century. It leapfrogged out of the eighteenth century into the twentieth century. This gave it much greater contact with the twentieth century than it could have had if, like Chicago, it had had a nineteenth century. For the nineteenth century was the great period of specialized and classified organization of human work and experience.

It is therefore very difficult for nineteenth-century countries to come to terms with the twentieth century. The Establishment, whether in law, education, or politics is highly bureaucratized and fragmented in accordance with the best achievements of literacy and mechanical industry. The Establishment, as it encounters the twentieth century, forms a splendid image of the nineteenth century by which to orient itself.

It is a kind of alluring and *Bonanza* image that the Establishment has of itself. David Dortort, the producer of *Bonanza,* told me recently that in Nairobi, in Africa, a strange little boy was found wandering. When questioned as to where he had come from, he said that he had come to Nairobi to meet Mr. Cartwright, and he said: "Because I have seen him on TV and I know he can help me." Dortort thought this was a very touching testimonial to his program and I think it is, too.

After all, *Bonanza* is not a depraved or misbegotten type of entertainment. It is a compassionate and noble program. But it is unmistakenly a rear-view mirror image, not specially suited to a fast-moving situation.

In many of the countries of the world *Bonanza* must look like a science-fiction promise of great things to come. But its meaning for twentieth-century Americans is rather different. The advantage of leapfrogging out of the eighteenth century into the twentieth century, as the West Coast has tended to do, lies in the greater flexibility of imaginative approach to problems. People who merely moved out of the nineteenth century into the twentieth century are less endowed with imaginative power. Moreover, as *Bonanza* shows, we, who have gradually emerged from the nineteenth century into the twentieth century, are strongly inclined to leapfrog back into the nineteenth century. The astronaut in Bonanzaland may feel friendly and at home, but he can't avoid looking like a great big teenager doing a makeup year in elementary school. It is not a course fraught with exciting opportunities.

Today we would seem largely to be reading childish forms of science fiction, occupying a childish world that is in actual fact far more fantastic and exciting than anything dreamed of in science fiction. In our rapidly changing environment, one technology succeeds another every few months almost. For example, colour TV is a new technology, a new medium, not just the adding of colour to the old TV medium. One way in which we can recognize the arrival of a new form is to notice its power to revive old forms.

Batman is an example. As a revival of a comic-book entertainment of a few years ago, it would seem to be a response to the novelty of colour TV. The mask-like image of Batman has many of the iconic characteristics of Byzantine art. Colour TV gives new stress to the tactile sciences, as compared with black-and-white TV. The world of the comic book and of the cartoon is multi-sensuous rather than just "visual."

If new environments, created by new technology, tend to turn the old one into an art form, it would be well to think about the meaning of our satellite environment today. When TV went around the old movie environment, it turned the movie into an art form. When the satellites went around the earth, they not only began to turn TV into an art form, in its turn, they began to turn the planet itself into an art form. In the years that lie ahead, we shall see the old nose cone, the earth, being given all the care and grooming that we have accorded to Williamsburg. This is always the fate of old environments when surrounded by new ones.

Eventually people will return to the planet as the old "stamping ground," the place where all began. Was it not Adlai Stevenson who said Plymouth Rock should have landed on the Pilgrims? When TV ceases to be the latest environment and becomes the content of the new satellite environment, we will stop trying to deal with it as if it were some kind of movie. The world of so-called Pop Art has been handed to us, as it were, by the new satellite environment. Pop Art is not a new environment of electric information, but the old mechanical environment suddenly observable as an art form.

But Pop Art is an indication that as the whole planet goes inside a new satellite-and-information environment made by man, we can no longer afford to deal with the human habitat as something given to us by Nature. We have now to accept the fact and responsibility that the entire human environment is an artifact, an art form, something that can be staged and manipulated like show biz.

In his book *Propaganda* Jacques Ellul has pointed out that propaganda does not consist of little separate messages moving through any one medium. Propaganda is the entire way of life in action. "Propaganda ends," he said, "when dialogue begins." We have now to begin a dialogue with our total human environment. We have now to arrange for the media to talk to each other instead of complaining about people who put the programs on the media.

Current discussion of media programming seems to take the line that it is the hot-dog vendors at the ball game who decide what kind of ball game we are going to see. We now have reached the stage when we must cope, not with the content of environments, but with the environments themselves. The James Bonds of our time are mythic ways of telling us that in the electric age man has returned to the status of the hunter. The hunter had been succeeded by the planter, the stationary specialist. The hunter had dealt with his entire environment as a totally unified thing. This is a natural course for man in the electric age.

Man, the hunter—the new electric man—is a man who crosses boundaries and who tries to deal with the total human environment as a single unit. This is a new strategy born of a new situation. It helps to explain the strange importance of the newspaper reporter, the man who tries to deal with an entire world. It also helps to explain something that has mystified me for years. Why does real news have to be bad news? Advertising is all good news, and it doesn't seem to serve the function of news at all.

Is not bad news the order of interface and encounter? On the other hand, is not good news simply a one-way flow, lacking all encounter and interface? May this not help to explain why the poet and the artist, those who sharpen our perceptions, tend to be antisocial types who refuse to go along with the main currents and trends?

4. The Road to Responsibility

by William S. Paley

Mr. Paley is Chairman of the Board, Columbia Broadcasting System, Inc. The following address was delivered before the National Association of Radio and Television Broadcasters in Chicago, May 25, 1964. It is reprinted here by permission of Mr. Paley and the CBS network.

In our turbulent industry today I have no lack of problems to choose for the topic of my talk this morning. Broadcasting gets broader all the time. The problems we have before us are numerous and complex and touch almost every phase of the broadcaster's world: his business and economic preoccupations, the technological developments to the medium; his relationship with government; his responsibility to the public at large. But you will permit me, I am sure, after acknowledging their very great importance, to put all but one of these categories aside and direct your attention to one significant aspect of our responsibilities to the public at large. And so, I have chosen as my topic: the broadcaster's role and responsibilities in the field of news and public affairs. This problem is neither transitory nor peculiar to any other segment of the industry. On the contrary, the proper exercise of a broadcaster's functions in this field is a responsibility which every broadcaster must face and the problems relating to it are common to each of us.

Another reason I want to discuss news and public affairs is because this part of a broadcaster's operation has long been close to my heart and uppermost in my mind; also because I believe that discussion and exchange of ideas on this topic may well lead to newer and more significant plateaus for every segment of broadcasting.

It is my belief that if we know what we are doing in the world of news and public affairs, we are *secure;* if we do not know, we are in danger—in danger of encroachments from government, in danger of criticism, destructive and deserved from other powerful organs of opinion, in danger of criticism from the thinking and leading citizens of our Republic.

CRITICAL ROLE OF NEWS

Let me say that when I talk about news and public affairs this morning, I mean more than just what we call, in this business "the hard news." I mean also all the related fields—the opinion broadcasts, the debates, the feature projects, the documentaries, the panel discussions—as well as the direct on-the-spot coverage of news and public events as these occur. So here I am concerned not just with the raw news, but with the ideas, issues, and controversies that concern the public.

At no period in our history has the function of news and public affairs broadcasting been so critical and important to our national life. The movement of world events on both the national and international scenes takes on increasing significance each day for the welfare and security of each citizen.

These events not only affect *how* he lives, but, in some instances, *whether* he will live at all. Issues have become extremely complicated, giving rise to intense emotion, to a deep longing for answers, and hence demanding greater knowledge and—above all—understanding. It is part of our democratic tradition that facts and exposure to other people's views and opinions have a way of driving out emotional prejudice and of leading the way to answers which are more nearly correct.

These conditions and circumstances provide the broadcaster with an unprecedented opportunity to move ahead in this field of news and public affairs. We have today within our grasp the opportunity to provide an extraordinary public service in a troubled world and, at the same time, to increase our stature and strength as broadcasters. However, one does not receive positions of public trust and of strength on a silver tray. They must be won by resolution, courage, and performance. In these areas, I think, we still have quite some distance to go.

BROADCASTING VIS-À-VIS THE PRESS

Let me give you my reason for thinking so. Consider for a moment some of the significant differences between a broadcaster and the publisher of a newspaper, or, more broadly, the difference between what a radio transmitter does and a printing press does. The printing press came into being in a rudimentary, slow-moving society, in an excessively limited world. Even so, it needed several centuries of struggle to become as free as it is today and to be placed at the service of *all* the people, not just at the service of scholars or priests or governors or public administrators. In the vast present-day diffusion of printing we have undesirable comic books—but we also have great newspapers, an incredible world of magazines, and big books for a quarter. Editors today come from a line of men that have centuries of experience behind them in getting news printed and diffused against the pressures of persons who thought it shouldn't be. The press may not be as free as it likes, but it's plenty free. Through the years it has learned how to keep a manageable degree of freedom against the pressures of readers, advertisers, and powerful critics in government and other high places.

Now, by contrast, consider broadcasting. Against the centuries of hard-bought experience of the press, what have we? Why, we have about a quarter part of one century. Considering this pitiful little span of time, I think we've done rather well. I think it's remarkable that we have been as successful as we have—that we have learned things that have taken other people centuries to get the hang of: the subtleties, traps, and pitfalls that lie in the path of the unwary.

Another thing we had to learn in our beginnings—the hardest possible thing to have to learn fast: we had to learn what our new broadcasting medium was for. Some people thought broadcasting would be for education. Some thought merely that it would replace the phonograph. Some thought it would remake the world, which it hasn't. Some thought it would revolutionize politics, which it has. Some thought it would put newspapers and magazines out of business; instead, it has joined them in an intense but friendly—well, sort of friendly—competition to carry the messages of editors and advertisers to greater areas than ever before.

THE BROADEST MEANS OF COMMUNICATION

But, of course, broadcasting did not limit itself to any *one* of these. Instead it became almost *all* the things that were imagined for it. And when television added broadcast sight to broadcast sound, broadcasting then became, and I am sure will always remain, the broadest means of interchanging, communicating, and diffusing ideas, moods, sights, emotions, facts, images (should I say color?)—and also confusion—in the history of man. We now, in this industry, partake of the newspaper, of the magazine, of the stage, the movies, the concert hall, the lecture platform, the museum, the medical center, the university, and the battlefield. To say nothing of the town meeting, the Senate committee room, the whistle stops of political campaigns and the auditoriums for great debates. Indeed, I think we can say, "Name it, and we are *of* it."

No wonder we are all a little confused. A grant of opportunities has been offered to us of this industry wholly unlike, in range and scope, any grant of opportunities to any other group of modern men. And it happened in a democracy! Which must make us eternally wary of abusing it.

THE NEED FOR STATUS

Since by relative time we are newcomers to the field, we are short in tradition and we still have a lot to learn. The old-time newspaper editor, with his centuries of professional experience, was accustomed to letters saying "Dear Sir, you cur." He was accustomed to being horsewhipped by outraged ladies, and shot at, or indeed shot, by those who considered themselves libeled by some small, innocent item. He was also sued frequently; another painful form of the expression of displeasure.

I must say that I have no desire to stir up any additional public violence today. In this industry we should continue, as in the past, merely to shoot one another and not encourage the public to think that this pistol range is open to them, too.

Old-time editors used to take such threats and actions in their stride, as a part of their occupation. I think we broadcasters can afford a certain amount of the same stride in the face of our letter-writing, telegraph, and telephone critics. If we are fair and responsible in our decisions, we will gain the approval and the respect of the large majority of the people.

But, can we in all fairness and good conscience ask our medium to be as free of threats of boycott, as free of political pressures, as respected as the great newspapers of our country if we shirk our responsibilities in this field? We claim for ourselves, and quite properly I believe, the great historical and constitutional rights and privileges which have been maintained by the press of this country. We recognize the plain fact of our power for good or evil, through the enormous force of our media. Yet the question remains whether we enjoy in the public mind the status which is a natural corollary of our rights and privileges. I respectfully submit that we do not and that we will not until we have shown through clean performance that we have faced up to our opportunities and to our responsibilities.

But it is one thing to resolve to expend greater effort and courage in the field of news and public affairs. It is quite another—and far more difficult thing—to determine how to do it, and do it well.

For one thing, we cannot just imitate the press. With all the similarities between the press and broadcasting, there are still powerful differences between us. And so, with all due deference to the press and its long-established traditions, it can only set us on the road. We will have to do our own driving and find our own right way.

I would not be so presumptuous as to say that there is any *single* right way. Each broadcaster will have to find his own. But I would like to suggest certain steps which might be taken by the broadcaster who finds some truth in my words and who, having not yet done so, wants to build a responsible and respected place for himself in this field.

THREE STEPS TO RESPONSIBILITY

First and perhaps most obvious, the operation of news and public affairs in a broadcasting organization should be given great emphasis and attention by top management. Top management should adjust itself to the fact that this area of the broadcaster's operation is at least as important as any of the other areas coming under his jurisdiction and supervision.

Second, a broadcaster must build a strong news organization—not in numbers, which is not in itself controlling, but in quality. Certainly he should choose the man to head his station's news operation or his public affairs operation, or both, as carefully and as thoughtfully as he would choose the head of his most important departments. Having been selected, this person must be invested with status and the proper authority.

Third, a broadcaster who wants to exercise his responsibilities in this field cannot go into it blindly if disaster is not to overtake him. For his own protection, as well as for the protection of his listeners, he must work out in advance well-defined and clearly stated general policies that will govern his operations.

I make no pleas for uniformity in the ground rules which each broadcaster establishes in this field. Only one basic tenet must be observed: There must be fairness and balance. No matter what the station owner's personal predilections—and he is bound to have them—there must be fairness and balance among all viewpoints.

Beyond this fundamental principle, each station will doubtless work out its own ground rules in the exercise of its vital functions in the area of news and public affairs. The more different approaches there are—the more searching and experimentation there is—the better off all of us will be, for good new ideas will stimulate the competitive forces in this field as they do in other fields.

CBS POLICIES AND GROUND RULES

Reminding you, then, that ours is only one possible approach to a set of ground rules, I would like to outline the policies in this area which we at CBS have worked out over the years. I do so as an example of the thought processes that a broadcaster must go through in order to determine how to exercise the responsibilities in the area of his activities. Our policies, briefly, are these:

In news programs there is to be no opinion or slanting. The news reporting must be straight and objective.

In news analysis there is to be elucidation, illumination, and explanation of the facts and situations, but without bias or editorialization.

In both news and news analysis, the goal of the news broadcaster or the news analyst must be objectivity. I think we all recognize that human nature is such that no newsman is entirely free from his own personal prejudices, experience, and opinions and that accordingly, 100 per cent objectivity may not always be possible. But the important factor is that the news broadcaster and the news analyst must have the will and the intent to be objective. That will and that intent, genuinely held and deeply instilled in him, is the best assurance of objectivity. His aim should be to make it possible for the listener to know the facts and to weigh them carefully so that he can better make up his own mind. Our policies also provide that significant viewpoints on important controversial issues are afforded the opportunity of expression—largely through time periods which are allocated free of charge to outside people and organizations representing opposite viewpoints. The programs in these periods take the form of straight talks, debates, and panel discussions.

Opinion broadcasts must be labeled for what they are. In particular, opinion must be separated from news. The listener is entitled to know what he is receiving, news or opinion, and if it be opinion, whose opinion.

In other types of information programs, such as the feature or documentary program, produced by us, the expression of opinion might properly take place. When it does take place, it should be by the decision of management or through the delegation of authority to a member of the staff producing a particular program. Such delegation, however, must be to one who is trained and responsible and in whose integrity and devotion to democratic principles we repose complete confidence.

When opinion is expressed in any type of information program—excluding news and news analysis where opinion is not allowed—opportunity for reply is given to the person with whom issue has been taken, or to a responsible spokesman representing an opposite viewpoint.

An advertiser who sponsors any type of information program produced by us does not thereby purchase, or in any way gain, any rights to control the contents of the program.

Programs presenting news, news analysis, discussion, debates, feature, or documentary material, or on-the-spot pickups are under our direct control and supervision and we bear full responsibility for such broadcasts.

THE CORNERSTONE: FAIRNESS AND BALANCE

I have not as yet touched on the question of editorialization by the broadcaster—that is, opinion which might be expressed by the broadcaster in his own name or in the name of the station or network. The broadcaster has the same right to editorialize and the same right to independent expression as the free press. I am not urging anyone to exercise this right. It is and should be a matter of personal preference. However, I would urge that we fight to preserve this right should it ever be threatened. In this connection, I would like to say that it is not the act of editorialization which puts the bite and the backbone into a news and public affairs operation. It is rather the fact of having a rounded and vital schedule which does the trick; of having a schedule of unbiased news which covers all categories, national, regional, and local; of having a fair and objective analysis of that portion of the news which calls for background and interpretation; of having a schedule of controversy in which the issues of concern to the listeners will get full and responsible airing; of having public feature programs and documentaries which put the spotlight on conditions worthy of the listener's attention.

As has become apparent, the principle of fairness and balance is the cornerstone of our news and public affairs policy, as it should be of any broadcaster. I want to point out, however, that this principle cannot be reduced to a mathematical formula or even to a set of rigid rules which are self-executing and will cover all the possible circumstances.

In the free American broadcasting system, fairness and balance must be maintained through the exercise of fair and courageous judgment by the station or the network management. And it must be recognized that there *is* a difference between men, ideas, and institutions: some are good and some are bad, and it is up to us to know the difference—to know what will uphold democracy and what will undermine it—and then not to do the latter!

Some people may question the desirability of placing in the hands of the broadcaster this important element of control. To this point I would say that undoubtedly there may be abuses, as there are in other media. But I for one have enough faith in the vitality of the democratic process, in the intelligence of the American people and in the freshness of the competitive climate to believe that the good will and the determined intent of broadcasters to be fair, coupled with the powerful voice of the people, will provide far better protection against abuse than any other form of control.

And let me remind you that those who would take this control away from the broadcaster are the ones who would put it in the hands of government.

Here, then, are our thoughts and our ground rules, evolved through the years, on how best to exercise our responsibilities in the area of news and public affairs. You may disagree with some of them; you may have or develop better ones from which we at CBS may learn much.

TIMIDITY IS SELF-BREEDING

I would like to say that too often public officials, legislators, and other people in public life look upon the broadcasting organization primarily as an instrument created to serve their own purposes, whatever these may be. They do not sufficiently regard the broadcaster as a free and autonomous institution exercising to the best of his ability an influence and responsibility dedicated to the interests of all the people. The fact is—our own timidity in the vital areas of public information is self-perpetuating; it breeds pressures which in turn breed further timidity. Our excursions into the responsible exercise of our functions in the field of news and public affairs are often too spasmodic, too tentative, or too sensitive to permit us to realize our own independence and stature. This must be corrected.

The important thing, whatever the ground rules you set for yourselves, is to consider carefully whether you should not increase your activities and your emphasis in the significant field of news and public affairs. If you do so—if you develop an active, responsible, and eager organization—if you move vigorously into this area, you will, I am convinced, do the country and broadcasting an enormous service.

I have made these suggestions today because I am proud of broadcasting and I want to see the broadcaster seize the extra opportunities which are within his grasp and thereby raise his status among the people he is serving. If he does so he will be putting himself into the stream of life which brings meaning, satisfaction, and a sense of achievement which cannot be matched by any other kind of reward.

5. Remarks Before the International Radio and Television Society

by *E. William Henry*

Mr. Henry, former Federal Communications Commission Chairman, has been perhaps one of the most articulate spokesmen and critics of the broadcasting industry, its practices, and its responsibilities. Many of his speeches and addresses have become, and are, virtual landmarks in the recent literature of broadcasting, and deserve wide dissemination. This speech was originally delivered before the International Radio and Television Society on October 2, 1964, at the Waldorf-Astoria Hotel, New York City. It is reprinted here by permission of the author.

Three years ago, almost to the day, Newton Minow stood before you and urged a concerted effort on your part to produce more and better children's programs. Your group proceeded forthwith out of business—reorganized itself, and adopted a new name. The next year Chairman Minow returned, delivered a stimulating oration, and three months later the trade press rumored that he was going out of business.

Last year I appeared for the first time before you and made a modest proposal concerning the limitation of radio and television commercials. That got about as far as Jonathan Swift's proposal in 1729 to relieve the Irish famine by killing all newborn babies. But as his proposal eventually prompted a few souls to combat starvation, I'm hoping Commission history will eventually reflect action to combat a further rise in commercial feasting.

At any rate—I approach this particular podium with some trepidation, for it's obviously loaded, and I'm not sure whether it's pointed in your direction or in mine!

But one thing we know for sure—the IRTS podium is a sturdy one, for each year it holds three score or more of the most important figures in the broadcasting business. It's an honor to share again this platform with them, and a pleasure to have the opportunity to try out an idea or two on such a knowledgeable audience.

Since we last met with you a year ago the Commission hasn't had a dull moment. We have stirred up some controversy, re-surveyed old lands, and plowed a new acre or two.

In the area of the fairness doctrine we have issued our long-awaited primer, and we hope it has, in addition to making you write your lawyer, been of some guidance.

In the *Voice of Cullman* case, we held that a broadcaster's obligation to be fair—to deal with both sides of a controversy—was not ultimately dependent on his ability or inability to obtain sponsorship for either or both sides. In so doing, we simply reaffirmed the paramount right of the public to be informed—the right to hear spokesmen who could not afford to purchase time, as well as those who could. From the public's point of view nothing could be fairer than that, and I think if you examine the ruling closely, you'll agree with our unanimous decision.

During the last year we have also worked continuously for the promotion and development of additional television stations in the UHF band, and to foster the goals of the all-channel law. As you know, the VHF theater is packed as solidly as a Saturday matinee of *Hello Dolly*—and several late applicants are fighting for standing room only. So UHF alone can provide the multiple outlets—and the potential for program diversity—that our system now lacks.

We think we are going to be successful, and that UHF—although in its second appearance on the track after falling by the wayside in the first heat—is off to a good start and will finish the race.

In our continuing efforts to enact rules and establish policies that will be conducive to good broadcasting, we are engaged in a number of important studies—some new, some continuing. Chief among these are:

. . . Our study of proposals for the regulation of community antenna television systems and pay television;

. . . Multiple ownership of broadcast facilities, and the over-all question of undue concentration of control in the ownership of mass media;

. . . Control by networks of the sources of program supply;

. . . And of course, that perennial favorite of mine, the problem of loud commercials.

The Commission just yesterday took to the field to do some on-the-spot investigation of the practices and techniques of audio control for both live and recorded commercials, and was, I believe, much enlightened. However complex this problem may be, mutual effort on your part and ours should solve it. Some of you have been most helpful to us in our efforts to tackle it, and we are indeed grateful.

Incidentally, I ran across a poem the other day which was listed as being of anonymous American origin, and which obviously antedates both radio and television. It is reportedly a great favorite of the Duke of Windsor, and goes as follows:

He who whispers down a well
About the goods he has to sell
Will never reap those golden dollars
Like him who shows them round and hollers.

Now I realize that most broadcasters deny that there's any "hollering" in radio and television. To them, I would only repeat the story of the sophisticated resident of New York City who was asked whether he believed broadcasters were experimenting with subliminal advertising. "I didn't at first," he replied. "Then one day I went out and bought a tractor—for my wife."

There are many other matters vital to the healthy growth of broadcasting, and to your special interest: television. But today I want to talk about only two: educational television and money.

I know that only one of these holds a burning interest for you—and that, of course, is educational television. But I want to assure you that I have a good reason for talking about both.

I raise the subject of ETV with this particular audience to emphasize its overwhelming importance to the future of the television medium in this country. And later on, I want to suggest to this audience that you have a special stake in this subject.

I raise the question of money simply because educational television programs are not produced in a vacuum. In that medium, as in commercial television, if you want to dance, you have to pay the fiddler.

Demands for a supplementary, educational broadcasting service are far from new. Three decades ago, in 1934, the Congress was considering the bills that became the Communications Act of that year. Father John B. Harney, Superior of the Paulist Fathers, then proposed an amendment requiring the government to allocate one-fourth of all broadcasting facilities to "education, religious, agricultural, labor, cooperative, and similar nonprofit making associations."

Senator Fess of Ohio, supporting Father Harney, said: "Ever since the radio has been an agency of communication there has been complaint about the slight attention given to matters of an educational character, cultural, as well as religious."

The then young commercial broadcast industry fought Father Harney's proposal. Acknowledging the "manifest duty" of the FCC to require public service programming from commercial stations, industry representatives argued that nothing more was necessary.

And the industry won its fight.

Meanwhile, its technicians were constructing the world's finest broadcasting system, which was to produce some of the world's finest programs. Yet, to paraphrase a characterization once made of Hollywood, broadcasters sat down at their magnificent Steinway and—more often than not—played "Chopsticks."

As a result, one of the characteristic aspects of the American broadcasting scene has been the tension between those who sit in your chairs and those who sit in mine. Trying to carry out our "manifest duty," we push for public service programming which often costs you time, effort, and money; and you—not unnaturally—are often reluctant dragons. We shake the finger and lift the eyebrow. You holler "censorship!" and wave the Constitution.

There are times when tension ebbs—sometimes for years. But the troops regroup, particularly when public support for either side appears, and the battle rages anew.

Meanwhile, the Commission has come to accept the principle behind Father Harney's 30-year-old proposal. It has reserved a substantial portion of the available assignments for the specific use of noncommercial, educational organizations. We have 293 radio and 93 educational television stations now on the air. The federal government's program under the Magnuson Act, of matching grants for the construction of educational stations, is well under way. That far-sighted piece of legislation has been a tremendous impetus to ETV, and predictions that we will have over two hundred educational television stations by 1970 are not idle.

"Well," you may say, "What's the problem? If the commercial system is inadequate, we have a noncommercial system. What more can you ask?"

The problem, ladies and gentlemen, is money.

Over three-fourths of the 93 operating educational television stations are on the air 5 days or less per week, and many of these operate only a few hours each day. Even well-known stations in large communities, such as WNDT in New York City, have recently been forced to curtail their operations for want of money.

According to a recent survey, the median educational television station had about $100,000 per year to spend on all of its programs—both for in-school instruction and for the general audience. Again speaking in very rough but indicative terms—this amounts to some $100 to $200 per broadcast hour. I leave it to your experience and judgment to determine what kind of programs can be provided by individual stations on a budget of $100 to $200 per hour.

And what about the cost of programming for nationwide distribution?

In the current season, by cutting its evening offerings from 10 to 5 hours per week, the National Educational Television Center has been able to spend a little over $19,000 on each hour of nighttime programming. In the same season, the three commercial television networks are spending an average of about $125,000 on each hour of prime time programming—over six times as much. This disparity is compounded by the fact that each commercial network produces about five times as much evening programming as does N.E.T.

Those of you who dismiss educational television programming as dull and unimaginative—who point out how often people with college educations prefer commercial programs to those of educational stations—should keep these rather startling cost differentials in mind. Available money, of course, doesn't guarantee success, nor does good programming in every instance require it. Despite the vast sums poured into commercial television programs, including the enormous price that talent and talent agencies bring in today's market, a hefty percentage of these programs go off the air after the first season. But there comes a point at which it's futile to expect brilliance, imagination, artistry, and boldness to be long supported by a bootstrap tied to a shoestring.

The truth is that educational television has accomplished wonders with the resources at its command. But the time has come to say that it will never realize its full potential until its financial base rises to a radically new level. While that level does not have to be anywhere near the posh plateau inhabited by commercial television, it must provide support for good programming on a realistic basis. That educational television should permanently struggle for subsistence is intolerable.

Let's look at some informal estimates on programming costs.

For educational programming on a nationwide basis, and programming alone, most estimates are in the range of 20 to 25 million dollars annually. If a national, interconnected network of educational television stations is desirable (and respected educational broadcasters differ on this question) an additional 6.5 to 7 million dollars per year should be added. These estimates do not include research, administration, and a host of other overhead costs. For an additional 30 hours per week of local programming, estimates range from $150,000 to $200,000 per station per year.

So, the production of television programs is no dime-a-dance proposition. Sums of the size just mentioned bring into sharp focus the gluttonous appetite of the fiddler.

The current production of many fine programs on educational television—at the national and at the local level—is a glowing tribute to their producers and station personnel. But we are deluding ourselves and the public when we think and talk of educational television as a broad alternative and satisfactory supplement to the public service programs of commercial stations—when we paint shining word pictures of a different kind of television, free from the pressures of a commercial system and open to the creative talent of a growing America—if at the same time we claim it can grow and prosper within its present extremely limited financial framework.

So where does the money come from?

Many different solutions have been suggested from time to time.

While we now all agree that a public service program is not necessarily sullied by sponsorship, the proposal that educational television sell time to advertisers has some obvious drawbacks.

Program underwriting—where credit is given to the commercial producer, without advertising—is permitted. But it is self-evident that underwriting is a program-by-program approach which cannot be the major answer to the question of money for educational television.

Letting educational stations use part of their time for pay television has also been suggested. But again, if we are talking about a "television correspondence school" type of venture, it is unlikely that massive funds would be generated. And broader pay television ventures—such as first-run movies—pose a number of well-publicized problems, apart from the complications that an educational setting would add.

On the whole, then, the hope that educational television might find the needed support by selling a service in the marketplace offers little promise of filling the till. We are thus left with two alternatives: (1) the generosity—charity, if you will—of the public, of foundations, and businesses—including the broadcasting business; and (2) the taxing powers of the local, state, and federal governments.

Let us quickly agree that the greater of these is charity.

For this reason, we need more effort—not less—in the sphere of voluntary donations. Many people in the broadcasting industry—many, in fact, in this room today—have given generously. They need to give more, and others should join them.

The Ford Foundation has single-handedly performed a giant task in this field—one that has put us all in its debt. Hopefully, it will continue and broaden its efforts, and other foundations will find new and distinctive ways to make their contributions.

In May of 1963, a national conference on the economics of educational television was held at Brandeis University. That conference recommended the establishment of a private, nonprofit corporation which would—among other things—coordinate national fund-raising efforts on behalf of educational television. We need that agency.

In the same month (May 1963) Frank Stanton proposed "a mighty annual campaign" on behalf of educational stations, one that "ought to involve not just the educational and television worlds, but the churches, the civic organizations, the business community—the people." We needed that campaign when he suggested it, and we still need it.

The National Educational Television Center under the able guidance of Jack White has done a tremendous job. Local stations such as San Francisco's KQED and others, have had astounding success in local fund-raising campaigns. We need them too.

In short, the charitable impulse can be productive as well as noble. The number and variety of worthy activities that the people of this country support on a voluntary basis constitute one of the best refutations of the often-heard remark that American civilization is wholly dominated by crassness and materialism.

Indeed, no one can seriously quarrel with the proposition that private sources of funds are the best sources for educational television programming—if they can do the job.

And the multiplicity of financial sources for educational television will always be the best guarantee of its independence.

But we are a long way from the goal. While the sums required for ETV are small in comparison with our nation's wealth, they are large in the world of charity and voluntary giving. We simply cannot leave other alternatives unexamined and unexplored.

One source of supplemental funds for educational television programming is local and state governments, which already provide the bulk of its support. If we must have government in the picture, the natural impulse is to turn to government at the level nearest the communities, and nearest the educational institutions most directly concerned. Moreover, local and state governments are already in the field.

This approach has considerable appeal to me, as I'm sure it does to you. Certainly, state and local governments will always be the primary source of funds for strictly instructional television programming.

On the other hand, it is difficult to quarrel with a recent article in *Business Week* which stated:

> State-local expenditures doubled between 1950 and 1960—to about $51-billion—and they are expected to double that by 1970. This includes spending under federal grants. * * *
>
> The trouble is, state and local governments are bumping against the ceiling of their present powers of taxation, and there are limited sources of revenue open to them. * * *
>
> Anyway it happens . . . a solution cannot come too soon for embattled state and local authorities.

Statements such as this force us to examine ways in which the national government might finance that part of an educational television system which is truly national in scope.

It has been recommended, by Hartford Gunn of WGBH, as I indicated earlier, by the Brandeis Conference, and probably by others, that a private national agency be created to coordinate national fund-raising for educational television from both private and public sources. Presumably, members of such a board could be elected by the local educational stations themselves—leaving open, for the moment, the question of whether to apply the principle of one station—one vote! And as you know, Walter Lippmann and others have suggested a federally endowed educational television system, along the lines of land grant colleges.

Other suggestions include annual appropriations by Congress, either directly to the local stations, or to a national agency of the kind mentioned.

Still other sources of federal funds have been proposed from time to time. Senator Dill and, later, John Fischer of *Harper's Magazine*, have proposed a rental or fee system for the use of commercial broadcasting frequencies, with the proceeds to go for noncommercial programming. It has also been suggested that the fees the FCC now charges for the filing of broadcasting applications be automatically turned over to educational television.

Underlying all these possibilities for federal support are two basic questions: First, would the federal government be mature enough, and wise enough, to support a broad educational and cultural service without placing restrictions on the nature of the product? Second, would a strong commercial television industry be ready at all times to defend against governmental interference the right of its noncommercial brothers to be daring—to be free-wheeling and free-thinking.

These are the major alternative sources of ETV's finances. All have their drawbacks and difficulties. It is for this very reason that we have a problem. Nor would I claim that I have the answers.

But this I do know—we must not fail to supply educational television, one way or another, with sufficient funds. The result is far too important for the future of this industry and this country.

In a nation committed to eliminating poverty in all its forms, we should not overlook what might be called "cultural poverty." For the vigor of a democratic civilization depends upon the availability of knowledge and enlightenment, culture and beauty—not to an elite, but to all who want to learn—to all whose minds are undernourished. Our public school system, our free libraries and museums—all these are monuments to our belief in this proposition. And a critical function of educational television is to serve those who seek enlightenment on all subjects from the beauty of a line in a painting to the fascinating story of mankind's rise from caveman to astronaut. The elimination of cultural poverty is surely one of broadcasting's major responsibilities. And I would suggest that you who are familiar with the awesome power of both the atom and the electron tube might look with new conviction at the sobering thought once expressed by Thomas Carlyle: "That there should one man die ignorant who had capacity for knowledge, this I call tragedy."

The time for critical and broad-range thinking on this subject is now. The next decade will see the basic physical plant of our educational television system substantially completed. We must plan now if that plant is to be put to fullest use.

In these circumstances we cannot let our thinking be paralyzed by simple fear of the future. In dealing with the problems raised by each of the alternatives I have mentioned, we must not succumb to what Larry Laurent has called the habit of "manic extrapolation—the reckless projection of present trends into future catastrophe." Rather, having recognized potential dangers and diffiiculties, we must look for ways to overcome them. For as Justice Brandeis once wrote: "If we would lead by the light of reason, we must let our minds be bold."

Finally, and most importantly to this audience, many of the answers lie with all of you in this room today.

There is nothing startling in the thought that those who reap a profit from the use of broadcasting frequencies have a special obligation to contribute to the support of educational television. If you have already done much, you should do more. Educational television is entitled to look to you for a portion of its financial support.

Nor is there anything unusual in the idea that you have a special ability to help. You operate the most powerful selling instrument ever invented. If it can sell soap, automobiles, and potato chips, it can certainly sell an idea with the intrinsic merit of educational television. You have the know-how and you have the facilities.

So, why should this organization not take the lead in carrying out Dr. Stanton's proposal of May 1963 that commercial broadcasters help educational stations plan and launch an annual campaign for funds? And if not this organization, why not the NAB or an *ad hoc* committee, with representatives from networks, stations, advertising agencies, program producers, and every other element of this great industry?

Beyond this, you have a broader capacity to help. This cause needs ingenuity and enthusiasm, and you have a plentiful supply of both these qualities—in behalf of causes close to your hearts. I have seen it close up. So I urge you: put this cause close to your hearts. With your active sympathy and support it can succeed.

Why should you do all this? What is your stake?

Well, there are some obvious reasons. You are part of American television, and television as a medium needs the contribution that only a noncommercial service can make. You have a need for competition from telecasters who are not bound by the inevitable pressures of the marketplace, for such competition raises the level of the entire medium. You also have a need for a place in the medium where new ideas and techniques can more easily be tested—where there is no automatic penalty for failure to attract a maximum audience.

There is another less obvious reason: For thirty years your public service obligations have been the subject of controversy and debate. The creation of a meaningful national educational system can have a real impact on the course of this debate. Not that it would take over your public service role. I think we can all agree that it would be unwise to create a situation in which the medium had only one source of education and enlightenment, one source of broad information on public affairs, and one source of cultural programming. And public service programs, as well as sponsors, should be able to take advantage of the strong lead-in a commercial television schedule can supply.

There is no doubt in my mind, however, that we would all be in a better position to understand commercial television's true public service role if there were a complementary national television service, devoted full time to the needs and interests which your primary mission often leads you to skimp or ignore. With such a yardstick, we would know more about what you should and should not be expected to do. Your responsibilities could be better defined, as could the tasks that only a specialized service can undertake. The problem of your proper functioning would lose, I think, many of its angrier and more difficult overtones. It would become less a field of battle and more a field for growth.

Until that day comes, we are all in the hot kitchen made famous by President Truman. Part of the FCC's job is to keep the temperature in the broadcasting oven at an appropriate level. That we are doing and shall continue to do. But the door to the kitchen isn't nailed shut. I invite you to help us open it to the fresh winds of the future.

6. Insights into Broadcast Industry Profits

by Emanuel Celler

The following is an excerpt from the hearings conducted before the Antitrust Subcommittee (Subcommittee No. 5) of the Committee on the Judiciary, House of Representatives, Eighty-Fourth Congress, Second Session. The hearings were conducted under the auspices of the Chairman, Representative Emanuel Celler, Chairman of the House Judiciary Committee. At the time of the hearings, George C. McConnaughey was Chairman of the Federal Communications Commission.

THE CHAIRMAN: First with respect to revenues, it is correct, is it not, that for 1955 Columbia Broadcasting System and the National Broadcasting Co. networks and their nine wholly owned television stations had broadcast revenues of $312,658,470?

MR. MCCONNAUGHEY: Yes, sir.

THE CHAIRMAN: That is correct?

MR. MCCONNAUGHEY: Yes, sir; that is correct.

THE CHAIRMAN: It is also correct, is it not, that these revenues constituted 41.99 per cent of the revenue for the entire television industry?

MR. MCCONNAUGHEY: That is correct, sir.

THE CHAIRMAN: It is also correct, I take it, that the revenue of CBS plus its 4 television stations excluding radio was $153,614,317 and was equal to 20.63 per cent of the revenue of the entire television industry?

MR. MCCONNAUGHEY: That is correct, Mr. Chairman.

THE CHAIRMAN: It is also correct that CBS revenue from network operations in 1955 was $121,953,917, and NBC revenue from network operations was $124,353,526?

MR. McCONNAUGHEY: That is correct, sir.

THE CHAIRMAN: And it is correct that the combined revenue from CBS' and NBC's network operations in 1955 was 87.2 per cent of the broadcast revenue of all television networks?

MR. McCONNAUGHEY: That is correct.

THE CHAIRMAN: Now turning to the net income before federal income tax of CBS and NBC operations, it is correct, I take it, that in 1955 the CBS and the NBC networks together with their 9 stations had net income before taxes of $65,050,186?

MR. McCONNAUGHEY: That is correct, sir.

THE CHAIRMAN: And that this is 43.4 per cent of the net income before taxes of the entire television industry?

MR. McCONNAUGHEY: Yes, sir.

THE CHAIRMAN: That CBS and its 4 stations had a net income of $34,-870,837 or 23.2 per cent of the income of the entire industry?

MR. McCONNAUGHEY: That is correct, sir.

THE CHAIRMAN: And that NBC with 5 stations had a net income of $30,179,349 or 20.1 per cent of the income of the entire industry?

MR. McCONNAUGHEY: Yes, sir.

THE CHAIRMAN: Now I should like to ask you some questions concerning ratio of 1955 income to total investment in broadcast property as of December 31, 1954.

First, it is true, is it not, that as of December 31, 1954, CBS and NBC, plus their 9 owned stations, had an investment in broadcast property of $50,067,737?

MR. McCONNAUGHEY: That is correct, sir.

THE CHAIRMAN: That is correct?

MR. McCONNAUGHEY: Yes, sir; it is.

THE CHAIRMAN: All of these figures are taken from your record?

MR. McCONNAUGHEY: Yes; they are.

THE CHAIRMAN: This is a reaffirmation. You have already testified, have you not, that in 1955 CBS and NBC, plus their 9 owned stations, had a net income before taxes of $65,050,186?

MR. McCONNAUGHEY: Yes, sir.

THE CHAIRMAN: This means, does it not, that in 1955 CBS and NBC recovered, both these systems recovered, 131 per cent of their total investment in broadcast property?

MR. McCONNAUGHEY: Yes, sir, that is correct.

THE CHAIRMAN: That is, in 1 year, in 1955, they recovered back 131 per cent of their total investment in broadcast property?

MR. MCCONNAUGHEY: That is correct, Mr. Chairman.

THE CHAIRMAN: It is also correct, is it not, that in 1954, CBS and NBC recovered 99 per cent of their total investment in broadcast property? I refer to the Bricker report, appendix chart 1; that is correct, isn't it?

MR. MCCONNAUGHEY: Yes, sir; it is correct.

THE CHAIRMAN: And it is correct that CBS and NBC is 1953 recovered 53 per cent of their total investment in broadcast properties?

MR. MCCONNAUGHEY: Yes, sir.

THE CHAIRMAN: Now I should like to direct your attention to the rate of return of CBS on its investment.

It is correct, is it not, that in 1955 CBS, plus its 4 television stations, reported a net income of $34,870,837?

MR. MCCONNAUGHEY: Yes, sir; that is correct.

THE CHAIRMAN: And that, as of December 31, 1954, CBS had a total investment in broadcast property of $26,958,279?

MR. MCCONNAUGHEY: That is correct, sir.

THE CHAIRMAN: So therefore, in 1955 CBS in its network and station operations earned 129 per cent of its total net investment?

MR. MCCONNAUGHEY: That is correct.

THE CHAIRMAN: In 1 year it got back 129 per cent of its total investment, and that year was 1955?

MR. MCCONNAUGHEY: That is in tangible broadcast property; yes.

THE CHAIRMAN: That is what I mean, tangible broadcast property. In 1954 CBS and its television stations had a return of 108 per cent of its total net investment; is that right?

MR. MCCONNAUGHEY: That is correct.

THE CHAIRMAN: It is also correct that in 1953 CBS earned 54 per cent return on its net investment?

MR. MCCONNAUGHEY: That is correct, sir.

THE CHAIRMAN: Now turning to NBC, it is correct, I take it, that in 1955 NBC, plus its 5 television stations, had a net income before taxes of $30,179,349?

MR. MCCONNAUGHEY: That is correct.

THE CHAIRMAN: It is also a fact that as of December 31, 1954, NBC had a total investment in broadcast property of $23,109,458?

MR. MCCONNAUGHEY: That is correct.

THE CHAIRMAN: So that in 1955, this same year, NBC and its owned stations had a return before federal income taxes of 133 per cent on its investment.

MR. MCCONNAUGHEY: That is correct.

THE CHAIRMAN: And in 1953 that same company had a return of 52 per cent of its investment?

MR. MCCONNAUGHEY: That is correct, sir.

THE CHAIRMAN: Now I want to ask you some questions on investment of network-owned television stations, stations owned by CBS and NBC. First, it is correct, is it not, that the 9 stations owned by CBS and NBC earned an income in 1955, before taxes, of $30,081,992?

MR. MCCONNAUGHEY: That is correct, sir.

THE CHAIRMAN: And that as of December 31, 1954, CBS and NBC had an investment in their stations of $9,973,056?

MR. MCCONNAUGHEY: That is correct.

THE CHAIRMAN: So, therefore, it is correct, is it not, that in 1955 the television stations owned by CBS and NBC earned 307 per cent of their total investment in broadcast property?

MR. MCCONNAUGHEY: That is right, sir.

THE CHAIRMAN: It is correct, too, is it not, that in 1954 the TV stations owned by those 2 companies earned 330 per cent on their investment?

MR. MCCONNAUGHEY: That is correct.

THE CHAIRMAN: And in 1953 those same companies earned 239 per cent of their investment?

MR. MCCONNAUGHEY: That is correct, sir.

THE CHAIRMAN: Turning to the return on investment of television stations owned by CBS, it is correct, is it not, that in 1955, the same year, the 4 television stations owned by CBS earned 282 per cent on investment?

MR. MCCONNAUGHEY: That is correct, sir.

THE CHAIRMAN: It is also true that in 1954 stations owned by CBS earned 370 per cent on investment?

MR. MCCONNAUGHEY: That is correct.

THE CHAIRMAN: And in 1953 the return was 226 per cent on investment?

MR. MCCONNAUGHEY: That is correct, sir.

THE CHAIRMAN: Now as to NBC, in 1955 NBC's 5 television stations earned 335 per cent on investment?

MR. MCCONNAUGHEY: That is correct, sir.

THE CHAIRMAN: And that in 1954 television stations of NBC yielded 297 per cent on investment?

MR. MCCONNAUGHEY: That is right, sir.

THE CHAIRMAN: And in 1953, 251 per cent on investment?

MR. MCCONNAUGHEY: That is correct, sir.

THE CHAIRMAN: Now at this point I offer in evidence charts showing 1955, 1954, and 1953 broadcast revenues, expenses, and income of the entire television network industry, broken down by television networks, television-network-owned stations, and television networks plus their 16 network-owned stations.

MR. McCONNAUGHEY: You refer of course, when you say "investment," to tangible broadcast properties?

THE CHAIRMAN: That is right; I limit myself to that, naturally, which is their investment of course.

MR. McCONNAUGHEY: That is correct.

THE CHAIRMAN: That does not take into consideration whatever investment they have, if any, in talent and things of that sort. We are speaking of net returns; and, in those net returns, expenses for talent and other costs of operation are considered?

MR. McCONNAUGHEY: That is right, sir.

THE CHAIRMAN: I would like to ask you about financial data with regard to the network-owned stations. You have made available to this committee at its request, certain financial data on network-owned television stations.

Now in 1954 CBS owned three television stations, namely, KNXT in Los Angeles, WBBM in Chicago, WCBS in New York. That is correct, is it not?

MR. McCONNAUGHEY: Yes, sir.

THE CHAIRMAN: The net income of these three stations in 1954 before federal income taxes was $12,276,443?

MR. McCONNAUGHEY: Yes, sir.

THE CHAIRMAN: The net investment in tangible broadcast property of these three stations as of December 31, 1953, was $3,332,023; is that correct?

MR. McCONNAUGHEY: Yes, sir.

THE CHAIRMAN: This means, does it not, that in 1954 the 3 CBS television stations recovered 370 per cent of their total investment in broadcast properties?

MR. McCONNAUGHEY: That is correct, sir.

THE CHAIRMAN: In 1955 CBS owned four television stations, KNIT, Los Angeles, WBBM, Chicago, WXIX, Milwaukee, and WCBS, New York. That is correct, isn't it?

MR. McCONNAUGHEY: Yes, sir.

THE CHAIRMAN: The net income of these 4 stations in 1955 before federal income taxes was $14,505,459?

MR. McCONNAUGHEY: That is right.

THE CHAIRMAN: The net investment in tangible broadcast property of those 4 stations as of December 31, 1954, was $5,146,981?

MR. McCONNAUGHEY: Yes, sir.

THE CHAIRMAN: This means, does it not, that in 1955 the 4 CBS television stations recovered 282 per cent of their total investment in broadcast property in that 1 year?

MR. McCONNAUGHEY: Yes, sir; that is right.

THE CHAIRMAN: Let me ask you about just one CBS television station particularly; namely, WCBS in New York.
Isn't it a fact that WCBS in New York had a net income before federal income taxes in 1953 of $5,571,777?

MR. McCONNAUGHEY: That is correct, sir.

THE CHAIRMAN: And that WCBS, the same station, had a net investment in tangible broadcast property as of December 31, 1952, of $528,911?

MR. McCONNAUGHEY: That is right, sir.

THE CHAIRMAN: Which means, does it not, that WCBS recovered 1,053 per cent of its total investment in broadcast property in 1953?

MR. McCONNAUGHEY: That is correct, sir.

THE CHAIRMAN: In 1954 WCBS had a net income of $8,206,416 before federal income taxes?

MR. McCONNAUGHEY: Yes, sir.

THE CHAIRMAN: And that WCBS had a net investment in tangible broadcast property as of December 1, 1953, of $447,420?

MR. McCONNAUGHEY: That is correct, sir.

THE CHAIRMAN: This means, does it not, that in 1954 WCBS made 1,834 per cent on its total investment in broadcast property?

MR. McCONNAUGHEY: That is correct, sir.

THE CHAIRMAN: In 1955, WCBS had a net income before federal income taxes of $9,375,339?

MR. McCONNAUGHEY: Yes, sir.

THE CHAIRMAN: And that WCBS had a net investment in tangible broadcast properties as of December 31, 1954, of $409,484?

MR. McCONNAUGHEY: That is correct, sir.

THE CHAIRMAN: This means, does it not, that in 1955, and I give emphasis to these figures, WCBS recovered 2,290 per cent on its total investment in broadcast property?

MR. McCONNAUGHEY: That is correct, sir.

THE CHAIRMAN: You would say, would you not, those are high profits?

MR. McCONNAUGHEY: Extremely high profits.

7. Television's Role in the American Democracy

by Robert W. Sarnoff

Mr. Sarnoff, now President of RCA, was Chairman of the Board, National Broadcasting Company, when this address was given. The address reprinted here was delivered before the Chicago World Trade Conference, Chicago, Illinois, on March 5, 1963. It is reprinted here through special permission granted by Mr. Sarnoff and NBC.

It is a high honor to be asked to speak before this knowledgeable audience, and I am grateful for your challenging invitation.

In approaching my assignment this evening, I am mindful of the many eminent men of government and industry who have occupied this rostrum in the past. This fills me with a sense of modesty, and brings to mind a remark made by another speaker on a different occasion. That, as you may recall, was when Sir Winston Churchill said of his personal friend and political foe, Clement Attlee: "He is a modest man with much to be modest about."

Since your last annual conference, the world has experienced both change and stalemate, and from the vantage point of the West, a normal complement of frustrations. There are fresh Indian graves in the Himalayas. Draining jungle wars continue in Southeast Asia. The Wall still stands in Berlin, and Cuba remains a Communist fortress in our hemispheric seas. Even the rupture between the two goliaths of world Communism was prompted by how—not whether—to bury us.

Yet, the past year has also seen resolute strides by those nations with a commitment to freedom. In no section of the earth did we yield peoples or principles to Communism, and we recently passed the eyeball test without a blink. In such critical outposts as Formosa and West Berlin, our posture is stronger, not weaker, than a year ago. In the contest beyond the earth's atmosphere, our astronauts brought us nearer to competitive parity with Russia; and our unmanned satellites, such as Telstar, Tiros, and Relay, gave us clear leadership in global space communications.

But if I were to single out one event of paramount significance in the last year, it would be the performance of the American economy. Its continued resilience and strength, its ability to weather the worst market collapse in thirty-three years and then resume its forward progress, were more meaningful than any political event. There would be no free Berlins, no pacified

Congos, if this powerful machine of individual and competitive enterprise were to falter and to fail.

In a recent, eloquent statement President Kennedy said:

> We shall be judged more by what we do at home than what we preach abroad. Nothing we could do to help the developing countries would help them half as much as a booming United States economy. And nothing our opponents could do to encourage their own ambitions would encourage them half as much as a lagging United States economy.

FREE SOCIETY—FREE ENTERPRISE

It is to the indivisible goal of keeping our economy strong and our society free that I would like to address myself. And I hope it will not be regarded as immodest of me to suggest that the industry I represent—television—plays a role of decisive importance in stimulating economic growth and in reinforcing the strength of our democratic process.

Essentially, democracy is a union of two concepts. Television was born of both and supports both.

One is the concept of free expression which the late Judge Learned Hand characterized as "brave reliance upon free discussion." Rooted in tradition and sheltered by law, it holds that citizens of a democracy, given free access to knowledge, and freedom to discuss issues and views, can best judge their own interests and best guide their own destiny.

The other is the concept of a competitive free enterprise economy as best calculated to meet the needs of the individual and the nation. It has a dual premise: that open competition for public favor spurs the constant improvement of goods and services; and that the encouragement of mass demand sparks mass production, which, in turn, decreases the cost and increases the availability of these goods and services.

Both of these principles center on the individual as the master, not the servant, of the state; and both support the conviction that he can best realize his aspirations through ways of his own choosing. The opposite is true of the closed society, where the state is the master, controlling personal expression, political choice, and all economic activity. The combat between the two systems is waged at every level—not only as a war of ideas, but as a war of economic strength.

Historically, free enterprise and democracy have nourished one another. The revolutions that led to the modern political systems of the West also fostered the rise of mercantile enterprise, the forerunner of the modern competitive free economy. Up to this day, those nations achieving the highest degree of consumer-oriented industrialization have also attained the most effective self-government.

Television's role in supporting this economic and political process is often obscured by the pervasive yet intimate nature of the service it offers. Most people have strong and subjective programming likes and dislikes.

They might love the *Beverly Hillbillies* and be bored by the NBC *Opera*, or vice versa. They might become irritated by a commercial, or by a newscaster's comments on a subject where they have a preconceived judgment. The net effect—and this is perfectly natural—is that their personal preferences tend to eclipse a broader understanding of the medium's catalytic function in our society. I suggest the time is overdue for thoughtful Americans to begin evaluating the total dimension of the television service.

Its physical dimension is that of a service meeting so many needs and demands that in the United States in the last dozen years its circulation has grown from 10½ million sets to nearly 59 million. The number of television stations, both commercial and noncommercial, has increased from 107 to 647. And television advertisers have expanded their annual expenditures from $332 million to $1¾ billion.

Our technology and programming have also provided substantial impetus to the growth of television abroad, in both the established and the emerging nations. From 1951 through 1961, the last year for which figures are available, the number of sets outside North America grew from 1.2 million to 54 million. At the current rate of growth, the total will probably exceed 74 million by the end of this year.

To understand television's economic role, one must first relate it to the nature of our economy. Economic growth, as you who live by trade are well aware, hinges on mounting production and a high level of employment, both stimulated by increased consumption. In a free economy, production expansion depends primarily on rising consumer demand; and in the mature American economy, rising demand requires, in addition to population growth, the continuous stimulation of consumer desires.

AN ECONOMIC FORCE

The primary stimulant is advertising, and among all forms of advertising, television has unique capabilities that power the American economy. For television is more than an advertising tool; like advertising, it creates demand, but with sight, sound, color, and demonstration, it goes further and functions as a direct selling force. Its sales messages reach millions simultaneously, yet with the personal persuasion of one individual speaking to another in his home. With its ability to show not only what a product is but what it does, television has given American industry a powerful means of sustaining traditional consumer demands and developing new ones. It also has speeded and streamlined the distribution process. This is a contribution of particular value to our economy, where distribution cost is so important an element of end-cost to consumers.

Television's sales impact has contributed to, and has been accompanied by, a marketing revolution in which the primary selling function has shifted from the dealer to the manufacturer. In the past, the dealer had the responsibility for developing the manufacturer's market. Today the manufacturer helps create the market for the dealer by speaking directly to his customers. He does this not only for consumer purchases but also, in increasing degree, for the sale of those products that are purchased for the ultimate user by someone else, such as plywood and plate glass for the home and aluminum for the automobile.

This ability of the manufacturer to engage in mass selling as well as mass production—whether of packaged goods on the shelves of supermarkets or automobiles or home appliances—has given our economy a highly effective means of continuous expansion.

It is against this perspective that criticism of an advertiser-supported television system should be considered—criticism which claims that the marketing function of the medium prevents it from properly discharging its program function in serving the audience. To my mind, there is no inconsistency, but a close parallel, between these functions. Both seek to engage the interests of large audiences, and this is a valid goal of a mass medium of entertainment and information, quite apart from its marketing role. Additionally, television recognizes minority interests and in doing so, it also serves the advertisers interested in such specialized audiences.

The debate over whether television strikes a proper balance between broad and specialized interests turns on a matter of degree. If there is such a thing as a perfect and ultimate balance, I will not claim that we have reached it. Yet the debate tends to lose sight of an undebatable fact—the basic contribution commercial television makes to the national economy; and the paramount need for national economic strength in preserving the institutions of our free society.

The premise of growth in the American economy is consumption—a principle underscored by the President in urging that billions of dollars be released to stimulate spending by private consumers, private investors, and corporate enterprises.

Is the stimulation of private spending incompatible with meeting our public responsibilities? History argues otherwise, for as our consumption has increased, so has our allocation for essential services: billions of dollars for education, for social security benefits, for public welfare and old-age assistance, for highways and police and fire protection. While consuming more, we have paid the highest taxes in peacetime history, fought a war in Korea, given billions in foreign aid and maintained and strengthened our global defense structure, so vital to the survival of the free world.

PERSONAL DESIRES AND PUBLIC OBLIGATIONS

Fortunately, we Americans need not choose between satisfying our personal desires and fulfilling our public obligations. We are spared such a choice by a rare, perhaps unique combination of blessings: our vast natural resources, our unsurpassed technological skill, and a free and expanding economy based upon prosperity through consumption. Thus we can accomplish both goals. We can enjoy all the things that make work easier and leisure more fun—and at the same time meet the needs of society and the demands of security.

But we can sustain this formula, I believe, only if we maintain a protective and jealous attitude toward those institutions that make it possible. To do so, we must understand the nature of the political and economic forces that shape our environment. This is the function of free media of communications in a free society, anchored in the Jeffersonian conviction that men are inherently capable of making proper judgments when they are properly informed.

In this dimension of its service, television—alone of all media—is capable of bringing the sight and sound of great events of our time directly and instantaneously to nearly every man, woman, and child in the nation, whether the occasion is a national political convention or the tense drama of a manned space shot. It can and does place viewers in direct contact with the pressure of diplomatic crisis in United Nations debate and the violence of controversy over segregation on the University of Mississippi campus. And beyond showing and describing events as they occur, television has pressed the nation's search for truth through its documentaries and its debates on major public issues, such as social welfare, state legislative processes, our diminishing water resources, and legalized gambling.

As it has developed, television has properly intensified concentration on its journalistic function. For example, news and information programs account for more than 25 per cent of the total broadcast schedule of the NBC television network, and other networks and independent stations are also devoting increasing air time and creative effort to such presentation.

In addition to this concentration on equipping the citizen for more useful participation in society, the medium has forever altered the American political process. It has done so by presenting political candidates directly to the voters, culminating in the Presidential campaign of 1960 and "The Great Debate."

THE EQUAL TIME RESTRAINT

This unrivalled opportunity to assess the two major candidates took place only after the nation's broadcasters had won a Congressional respite from the equal-time law. Ironically, once the 1960 campaign was concluded, they were forced back into the legal straitjacket that makes the debates impractical by requiring equal time on the air for all Presidential candidates, no matter how quixotic their intent or meager their support.

I am hopeful that before the 1964 election campaign begins, the Congress will relieve the public and the broadcasters of this restriction. An early start has already been made in this direction. Chairman Harris of the House Interstate and Foreign Commerce Committee has introduced a

Resolution for suspension of the equal-time restriction to permit the 1964 Presidential and Vice-Presidential candidates to meet in face-to-face debate on the air. Yesterday, I testified in Washington in support of this proposal. However, I strongly urged that the Congress go further by eliminating, completely and permanently, the equal-time restraint which operates against the free flow of information in the crucial area of political judgment and choice. Given this freedom, and the responsibility broadcasters have already demonstrated in providing full and fair coverage, television could serve the voters at the state and local levels as it has served them nationally.

While conceding the immense value of the four confrontations of 1960 in exposing the Presidential candidates to intimate public scrutiny, some thoughtful analysts hold that they were neither "great" nor "debates." They argue that the format—which the candidates themselves helped develop—did not permit sufficient analysis of the issues for the guidance of the voter or, in fact, expose adequately what the issues were. This point of view is far from unanimous and addresses itself to method rather than principle, but I believe it is worthy of serious consideration.

Accordingly, in the expectation that the law will again be changed to permit debates between Presidential candidates, we should start now to refine the format of these televised encounters, seeking even more effective ways of assisting the American voter to make an informed choice.

A PLAN FOR 1964

As a major step in this direction, the National Broadcasting Company has enlisted the aid of the American Political Science Association, the nation's foremost professional organization devoted to the study of government, politics, and public affairs. I am pleased to announce that this distinguished organization has agreed to conduct an independent study, under a grant from NBC, to devise the best possible forms and procedures for televised political debates.

The Association has made many significant contributions to more effective government—its most recent, a widely acclaimed orientation course for new members of Congress, an innovation that is likely to become a Washington tradition.

It has selected a seven-man study group of distinguished political scientists and communications experts to carry out the project proposed by NBC. The group will be headed by the Association's president, Dr. Carl J. Friedrich, Eaton Professor of the Science of Government at Harvard University.[1]

---

[1] The other members of the study group are: Evron M. Kirkpatrick, Executive Director of the American Political Science Association; Harold Lasswell, Professor of

NBC's only participation in this study will be to underwrite the cost and provide necessary and basic information, including tape or film recordings of the 1960 debates. Whatever recommendations are arrived at will be the group's own, the result of careful, scholarly deliberation. By starting at this early date, the group will be able to present its findings well in advance of the 1964 Presidential campaign. I am confident that its proposals will be a major contribution to our democratic process.

Beyond equal time, we face the broader issue of whether any communications medium can effectively serve as an instrument of democracy if its freedom is curtailed. Today, television is fettered in many areas of journalistic enterprise. It cannot go wherever the public goes—into the halls of Congress, into public committee meetings of the House of Representatives, into most courtrooms.

Television does not seek this right in order to make a theatre of serious forums, and it recognizes the need for care and restraint. I emphasize, however, that wherever the public can attend, television should also be permitted to attend, so that it can serve as the eyes and ears of ALL the people. The right to witness public business should not be confined only to those whom the hearing room will hold, when television can bring the public business to everyone.

INSEPARABLE FREEDOMS

Whenever it serves, whether entertaining or informing, television functions best in a climate of freedom. It is paradoxical that in the area of news coverage, where television's need for freedom is recognized by all, restrictions on coverage should be placed through the equal-time penalty and the limitations on access. And it is even more paradoxical that among the strongest champions of television's freedom to report information and controversy without restraints are those who urge government restrictions on television entertainment. They would erect a double standard—one for information programs, another for entertainment programs—failing to recognize that freedom is indivisible. Would magazines and newspapers be free if only their news columns were unmolested, if the choice and content of features and fiction were subject to government influence, direct or indirect? But this is all part of that democratic process in which television was conceived. Often we who have the responsibility for guiding this service are accused of excessive sensitivity toward criticism. I assure you we welcome responsible criticism and take it seriously. It would be fatal to

---

Law and Political Science, Yale University; Richard Neustadt, Professor of Government, Columbia University; Peter Odegard, Professor of Political Science, University of California at Berkeley; Elmo Roper, Senior Partner, Elmo Roper and Associates; Gerhart Wiebe, Dean of the School of Public Relations and Communications, Boston University.

television's development if it were to operate in a vacuum of indifference and ignorance.

What we seek is understanding of the total dimension of our television service—its contribution to the political processes that keep us free, its impact upon the economic forces which keep America strong.

These essential and inseparable functions of political and economic freedom are the source of this nation's vitality and strength. Our capacity to support the arch of democracy both at home and around the world can be limited only if these freedoms are limited. It will grow only as we succeed in keeping them unencumbered. Thank you.

8. Forty Years of Progress

by Walter D. Scott

The following remarks are reprinted from "Forty Years of Progress," which was the keynote address given at the National Broadcasting Company's Annual Convention of Radio and Television Affiliates in Honolulu, Hawaii, May 11–13, 1966. The address is printed here in its entirety through special permission granted by Mr. Scott.

Although this is my first formal talk to you as NBC's Chairman I feel that the occasion is an extension of a long and happy association.

All of us sense that we are very much among old friends who have shared a common, fruitful experience and seek a common goal. The fact that this is a milestone meeting of the NBC affiliates is clearly evident, and it gives me great personal pleasure that we are all together, from the stations and the network, from radio and from television, for this fortieth anniversary celebration. We have come a long way to this gathering in our nation's newest state, just as we have come a long way together in broadcasting.

The nature of our unique association is hard to define for those who see it only from the outside. It is unlike any of the conventional relationships in business—for example, between manufacturers and distributors; movie studios and theatre owners; buyers and sellers; or principals and agents.

It is formalized by contract, but it is more intimate than the well-known parties of the first and second parts and needs no contract to bind it. Each network and each affiliated station is individualistic and independent, but in function they are interdependent; for the station cannot exist very well without the network, and a network cannot exist at all without affiliates.

Yet all of us also recognize that there are opposing, as well as common, interests between a network and affiliated stations. Both networks and stations seek increasing revenue to cover increasing costs and justify expanding risks, and a portion of this revenue must come from the same market. We share the time on your stations for network programming, as well as the revenue from the sale of this time, and sharing of a precious asset can cause strains, even among the closest of friends.

I mention these differences not to question them but to point out that they are, like the strains and stresses in even the best of marriages, facts of life. It happens, as in a good marriage, that the strains are small, compared with the satisfactions. If this were not so, the whole network enterprise would disintegrate, and divorce would destroy enormous values for broadcasting and for the public.

I take comfort and pride in the conviction that within this unique network-affiliate relationship, our association has a very special and distinctive quality, unmatched in the network business.

Perhaps it comes in part from NBC's tradition as a company that created and established the concept of national networking in order to furnish the first regular program service to stations and audiences.

Perhaps it is that while others have waited for the road to be marked out and paved, we have pioneered together and feel the comradeship of pioneers, first in radio, then in television, then in color—and always in service.

Perhaps still another explanation lies in the fact that NBC has always taken the long view in its operations and relationships, investing for the future and realizing that if we expect benefit from an association, we must contribute to it. So we have tried to identify and meet, as best we can, the real needs of those associated with us—affiliates, advertisers, and program suppliers. We believe we have attracted affiliates who also take the long view and judge our association for its meaning and substance, far beyond a day or a season.

The values created by the network system may be most evident to us because we confront them every day, either as station or network managers, and we have direct knowledge of the size and complexity of the effort needed to assure them.

We know that the broadcasting system of this country, like its political system, must live with the difficult problems of majority and minority, of stability and innovation. It also faces the special dilemma of a medium of expression operating under federal license.

We know that these influences are sometimes contradictory, sometimes harmonious. But we believe that such a system comes closest to reflecting and fitting the sort of country this is. And just as we accept and keep working at the imperfections of our social and political system, we support our broadcasting system against attack while we seek to improve it, because its strengths are greater than its defects.

We are by no means satisfied with the status quo of broadcasting and may never be. But its great developments have come and will continue to come from within, not from its detractors or its regulators.

We realize that the network operation itself is as singular as the network-affiliate relationship—a blend of forces, founded on enterprise, wholly supported by advertising, intensely competitive and living from hour to hour on popular acceptance.

In radio, it supplies an indispensable news service, with the most comprehensive news structure in existence. It stands in instant readiness to bring to everyone, wherever he is, information on any development of importance anywhere in the nation or the world.

In television, it fills an even broader role, providing a full service of entertainment, news, sports, and coverage of the great events.

The network undertakes the creative task of developing the programming and designing the schedule, the financial risk of underwriting it, and the entrepreneur's job of selling its advertising circulation to keep the entire complex moving. The stations furnish the facilities for bringing the programs to the public, provide essential local services to their individual communities, and compete at every level for audience attention.

This combination of skills and efforts has brought broadcasting to its present dimensions, a feat that has been wholly dependent on general public satisfaction with its service. Certainly a popular medium will not engage every member of the society in all it does. It will, in fact, infuriate some whose interests it does not match much of the time. But even they will find something of value in it if they look with any care. And upon reflection perhaps they will agree that tyranny of the minority over the majority is just as dangerous as disregard of the rights of the minority itself.

Broadcasting has grown in this country with a reach and momentum matched nowhere else in the world. Like many of you, I have been part of NBC through most of the cycles of this growth: during the late thirties and the forties when radio came to its supremacy as a national medium; through the birth pains of television; and now in the period when the spectacular growth of color recalls the explosion of black-and-white television 15 years ago.

I will not try to chronicle that history, but I think we can remind ourselves how far we have come by putting a spotlight on three points in time: the state of broadcasting when NBC was created 40 years ago; the broadcasting situation at the end of NBC's first 25 years; and its condition today. Perhaps this will give us a perspective on the future changes that we may expect to alter the face of broadcasting even more sharply.

In 1926, the new radio industry that showed such promise at its birth a few years before was desperately ill. The novelty of listening—just to hear something coming through a headset from far away—was wearing off. There was no regular, consistent program service to hold and build a national audience, and no systematic means of financing such a service. Radio stations were closing down. Their number, which had shot up to 1,400 in 1924 dropped to 620 by mid-1926, and the trend was down.

It was to reverse this trend that NBC was created. Commitments for network interconnections were made to the Telephone Company. Studios and facilities were built, and an organization was established to broadcast daily on a regular basis—all before a dollar of revenue was in hand, and before anyone could be sure that the whole enterprise would work. It was a risky undertaking, but it was necessary if radio broadcasting was to be converted into a national medium.

The initial NBC program was carried by a handful of stations scattered from Portland, Maine, to Kansas City. I was not there, but that's when it all started. And it was on the basis of this first network service that the upward spiral of broadcasting began. Six weeks after this inaugural broadcast, the NBC Radio Network had been extended from coast to coast.

Audiences grew, leading to increased advertising revenue, which financed expanded programming and led in turn to larger audiences. And all this supported a continuing growth in the number of stations. Within a year, there were three radio networks, and within five years, five million sets had multiplied to 20 million. So the anniversary we are celebrating today is not just an NBC event, but the fortieth anniversary of nationwide broadcasting itself.

If we turn the spotlight to 25 years from that beginning, we see another start underway—that of television. Again, a new cycle of growth was led by NBC and its affiliates, who together ventured into this promising but uncertain field while others sought to fend off the new medium as a threat. The risks were huge, although the enterprise soon became profitable for stations—long before there was any network profit. The pioneer's reward is evident from the fact that nearly 60 per cent of the pre-freeze television stations were on the NBC network.

In 1951—only 15 years ago—the total NBC Television Network consisted of about 60 stations, a fourth of them non-interconnected. Only 11 markets in the country had as many as three stations and the national set count stood at about 10 million. Today it is 70 million.

An hour evening program cost less than $50,000 compared with the average of about $150,000 today. The idea of spending a quarter of a million dollars on a single news documentary or pre-empting evening entertainment series for such news specials 40 to 50 times a year would have been considered preposterous. And the displacement of an entire evening's programming for a television analysis of a major subject would have been regarded as impossible. NBC News this year will operate on a budget of more than $70 million, and will be responsible for more than 25 per cent of our television schedule. And in the coming season, all of our nighttime programming, and almost all of the daytime schedule, will be in color.

The years have brought even more dramatic changes to network radio. Here again, NBC led the way in fashioning a new form that enabled network radio to maintain an essential service of continuing news and information features. It is a service that complements both national television and local radio, doing what neither can do. It is tailored to meet the demands of the modern radio audience—a mobile, busy audience that uses radio as a personal medium—and flexible enough to mesh with affiliates' local schedules.

The experience of 40 years has taught us that one constant in our industry is change. The principal difference we can see ahead is that the rate of change has quickened so that new problems and challenges will be on us almost before we can identify them. And throughout these accelerating changes in the economy and technology broadcasting there is one other constant that links the past with the future—the network-affiliate relationship.

If it is wisely administered, it will enable both stations and networks to harness the tides of change for an enlarged and improved service, alert to the developing needs of the audience and able to move with advertisers into the more sophisticated uses of merchandising and marketing. But if this relationship cannot maintain the resilience to meet these new opportunities, other forms of communication may dilute broadcasting's position as the nation's primary medium of entertainment and information.

The synchronous satellite, for example, has placed us on the threshold of a truly global system of instantaneous sight-and-sound communication. Satellite broadcasting directly to home or community receivers can reduce illiteracy and promote the advancement of underdeveloped areas. But if applied in countries with well-developed broadcasting systems, like the United States, it could cause many dislocations in station and network operations and raise serious social and economic problems. For several years NBC has been exploring the feasibility of a domestic system of interconnection to make network transmissions to affiliates more flexible, more complete, and more economical. NBC's plan, which we presented last month to COMSAT, envisions a workable satellite system to link networks and their affiliates, in radio and television, strengthening the relationship, not weakening it.

Other technological developments will multiply the information and entertainment service electronics can bring to every individual. One is the spread of audio-video communications by wire, already begun by CATV, to overcome the limitations imposed by frequency scarcity. Another is the increase in numbers of stations, as the UHF band is used more effectively. A third is the development of low-cost video recorders for homes, together with the use of television receivers for playback and print-out. And there will be further innovations that we cannot now predict.

We must remember that a television set, even as now designed, is technically capable of delivering scores of services, and we must expect that additional ones will arise to fill more of this capability.

The opportunity offered broadcasting is to keep its programming attuned to the interests of a constantly changing audience and to enlarge both its broad and specialized appeal. Only by doing so will it be able to compete effectively with all other forms of entertainment and information that can be brought into the home in the future. It has succeeded in this task through the changes of the years, but its mission will become more difficult in the years ahead. And more than ever, it will need forward-looking and imaginative management at stations and within network organizations. Speaking personally and for NBC, I want to assure you that your network will do all it can to plan constructively for our future together and to contribute to the productivity of our association.

Through four decades of evolution, we have pursued certain objectives that remain valid and intact. I believe it would be appropriate, in closing, to outline how your NBC management views its own obligations and those of our organization.

First and foremost, we acknowledge that ours is a dynamic, progressive society, and it is our first responsibility to be alert and sensitive to the variety of its needs and demands.

We will continue conscientiously to seek a fair balance between the broad interests of the mass audience and the range of specialized preferences that also exist within the population.

We will encourage our producers and our outside suppliers to seek fresh, creative approaches so that we may constantly invigorate all aspects of our service—popular entertainment as well as programming that seeks to enlighten and stimulate.

Although we must maintain the strongest possible economic support for our operation, we will not hesitate to take sound and reasonable risks in the interest of continued improvement of the network service.

We are determined to retain and expand NBC's leadership in the field of news and information, and thus enforce television's journalistic role, by devoting the time, money, talent, and energy needed for the proper coverage of the great public events and issues of our time.

We will continue to campaign for full parity with the print media in the coverage of legislative bodies and the courts, in the interest of improving the public's view and understanding of public affairs.

We will take prompt and full advantage of technological advances, to improve television's service to the public as a medium of entertainment, information, and education.

I am well aware that our attainment will not always match our aspirations. Nor will we fulfill all of these intentions to the total satisfaction of our critics or our friends. But we will try.

We have reached the age of 40 in good health and good spirits, and, I think, with our successes far outweighing our failures. We—NBC and its affiliate family—have long been in the forefront of our industry and we have won the acceptance and respect of a public that needs and expects the best service we can deliver. We will retain and deepen this acceptance and respect in exact proportion to the quality of our performance and the degree to which we strive to meet our responsibilities.

This, it seems to me, is our greatest challenge, and I am confident that working together we will meet it squarely and fully.

II. Criticism

RESPONSIBLE CRITICISM is essential to any art form or institution if it is to remain sound and healthy. By many of its practices, broadcasting has invited criticism from a variety of directions.

Some of this criticism, like that of E. William Henry, is from government. Some is from practitioners themselves, within the institution's walls, like that of Fairfax Cone, one of the advertising industry's most articulate statesmen. Some is from social critics who see broadcasting's dilemma as a part of a problem common to all media: the temptation to lower standards in order to reach "broader" audiences. And, finally, some comes from former writers like Dalton Trumbo and the late Shelby Gordon, sadly or angrily, with explanations for their sorrow and anger.

9. The Retreat From Excellence

by Claude M. Fuess

Mr. Fuess was headmaster emeritus of Phillips Academy, Andover, Massachusetts. The essay originally appeared in The Saturday Review *for March 26, 1960. It is reprinted here through special permission of the author's family and estate.*

A worried mother recently called me by telephone to ask me to intervene for her son, who had failed to win a scholarship at a well-known private school.

"How did he do on his tests?" I asked.

"I don't know," she confessed plaintively. "But what difference does that make? He's a lot better than most boys who get scholarships!"

I tried to explain to her that this lad, in our democracy, could in due season marry and vote and run for office and qualify for social security, but that when he applied for financial aid at a school, he entered into a contest in which he had to match his brain power against that of others. But the mother just didn't understand that the mere fact of her son's existence doesn't make him the equal of anybody in competition.

Her attitude is a significant illustration of a pattern which is recognizable throughout our society. The pattern is simply that we are increasingly ignoring the important differences among people. In New England a school custodian (a modern euphemism for "janitor") often receives a higher salary than the teacher whose room he cleans. This has always seemed to me something absurd, and once when I said so at a meeting a man rose and asked indignantly, "Why not? He works just as hard, doesn't he?" The prolonged and intensive training of a mathematics instructor seemed to this critic entirely irrelevant. After all, one citizen is as valuable as another.

We live in an age of the average, and even the average is not as high as it should or could be. I have spent some time in hospitals and know how reassuring it is to be attended by a physician who knows his business, who thinks and speaks and acts with the confidence acquired by thorough training. After you have been examined by a nurse who barely differentiates the patella from the esophagus, how cheering it is to come under the tender care of a professional who doesn't wield a hypodermic needle as if it were an ice pick. The current incompetence is due, I am told, largely to lack of competition. Nurses are needed so badly that hospitals take what they can get—and the "take" is often astonishing.

That one citizen is as good as another is a favorite American axiom, supposed to express the very essence of our Constitution and way of life. But just what do we mean when we utter that platitude? One surgeon is not as good as another. One plumber is not as good as another. We soon become aware of this when we require the attention of either. Yet in political and economic matters we appear to have reached a point where knowledge and specialized training count for very little. A newspaper reporter is sent out on the street to collect the views of various passers-by on such a question as "Should the United States defend Formosa?" The answer of the bar-fly who doesn't even know where the island is located, or that it is an island, is quoted in the next edition just as solemnly as that of a college teacher of history. With the basic tenets of democracy—that all men are born free and equal and are entitled to life, liberty, and the pursuit of happiness—no decent American can possibly take issue. But that the opinion of one citizen on a technical subject is just as authoritative as that of another is manifestly absurd. And to accept the opinions of all comers as having the same value is surely to encourage a cult of mediocrity.

Recent prognostications for the 1960's over radio and television, and even in legislative halls, have been preponderantly optimistic. The orthodox American mood today is confident and complacent. Very little mention has been made of significant changes which, whether we approve of them or not, have taken place since the close of World War I, and which have had and are having far-reaching consequences. The current tendency to reduce everybody to the same level is illustrated not only in the broad results of our taxation policies, but also in such matters as the adjustment of the school curriculum to the abilities of the average or below-average pupils. Having moved almost unconsciously toward achieving the greatest good for the greatest number, we have neglected that small group of potential leaders through whom human happiness has in the past been attainable and attained.

The prevalence of a cult of mediocrity is evident in many phases of our national economy. Out of many trends a few samples will prove this point. One of the more obvious is the system under which all workmen, regardless of skill or attitude, receive the same pay for the same job. Under the existing policy nobody has any incentive for toiling harder or longer or more effectively than his less aggressive neighbors. Everybody is familiar with union regulations forbidding a member to lay more than a specified number of bricks in an eight-hour day. Those who are able to lay more, and would like to lay more for more money, find themselves frustrated. I have argued this matter with labor leaders, who insist that it is more important to protect the normal, or below-normal, workman than to encourage the exceptionally eager one. It should be possible to do both. The present tendency, however, is to eliminate competition among employees and thus to restrain the ambitious laborer. Clearly this policy does not, indeed cannot, stimulate increased production. The principle of "leveling down" is here perfectly illustrated in one of its worst aspects.

The widespread acceptance in our country of the doctrine of promotion by seniority alone has had much the same consequences. Let me illustrate what often happens in education. A teacher of average ability, once appointed to a school or college staff, can be sure of moving up automatically as retirement or death removes his older colleagues; and because of this guarantee, ambition frequently ebbs away. The amount of dead wood among faculty members of forty-five or older is considerable, as any honest dean will testify. Yet to advance a younger but much abler associate ahead of one of these comfortably situated elder professors is virtually an impossibility unless the old-timer strays from the marital fold or becomes a Communist suspect—which he is unlikely to do. Again I am aware that the existing system has its staunch defenders, and it unquestionably does favor some entirely respectable men and women. But of its contribution to the lowering of quality there can be little doubt.

The forces promoting mediocrity in our schools are too well known to require extensive documentation. The Carnegie Foundation for the Advancement of Teaching, summarizing some of the problems created by these forces, says: "Very large numbers of superior students are still not working up to their capacity, not being challenged to their best performance, not pursuing the hard intellectual programs which will develop their talents."

George F. Kennan, writing as a historian observing current events, refers scathingly to an educational system in which "quality has been extensively sacrificed to quantity." In a majority of secondary schools, the emphasis is not primarily on scholastic attainment, the vacations are too long, the standards are too low, and too much emphasis is placed on games. Parenthetically, it may be observed that athletics are not mediocre—at least not in the United States.

Nor is this deadening effect confined to educational areas. *Look* magazine, in an article entitled "Our Military Manpower Scandal," quotes the Cordiner Committee as saying that the army has "no adequate incentive system." It adds, "There is no extra pay for doing your job well." One officer who left the Air Force to enter business said, "Those people were ahead of me and were going to stay ahead of me. This kills aggressiveness in a young officer." It has been suggested as a basic reform that the armed forces should break with the traditional promotion system, so that the best men could advance more rapidly.

Perhaps recognition of the philosophy underlying such practices has damaged the competitive spirit of young Americans. College graduates entering business are reported to be more interested in security than in competition. The inclination to "take a chance" or indulge in a "calculated risk" is said to be not so prevalent as it was in our pioneer days. Recently I spent an afternoon with the chairman of one of the most enterprising manufacturing companies in America. He was lamenting that on his team, as he called it, he could discover few with the originality and resourcefulness required for dynamic leadership. We have, he said, dozens of fine-looking, well-groomed fellows for routine assignments. What he wanted was a few restless men with imagination, ready to break loose from conventional procedure and move into untried fields. Desire for security and mediocrity belong together.

Even in politics specialized knowledge is far from obligatory. Where I live, aspirants for office stress the fact that they are good family men, with a covey of children, as if demonstrated fecundity made them better lawmakers. In certain instances candidates are actually hampered by an acquaintance with history or economics. And if the results of elections seem to offer an advantage to mediocrity, whose fault is it? A considerable proportion of voters is prepared to accept garrulity, back-slapping, vague enthusiasm, and expressed good intentions as substitutes for talent.

As a practical asset, demonstrated skill is unimportant when community appointments are concerned. Recently in Cambridge, Massachusetts, the issue of veterans' preference for public school teachers was vigorously debated. The heads of local veterans' organizations argued that any former serviceman, no matter what grades he had received on his qualification tests, should be given an appointment ahead of non-servicemen who had passed with higher marks. The commander of the local American Legion post was quoted as asking, "What difference do a few points on an examination make anyhow?" The assumption was that virtually anybody could teach school. Perhaps this feeling is responsible for some of our present difficulties in education.

Much the same criticism may be made of the policy of giving "home town" applicants preference for salaried jobs, no matter what their ability. In this case, as in many others, some attribute other than professional excellence is used for the decision.

The American distrust of cleverness or quickness of mind, or indeed of conspicuous intellectual ability, has been noted by foreign observers. A young Korean, a graduate student at Harvard, wrote recently in the *Chris-*

tian Science Monitor some candid comments on the United States, in which he said:

> The American people want their President to be one of them. They would like to elect to the Presidency a man like themselves, they do not want the President to be the "great leader," "hero," or "superman," whose vision, outlook, and philosophy are remote from theirs. Instead they want their President's tastes, outlook, and philosophy to be similar to theirs.

It is significant that the number of first-rate statesmen in the United States of the post-Revolutionary period, when the country had a population of not much over three million, was greater than it is today when we have 180,000,000. One of the reasons, perhaps, is that our generation has reversed the pattern and established a cult, not of genius, but of mediocrity, by its approval of conformity and orthodoxy and the kindred colorless virtues which keep a social organism static. Various observers deplore the growing tendency to distrust those who are different, those who deviate in their thinking or writing or behavior from what we have been taught to consider normal. Uniformity of ideas seems to be regarded as a protection against the dangers of so-called radical thought. Although we concede that, in theory, heterogeneity is essential to progress, we give it little encouragement in our own vicinity.

Let me call new witnesses. Arthur E. Fetridge, predicting the course television will take during the coming year, says:

> I may be wrong, but certainly all the signs point to mediocrity and more mediocrity. The more commonplace a presentation the more people watch it. . . . The masses say they want mediocrity, and until the sponsor says he won't give it to them any more but instead forces them to accept something far better, mediocrity is what we are going to get.

In his latest book. *The American Conscience,* Roger Burlingame has summed all this up in one succinct paragraph:

> We are prosperous. We are complacent. Religion has become, for the most part, a social convention. . . . Skill is anonymous, thought is under pressure to conform, security has replaced venture as a dominant aim, intellect is in the discard, and politics are dictated by . . . mediocrity.

Fortunately, despite Mr. Burlingame's well-justified pessimism, some signs of a revolt are evident, especially in education. Thomas Jefferson, whose democracy cannot well be questioned, deliberately advocated an educational policy in Virginia which would gradually eliminate the unfit and open the way to brilliant students. Public inertia kept his program from being put into operation, but today even hard-boiled, thrifty taxpayers are advocating the early identification of talented boys and girls and their proper training on their way through school, college, and university. We have at last become fearful that the Soviets are producing more competent experts, particularly in the fields of science and modern languages, and fear may do for us what prosperity has been unable to accomplish. To meet this sudden challenge from a "barbaric" country we have already in some degree begun to sort out the young, using native aptitude as one of the criteria, and we are offering these gifted boys and girls the opportunity of going as far and as fast as they can. Everybody should have a fair chance, and the ablest should win.

Against this program for developing excellence are allied several evil forces: the sinister operations of politicians who thrive on public ignorance; the corrupting power of the press through its appeals to mass psychology; the decline of the pioneer spirit; the general relaxation of people seeking a fast and often a criminal buck; and the enervating influence of widespread luxury. These are no new phenomena in a democracy, but they have been spreading farther and reaching deeper in our times than in any American period with which I am familiar. And their combined effect is to promote the commonplace, the average, and the mediocre.

It can be argued that the United States and the Soviet Union, without either intent or realization, have been exchanging national philosophies. In our insistence that all men and women engaged in the same jobs should be compensated in the same amount, regardless of activity or quantity and quality of production, we have been moving toward the fundamental principles of Marxism. The Russians, on the other hand, convinced that exceptional energy and ability are national assets and should therefore be well rewarded, have adopted some of the stimulating practices associated with the capitalistic system. If they find a brilliant scientist, they apparently offer him unusual opportunities and privileges; and they now pay gifted writers, musicians, and even teachers, salaries which make their positions worth competing for. Richard L. Strout has reported that in the Soviet Union a university professor may receive from $35,000 to $50,000 at the current rate of exchange. He adds significantly, "Anybody who sees the sputnik as an isolated phenomenon is shortsighted. It is not a race between sputniks but a race between schools."

One truth is almost self-evident. Widespread mediocrity in key places, in business or education and even in the armed forces, can be deadening, even destructive, to any country. For Americans it is at the moment vital to provide a cultural climate in which exceptionally high intelligence is detected and subsidized accordingly. So much that is good is going on in these United States, and a small number of dedicated people are doing so much on so many levels, that our mass degeneration seems pitiful. Possibly we have been anesthesized by material prosperity and lack the vigor to move into action. Perhaps we should put up a sign "Wanted: A Satirist" and hope for his energizing appearance. But somehow our national leadership in religion, in education, and in statecraft should rise to the occasion and counteract the blight which mediocrity has been casting over our boasted culture.

10. The Two Publics of Television

by Fairfax M. Cone

This speech was delivered by Mr. Cone, Chairman of the Executive Committee of Foote, Cone and Belding, before the Broadcasting Advertising Club of Chicago, Sheraton Hotel, Chicago, Illinois, on October 10, 1961. It is reprinted here by permission of the author.

If the two publics of television were clearly what they seem to be—the mass and the minority—and if television were something that anyone could enter into as a broadcaster, there would probably be no more official controversy about programs and programming than there is about books and the movies. Criticism would undoubtedly still be with us; but criticism and controversy are two different things.

In the case of television almost any discussion of programs sooner or later comes down to a discussion of the physical rules of broadcasting that limits, and limits so drastically, the number of channels.

It is this limitation, that everyone here knows so well, that makes television so different from movies, books, magazines, and newspapers; or, for that matter, from any other means that we know for communication—either of entertainment or instruction.

Anyone who can afford it can make a movie, and there are various means and places available for showing it. Anyone can write a book, and even if he has to publish it himself, there are plenty of presses available to print it. Anyone can start a magazine or newspaper, and many people have. But only those people can get into television who can be allocated a channel for broadcasting that is not already in use.

These channels, or frequencies, are of limited number and, like other limited resources of our people—our forests, our lakes, and our rivers—these are necessarily supervised and controlled as to use by a government agency.

Just so, the question of programming becomes a matter of interest to the Federal Communications Commission, and thus different from the programming of movies, books, newspapers, or periodicals.

The extent to which this interest should *control* programming was the principal topic of discussion at the recent television symposium at Northwestern University.

I happen to have been a member of the panel (which included Newton Minow, Chairman of the FCC, Ward Quaal, General Manager of WGN-TV, and perhaps a half-dozen other station operators), and you know, I think, what the general conclusions were.

Government control seemed to every one of the station operators there, and equally to the several distinguished lawyers present, synonymous with censorship. And even Mr. Minow, who denied strenuously that censorship was in his mind, had to agree that this was implied in the licensing power of the FCC through which all stations operate.

It was Mr. Minow, as an attorney for the publisher, who freed the book, *Lady Chatterley's Lover*, from the censor's ban. And there can be no question of his distaste for censorship generally.

But neither do I think there can be any question but what television must come under some new conscientious, if not rigid control. For every step forward that television takes, and each with a considerable fanfare, it appears to take two steps backward.

The same little screen that last week gave us NBC's magnificent Ernest Hemingway story, *Life of a Giant*, and the immensely thoughtful and significant *Where We Stand: War or Peace*, with Howard K. Smith and CBS reporters from London, Paris, Berlin, and Moscow, brought us also a whole new series of tired rearrangements of old mystery and comedy routines, topped off with the nastiest, ugliest show I have ever seen on TV. This was the first installment of *Bus Stop*.

When it was argued at Northwestern that there is no other control of books and magazines and newspapers than the right of the Post Office to deny mailing privileges in defense of morals, it was pointed out that there this is ample control.

Personally, having read *The Carpetbaggers* and *Valhalla*, and seeing *Playboy* magazine, I doubt this. I hope that Simon & Schuster, who published *The Carpetbaggers*, and G. P. Putnam's Sons, who published *Valhalla*, will come to regret these lapses from taste and dignity and lapse no more; and I hope that some day *Playboy* will grow up. I am sure that there will always be people who think that looking inside girls' blouses is the height of sly and not very dangerous excitement; and *Playboy* takes the blouses off for them.

In the book business there have always been under-the-counter items. Who can forget *Fanny Hill?* Or the fact that Mark Twain wrote one of the world's most vulgar, most obscene little books. But these *were* surreptitiously published and distributed.

Books like *The Carpetbaggers, Valhalla, From Here to Eternity, Tropic of Cancer* are generally available, and available to everyone. Also, the movies have come to have a predilection for the seamy side of life, with the glamorization of the gangster and his sultry girl friend as standard, almost, as the slouch-hatted, cigar-smoking villain of TV, each with his busty moll. Only the *major* magazines have not entirely let down the bars. But this is no excuse for television.

It would be easy to say that TV's biggest trouble is the problem of numbers, and that these are not the operators' faults, but rather that they are a combination of mass interests on the one hand, and the meagerness of the minority on the other. Any study that anyone wants to make of the Nielsen or Arbitron, or any other television audience figures, gives aid and comfort to this point of view.

There is no doubt about it that a large number of people *are* getting what they want. Indeed, it is even possible that a *majority* of people are getting what they want, crummy as much of it is. But even if this be true, there is a sizable minority that rarely is satisfied.

When almost three hours pass on a Saturday afternoon in Chicago with no other choice for the viewer than a baseball game (between the Cubs and the Phillies) and *three* football games, something is wrong. Something is terribly wrong. And it needs to be corrected.

I have argued in the past, and publicly, that there is a good deal of television programming that is very good programming. And I can argue this again. And I will. But it is impossible to argue that there is not also a great deal of television that is stupid and dreary and ugly, and some that is vicious. Worse still, it is impossible to argue that you can *choose* your television, for you can't. You must take it, or leave it, for better or for worse. Once you take it you have had it, to the exclusion of everything else that was broadcast simultaneously. And once you have left it you have missed it forever.

This is a disadvantage in television that may eventually be remedied by automatic house-taping of shows to be rerun at the viewer's choice. Meanwhile, and this will probably be a long while, the disadvantage is real and the consequences are serious—at least in the views of the Northwestern conference. And in the view of Newton Minow and the FCC. And they should be to all of us. The public is getting its television according to the money that is in it for the owners and operators of television stations.

If you want to, I suppose, you can go back of the stations and say that it is the advertiser and his agent who *pay* the money that are responsible. And in a way you would be right. For if the advertiser were to withdraw *his* support from television there would be no support. Without the advertiser or subsidies from government, there would be no television.

Here, I would like to remind us all again how different broadcasting is from the printed media of news and instruction and entertainment and advertising. The large audience attractions which build magazine and newspaper audiences and attract advertising stand side by side in type with *limited* audience attractions.

There is no exclusivity here. The audience shifts within the pages of the magazine and between the columns of the newspaper. And the advertiser pays equally for it all. The huge difficulty in television is that it has always been sold on a selective basis; selecting the big numbers. The advertiser virtually has bought the audience he wanted. And he has "served" that audience and no other.

So we come back to the *two* publics; and the fact that, as I said, these are not *clearly* a majority and a minority. Network and station and advertiser operations have wanted to make them this; and had they been successful probably everyone would be happier. Including the FCC and Mr. Minow, and me.

We would have mass programs of soporific drama and comedy and sleazy mystery and (I suppose) mayhem, just as we do now, for what we call the majority. And we would have programs of somewhat more intellectual and stimulating character for what we call the minority.

The difference is that we would have them at the same time, and in competition. And all the silly soap drama and corny horse opera and naked city crime wouldn't fill the best hours for viewing, while *Wisdom* comes on at 4 o'clock on Sunday afternoons and Irv Kupcinet starts one of the most provocative and challenging programs in all television at 12:15 (fifteen minutes after midnight) on Sunday mornings.

Varying types of programs would be on in identical time periods: on *big* and *little* stations, as it were, just as *Life* and *Look* and the *Saturday Evening Post* are on the same newsstands with the *Atlantic* and *Harper's*, and the *Saturday Review* and *The Reporter*. Our *real* difficulty is that we have no little television stations. We have only giants, fighting each other. And this is understandable.

In a very different way, the *Chicago Tribune* was begun by Joseph Medill to be his voice in matters political and economic. The *Chicago Sun-Times* was put together by Marshall Field to be *his* voice in similar matters against the voices of the heirs of Medill. The *Saturday Evening*

Post was founded by Benjamin Franklin to be *his* voice. *Life* was established to bring a new kind of illustrated news to the public. And *Look* was planned to be somewhere in between those others.

Television's great networks (and there are no other kind) all were planned for a totally different reason; not as public servants, for there was no public service involved at the outset; but purely and simply as business enterprises, planned to make money.

Please let me be clear. There is nothing the matter with this. At least, there is nothing the matter with making money: Arnold Toynbee, Arthur Schlesinger, and John Kenneth Galbraith notwithstanding. The trouble, yours and mine (and the public's too) is the limitation in the number of television channels that makes all three networks and most independent stations fight for the same mass of viewers hour after hour after hour.

It is said by certain researchers, and I have recently come to believe this, that the public is composed of three broad groups of people:

> First, there is 20 per cent whose minds are made up about most things and cannot be changed;
>
> Second, there is another 20 per cent who will try almost anything and follow almost any fad without conviction. Together these are held to represent some 40 per cent of the population.
>
> Then there is a solid 60 per cent in the middle which is open only to sensible argument and moved only by satisfactory performances.

These are arbitrary percentages and the people within the groups surely change. However, as I said, the researchers believe these are real, discernible, workable groups, nevertheless. And smart business people, smart advertisers, work on this middle 60 per cent who in the long run represent success,

Television, unhappily, seems to be aimed primarily at the people whose tastes can't be changed and whose sights can't be raised. These are the people who represent the public's lowest taste. And this, I think, describes that bottom 20 per cent of our population at which television apparently is first directed.

Marya Mannes, who is the best writer on television anywhere around (and one of the most perceptive writers about anything) has a piece in a recent book about New York where she talks about "men in vicuna coats spitting on the sidewalk . . . and women in Dior dresses chewing gum"; the second target, I am sure, is that second 20 per cent that doesn't know what it wants and drifts along, dreaming, perhaps, of vicuna coats and Dior dresses, neither of which it will ever obtain.

What most television clearly is *not* aimed at is the 60 per cent in the middle that can be attracted and held only by sense and substance and good taste. Actually, the peak time viewing turns up about 60 per cent of

U.S. television sets; the average is around 55 and 60 per cent during prime nighttime hours.

So you see, if you are getting the bulk of the gum chewers and the lipmovers and the bulk of the no-opinion holders (who are not very apt to be at a PTA meeting or one of the English-Speaking Union, or a Great Books class), you are pretty apt to be attracting no more than a third of the great body of sensible and sensitive Americans whose tastes it is much the most important to satisfy. This, it seems to me, is something less than good business—for anyone.

Television is the greatest means of communication ever discovered, and our most important people, often times a majority *of the majority* are being largely left out of it, because networks and stations and advertisers alike, or seemingly alike, are out only for numbers—to beat the competition each one faces.

Nor is this the least bit hard to understand. If the network fails to produce programs that attract large audiences, the station can't sell its local time. If the network can't deliver the station, it has no reason to exist. And if the advertiser can't match *his* competition in advertising effectiveness for every dollar he spends, he cannot remain in competition. He must fail.

Now, the question is: Is there any way out, short of a system of UHF stations, like our FM stations in radio, and forty-odd million new television sets to bring in these new stations, to say nothing of the problem of building and operating such stations for limited audiences, for what is so vaguely referred to as better programming.

Let me say parenthetically that I haven't the foggiest notion what this better programming really is. I know that it is supposed to take children away from the cops and robbers and the Westerns; and I know that no books have ever done this successfully. If it is supposed to move adults from *The Untouchables* and *Andy Griffith* to opera and the ballet, I am sure that its promoters are wasting time.

The only course open to any of us, in my opinion (and us includes the FCC) is not to constrict television but to expand it; to broaden the choice within the present total time limits and to *keep* that choice broad, with the end in view to do more for the roughly 60 per cent of the public that by its television viewing habits is thought of as the minority, but which is actually the majority that is being so little considered today.

How would I do this? I first suggested a plan when I perceived that all television, or *almost* all television, would sooner or later be spot television. The sponsor had already become the alternate sponsor; and with his advertising distributed among three or four products over two weeks instead of concentrated on one, each week, he was soon to become no sponsor at all. He would be merely another advertiser on a purely spot basis. This he has become.

The magazine concept of television is now in effect with only two steps to be taken to make it what it should be. The first of these is to program as a responsible magazine would: to balance the weekly fare between regular and special entertainment features and regular and special features in the fields of controversy and ideas—even in the arts.

A Sunday afternoon "hidden feature" that I saw on WGN a few weeks ago had Heifetz and Rubenstein and Piatigorski in a classical jam session that made *Sing Along With Mitch* sound like a grammar school assembly exercise.

But let me not digress. Not only is such programming possible, it is also the means, and the only means, to experimentation. For implict in it, and the second step in this plan, is to revolve advertisers through the total week's programming; to cut out, as it were, preferred positions in the weekly schedule, and to open this up to experimentation, with every advertiser paying his share.

The method is as simple as the rotation of commercials by a single advertiser for six products through two weekly shows; only the network would rotate *all* advertisers through *all* except special shows.

There are people who oppose this. But I believe they are short-sighted. Because, if they hold to their preferred positions, which means positions in certain large-audience programs, their competitors must seek to equal them (they have no choice) and the level of television programming will remain precisely what it is—which is a national disgrace.

We are programming for what has long seemed to be the majority. But this is not the case. The majority is only the majority of viewers; it is not necessarily the majority of people. It is probably, in fact, a minority of the public. And *all* our people should be served.

It is in the province of the networks and the stations to do this next season. I have no doubt at all that most advertisers would agree to rotation through all kinds of programs—at a single rate for the nighttime period, and another single rate (for time and talent) for the daytime. For, after all, why should any but the very luckiest today not wish to remove the gamble they take alone when they, and I quote, "pick" shows.

American business became enamored of show business back in the heyday of radio. The devotion reached the stage of red-hot passion with television. But I think that disillusion has set in.

If it has, as I think it has, and most advertisers now view TV simply as another, even though greater medium of communication, that they are willing to leave to the communicators to operate, we can have fair sailing, together with all manner of just and pleasant rewards; not the least of which may be hands-off by the FCC.

11. Remarks Before the Annual National Broadcast Editorial Conference

by E. William Henry

Mr. Henry's address was delivered before the Annual National Broadcast Editorial Conference, Arden House, Harriman Campus, Columbia University, Harriman, New York, on July 7, 1964. It is reprinted here by permission of Mr. Henry.

The first sentence in most speeches, including my own, usually expresses the thought that the speaker is "privileged" to address the particular audience before him. Upon receiving the invitation from Peter Straus to speak to the Second Annual National Broadcast Editorial Conference, I looked up the report on your first conference last year at the University of Georgia. After studying that report, knowing full well that "editorializing"—in its broadest sense—touches the heart of public service broadcasting, and appreciating the quality of your commitment to this kind of broadcasting, I have no qualms in repeating the traditional opener. It is indeed a privilege to participate with you in this conference and to join your discussions on this vital subject. It is also an exciting challenge to try to add a stimulating thought to the many you have received from the outstanding speakers who have preceded me at this forum.

Speaking in New York State reminds me of a complaint recently made to me by an emigrant from New York City. "Washington," he said, "is a town with only one industry—government—and that's all anybody talks about!" As an emigrant from Tennessee, that strikes me as partially true, and the degree to which it is true may explain many of the games played in Washington to enliven leisure moments. You all know the one that is currently most popular: "Who's going to be the Democratic Party's nominee for Vice-President?" You may even have heard of the very latest one: "Would the proposed cut in salary raises for Supreme Court Justices violate the Fair Employment Practices Act?"

Lesser known, but even more intriguing, is the "Answer" game, probably originated by Kennedy speech writers, given widespread publicity by Theodore White's book *The Making of the President*, and modified for the masses by scores of local wits. The point of the game is to guess the question, having first been provided with the answer. It goes this way. The first answer is: "Washington Irving." What is the question that prompts this response? The experienced player will immediately reply: "Who was the first President of the United States, Morris?"

After that one, only the most fearless after-dinner speaker would proceed further on the same track. Nevertheless, here is another example.

Answer: "I'd rather be right than President."

Question: "What is your campaign slogan, Senator Goldwater?"

Now, the wonderful part of this pastime is that anyone and any number can play, and any subject is fair game. So, I brought along a couple for you tonight. See if you can guess the question.

Answer: "To err is human."

Question: "But Mr. Cronkite, suppose the computer makes a mistake next fall?"

Finally, here's one which has two answers (depending on your point of view), but only one question.

The answer may be either: "Industry statesman" or "Heavy-handed, narrow-minded bureaucrat."

Question: "What do you call a person who urges genuine adherence to high standards of public service?"

And so it goes. Maybe you can think of some more examples before the evening is over. If so, I wish you'd let me have them to take back home.

Tonight, however, I want to follow the more accepted procedure and discuss with you certain questions first—before we attempt to find answers.

Many years ago William Sumner wrote: "If you want war, nourish a doctrine." Although he wrote well before the formation of the Federal Communications Commission, some broadcasters will think he had our fairness doctrine in mind! And, of course, no speech by the Chairman of the FCC, at this time and at this place, would be complete without discussing some of the questions raised by this highly controversial, highly interesting 12-page policy statement.

As Sumner unknowingly predicted, the doctrine has indeed sparked a war. But it is a good war, for it is a war of words and thoughts fought on the ground of principle. And as such, it is an essential part of the democratic process.

Like all wars, this one has its partisans—led on the one side by "heavy-handed, narrow-minded bureaucrats" and on the other by "industry statesmen." But the matters at stake transcend the partisan interests of regulators and regulated alike. They go to the heart of the role played by a medium to which we are all dedicated in a country we all love. So let us tonight declare a temporary moratorium and discuss this subject simply in terms of the deep and abiding interest we share in American broadcasting, and our common goal of service to the public.

A year and a half ago one of our country's most distinguished journalists,

Mr. Harry Ashmore, told the California Newspaper Publishers' Association:

> In their moments of candor the proprietors of TV recognize their limitations. Not long ago the president of the American Broadcasting Company said bluntly that TV is primarily an entertainment medium, and only incidentally concerned with news and public information. There is an inherent reason why this is so. Broadcasting is the natural child of a union between show business and advertising. It was born as an impersonal corporate entity; responsibility for its performance is diffuse, and it is ultimately bound to the highest-bidder morality of the marketplace. Broadcasting does not have, and shows no sign of developing, a sustaining tradition of public service of the kind the best newspaper proprietors still recognize and act upon. It is not in the nature of the beast to serve the public interest at the price of its own.

Now I venture to say that few in this room would concede so much to Mr. Ashmore. Most if not all of us believe that broadcasting can achieve full stature as a medium of journalism, as well as a medium of entertainment. Your commitment to editorializing springs from a desire to reach that goal. And the government's allocation of huge portions of the radio spectrum to broadcasting rests as much on the medium's ability to provide information and ideas as on its ability to provide living room entertainment.

It is deeply troubling, therefore, to be told that the government's insistence on fairness in the treatment of controversy is hampering and even thwarting the growth of electronic journalism. It is disturbing to hear that networks seek a colorless "middle-of-the-road" point of view in their documentaries and news analyses. It is disconcerting to see flat statements that the integrity of news and editorial judgments has been impaired, that minority points of view are given short shrift or none at all, and that editorials are limited to safe topics or, in any event, to three-minute driblets.

Now I personally think that American broadcasting is a good deal livelier and more open to a spectrum of views on vital public issues than these criticisms suggest. But if the Commission's fairness doctrine, or Congress' approving reference to it in Section 315 of the Communications Act, are responsible for a tendency to play it safe, then there is still cause for grave concern. If boldness and daring get some chance, and minority points of view some exposure, only despite serious obstacles created by the policies of government, then the government itself should be the first to sound the alarm. After all, the Bill of Rights is a government document, not a corporate charter.

For the purposes of my talk tonight let us assume a working knowledge of the fairness doctrine on your part and mine, without going into its origins and specific applications in great detail. For such a discussion I refer you to the 1949 policy statement itself, and to our primer on fairness released yesterday.

There is an ancient legend that a skeptical young man once journeyed throughout the land seeking to have the meaning of religion summarized for him while he stood on one foot. If this were possible, he said, he would become a true believer. He was advised by many great and important men that he sought the impossible. Then one day a wise teacher said to him: "The true meaning of religion is 'Do unto others as you would have them do unto you.' The rest, my son, is commentary."

While far less important in the scheme of things, the essence of the fairness doctrine may be similarly distilled. Reduced to its basic terms, it holds that: If a broadcast licensee undertakes to present programming dealing with controversial issues of public importance, he must make reasonable efforts to present conflicting points of view on such issues. It is as simple as that. All the rest is commentary.

Some broadcasters claim that the principle is deceptive in its simplicity. In fact, there are some who charge it is hopelessly ambiguous. Its net effect, they argue, is that the broadcaster lives in daily fear that the FCC will disagree with his judgments and take action against him. Broadcasters, they continue, have a strong incentive to avoid controversy entirely by adopting an "impartial" approach in news programs, and by carefully avoiding sensitive issues in documentaries, actualities, and other public affairs programs. Recognizing that any effort to organize or analyze information requires some point of view, such critics conclude that the ultimate result is a predominantly bland, middle-of-the-road point of view, which provides room for the center of the political and social spectrum, but which ignores or downgrades anything that might conceivably be characterized as "extreme."

Like many curbstone judgments, this argument has a veneer of plausibility and therefore a superficial appeal. Admittedly, no one can define "fairness" for any and all conceivable situations, and reasonable people may disagree about its meaning in any particular circumstances. But the Commission doesn't require the broadcaster to make an arbitrary and subjective judgment about the treatment of controversial issues, at the risk that seven men in Washington may, in an equally arbitrary and subjective manner, disagree.

In an analogous situation, the law has been coping with the question of negligence for quite some time. No one can define "negligence" for any and all conceivable situations, and reasonable people often disagree about its meaning in particular circumstances. But the law defines negligence generally, and then allows juries to apply the general definition to concrete cases, upsetting their verdicts only if they are shown to be unreasonable.

The Commission has followed a similar course in dealing with "fairness." In its statements and opinions, it has enunciated only the most general principles—that the basic right to be protected is the public's right to hear both sides of a controversy, that a broadcaster has an obligation to respect that right, that he must make an affirmative effort to discharge this obligation over and above making his facilities available to contrasting points of view on demand, and so forth. It has committed the concrete application of these principles to the judgment of the licensee. He is the one to decide what issues shall be covered, who the spokesmen shall be, and what format shall be used. And the Commission has said time and again that it will not quarrel with any good-faith judgment.

In brief, the broadcaster who is a miser in dealing out fairness—who seeks to do only what gets him by—will find his legal obligation both troubling and ambiguous. The broadcaster who is willing to shoulder the responsibility for making decisions—even if the Commission should occasionally disagree with him—has nothing to fear. Indeed, as the Commission's 1949 Report on Editorializing expressly states:

> . . . the standard of public interest is not so rigid that an honest mistake or error in judgment on the part of a licensee will be or should be condemned where his overall record demonstrates a reasonable effort to provide a balanced presentation of comment and opinion. . . .

The Commission's actions for more than a decade have repeatedly demonstrated that it means these words. Time and again it has upheld the broadcaster's judgment against the complaints of those whose ideas of fairness differed. Without citing individual cases, let me quote a recent speech by Mr. Lawrence H. Rogers, President of Taft Broadcasting Company. Said Mr. Rogers:

> Since the first time I aired an editorial in 1956 I have probably received half a hundred letters from the FCC questioning a complaint or challenging my adherence to the Fairness Doctrine. Living dangerously? Perhaps. Yet I can state there has never been a single instance when the outcome of such a case was not assured, so long as the station had honestly attempted, in its own judgment, to live up to its license obligations. . . . It is my firm belief from all these experiences that an honest answer revealing an honest intent to serve the public interest will always result in a clean bill of health.

At this point, however, we reach another class of objections. There are some who think that specific Commission rulings in the area of fairness are wrong, or "go too far." There are many, for instance, who object to the requirement that a script be sent to someone who has been personally attacked. Requirements of this kind, they claim, are too rigid, impossible to administer, and unduly burdensome.

I know I speak for the Commission as a whole when I say that we welcome criticism of any ruling. In this instance, our reasoning has been as follows: Where the controversial issue involves the character, honesty, and integrity of a particular person and these have been directly and seriously attacked, it seems elementary—under the concept of fairness—that that person should have an opportunity to respond if he wishes. It seems equally elementary that he cannot respond if he does not know what was said about him. But there are various ways to accomplish this. A verbatim script cannot always be available. And so, the Commission has made it clear that the requirement is for the broadcaster to use his own best judgment and to do the best he can in the circumstances he faces. If a verbatim script is simply unavailable, a summary based on the knowledge at hand should be sent. If notification of the person attacked cannot reasonably be made at or prior to the time of broadcast, it should be made as soon as possible thereafter.

There are likewise some who object to our *Cullman* ruling. There we held that a broadcaster who has presented one side of a controversial issue on a paid basis, may not impose an absolute requirement that time be purchased for the presentation of opposing views. This ruling, [1] they claim, puts the representatives of opposing views in a position where they can demand free time, and makes the cautious broadcaster hesitate before accepting a sponsored program that espouses one side of an issue.

This reasoning leaves me unconvinced.

A broadcaster who accepts a sponsored program on a controversial issue makes a judgment that his community wants or needs exposure to points of view on that issue. If he can obtain sponsorship for the presentation of opposing views, well and good. If, for instance, there are two equally appropriate representatives of the opposing side and one of them offers to pay, the broadcaster may select the one with money. But to hold that he can make ability to pay the sole factor governing the public's right to hear opposing views would be to relinquish the fairness principle entirely. It is this reason, I believe, that made the Commission's ruling on this question a unanimous one.

---

[1] The actual ruling of the case (Public Notice 40775, September 9, 1963), while not lending itself to abbreviated treatment in a speech, may be summarized as follows: Where a licensee has chosen to broadcast a sponsored program which presents one side of a controversial issue, and has not presented (or does not plan to present) contrasting viewpoints in other programming, and has been unable to obtain paid sponsorship for the appropriate presentation of the contrasting viewpoint, he cannot reject a presentation otherwise suitable to the station—and thus leave the public uninformed—on the ground that he cannot obtain paid sponsorship for that presentation.

And as Mr. Lawrence Rogers has pointed out: ". . . it is only a coward who wishes to use the privilege of free speech to advocate only his own cause while he turns off the microphone to an opposing view."

So let me re-emphasize: If specific rulings in the realm of fairness are unreasonably burdening broadcasters, deterring their bolder efforts and causing them to adopt neutral, noncontroversial positions, let us discuss the matter and seek alternatives. Government requirements whose only justification is enunciation by government officials have no place in a free society. But we should realize that most of the rulings to which one hears objection logically follow from the basic principle of the fairness doctrine itself.

Here, I suggest, lies the key to the problem we are examining. Because the doctrine may be stated so simply, the usual conclusion is that the difficulty comes not with fundamental principle, but with the attempt to apply it to a myriad of different circumstances. Such, I believe, is not the case.

A member of the U.S. House of Representatives once summarized the general feeling on the subject when he told me: "I agree with the general principle of the fairness doctrine; I just disagree with the manner in which the FCC interprets and enforces it." At the risk of being thrown out of his office, I suggested to the Congressman that he had it backwards, and I suggest the same to you tonight (at perhaps the same risk). My belief is that the principle of the fairness doctrine is the tough part—that in practice our enforcement of this principle is reasonable and works as well as can be expected. If that principle is not accepted, we will continue to argue interminably over procedure. If that principle is accepted, 90 per cent of the arguments over FCC enforcement will vanish.

In the most basic terms of all, the fairness doctrine holds that a broadcaster who enters the field of controversy isn't perfectly free to propagandize as he sees fit. This is clear, unambiguous, and provides a well-delineated battlefield on which our war of words should be fought. Arguments over enforcement procedure are only guerrilla skirmishes in a swampy forest of rulings and precedents—often tiring, frustrating, and having little impact on the course of the conflict.

We must make our stand on the high ground, for it is here that the real decisions are made.

Having reached this ground, I have two basic things to say. The first is that I believe the fairness requirement is entirely appropriate.

Despite the growing number of radio and television stations, the frequency spectrum is a limited natural resource. Despite the growing sophistication of electronic technology, which permits the utilization of higher and higher frequencies, the demand increasingly exceeds the supply. The population explosion in the mobile radio services, for instance, is causing those users to look enviously at the large portions of desirable spectrum space now assigned to broadcasting.

The FCC is charged by law with the allocation of this resource to those who clamor for it with shining eyes and outstretched hands. It is thus the Commission, and not the marketplace, which determines how frequencies shall be apportioned between broadcasting, long-distance microwave, communications satellites, police cars, fire trucks, aircraft, ships, industry, amateurs, and a host of other useful services. It is the Commission, not the marketplace, which determines the maximum number of broadcast outlets any community may have, and the facilities that each may use. The FCC was created for precisely this purpose, and the Communications Act is based on the assumption that government must perform this function.

When government undertakes this kind of responsibility for a mass medium, it cannot avoid responsibility for the manner in which its licensed outlets are used. At a minimum, having excluded all but its chosen licensees from the medium, it has an obligation to see to it that those licensees do not themselves suppress free speech. This is the basic reason why Congress has written the fairness doctrine into Section 315 of the Communications Act.

And I would add that the fundamental fairness of the broadcasting medium—its openness to controversy, conflict, and dissenting views—is one of the most important reasons why our society can tolerate a disagreeable fact—the rapid decline in the number of competing daily newspapers.

At the same time, I recognize that no matter how reasonably the Commission may treat broadcasters who honestly try to be fair, the bare existence of the fairness requirement does impose some burdens on those who want to present programming on controversial issues. A letter from the FCC may not pose a realistic threat to a broadcaster's license, but it may take some work by top personnel to answer it. Planning for fairness will in any event require effort and imagination. Most important, an attempt to broadcast conflicting points of view will often require a greater allocation of broadcast time to controversial issues than might otherwise be the case.

But in the large view, the burdens imposed by the fairness doctrine are strictly secondary, and can neither deter nor prevent the success of a creative broadcaster who is seriously committed to provocative programming.

The creative broadcaster can reap good returns from programming directed to the important issues of our day—returns in prestige, in self-respect, and even in revenue. The fact that broadcasters usually rank high on the list of community leaders is not because they serve chunks of commercials between slices of old movies—it is because of their participation in community affairs and community issues.

The real difficulty lies with broadcasters who aren't seriously committed to the journalistic function or to the exposure of controversy.

Programming that represents a slight profit or even a loss does not interest them greatly. They carry it as part of a minimal public service effort, but they limit their commitment to the least that will pass muster with their community, with the FCC, and with their own conscience. They can be tempted into a venture beyond the minimum, but the least additional burden is enough to discourage them. For their primary interest lies elsewhere.

Their pole star is not the Peabody Award, but that idol of the airlanes, the latest Nielsen. If their program director's flirtation with his journalistic flame becomes too serious, the station manager quickly cuts his allowance. If the station manager gets similarly soft, or unduly enamoured of his own creativity, the appropriate corporate officer soon sets him straight. If this officer should by some oversight fail to react to this danger, the Board of Directors explains to him the facts of life. Finally, if the Board lets provocative programming frustrate stockholder demands for ever-increased earnings, the latter group eventually shows where the true power lies.

The obvious aim of this process is to keep foremost in everyone's mind one thought: Controversy may sell newspapers, but in this business it's the funny page that counts. Mr. Average Viewer will not consider buying your brand or brand X when an editorial has just made him apoplectic.

You may describe this process any way you will. But in our war of words, the ancient admonition still holds true: Know your enemy. Your enemy is not the fairness doctrine. If you find yourself confined within time segments so short that the only way to be stimulating is to be unfair—if you are allowed to be controversial only so long as the boss gets no letters from the FCC—your struggle is not with the fairness doctrine. Your struggle is with the forces in American broadcasting which were so colorfully castigated by Harry Ashmore.

Is journalistic broadcasting in America nothing but the tail on an entertainment dog? Is American broadcasting "ultimately bound to the highest-bidder morality of the marketplace"? And does it show any signs of developing "a sustaining tradition of public service of the kind the best newspaper proprietors still recognize and act upon"?

No government official can answer these questions. The challenge and the ability to respond are yours. But unless I much mistake the temper of this industry as a whole, and the aims of such conferences as this, your response will be a worthy one.

12. Hail, Blithe Spirit!

by Dalton Trumbo

Mr. Trumbo has won many honors as a screenwriter under his own name, and during the McCarthy era won a number of Oscars and Oscar nominations under other names. Mr. Trumbo says the title of the original manuscript copy for the article was submitted to The Nation *magazine under the title "Hail, Blithe Sprite," thus connecting one of the witnesses mentioned in the article who referred to herself as a "sprite" with the Shelley quotation. Apparently the allusion was not understood, and the editors changed the word to the present title. The article originally appeared in* The Nation *for October 24, 1959. It is reprinted by permission of the author.*

Who will ever be able to forget the days of our glory, before Sputnik I and Lunik II, when culture came to Madison Avenue, and eggheads drew better ratings than murderers? And who cares to estimate how many millions of savage little juveniles, denied their nightly beakers of blood, twisted sullenly while the old man roared his answers at the sweating slob on the TV screen, and mother filled the station breaks with gentle sermons on the cash value of education?

As for the characters who led this national assault on Parnassus—where could one find better proof of the richness, the bigness, the democratic variety of the American intellect? Illiterates and college professors, preachers and union organizers, club ladies and cowhands, actors and show clerks, cab drivers and jockeys, arthritic senior citizens, half-weaned babes, nubile females, pimply students, the dazed, the crazed, the lame, the halt, the grind: came they all to the booth, podium, board, or panel, and there worked they their steamy wiles. On July 16, 1956, Ringling Brothers and Barnum and Bailey Circus pulled down the Big Top forever, hopelessly outclassed by the grunting, twitching, stammering, grimacing herd of innocents and oddballs who prowled the American rumpus room from sea to shining sea.

The winners' pictures nested regularly on the front page of the national press. Editorial columns flowered with laudatory essays on the virtues of a country in which treasures of the intellect—or, at least, of memory—stood so high in public esteem, and paid off so handsomely. Sermons were preached. Lecture tours were arranged. Morals were drawn. School teachers looked nervously ahead to a time when conceivably they would be readmitted to the national community. Oceans of lotions, lathers, depilatories, and deodorants were sold. In Vegas they were laying eight to five on the renaissance by 1960.

And then they had to go and spoil it. Somebody hired this comedian as a "standby contestant," whatever that is. He stood by until a regularly employed contestant incautiously left her notebook in a dressing room. He filched the book. He read it. His heart stood still. Written there were questions *cum* answers for the next time around. Dear God, the show was fixed! Outraged conscience impelled him to legal threats.

The owners of the spectacle arranged for him to be paid $1,500, for which he signed a statement that the show was clean as a hound's mouth, and promised that he'd not talk to the law about anything. Shortly afterward, conscience nibbling once again, he lapsed. He began talking to all sorts of people until, in the lovely argot of his tribe, he "blew the whistle" on practically everybody. Asked why he hadn't stayed bought, he replied that the show was fraudulent, and he didn't regard the $1,500 as "hush money" any more. On this gamey note, the fat flew into the fan.

By the time the odor spread from Madison Avenue to the district attorney's office and the chambers of the grand jury, the TV industry, as it is called, was in movement: people investigating other people, people writing affidavits, people testifying boldly for free, people testifying shyly under subpoena, people dodging subpoenas, people leaving town, people coaching people, people lying, people being shocked, surprised, infuriated, betrayed, deceived, alarmed—and even indicted. Recusancy turned epidemic among former contestants, some of whom were cruelly trampled in the general rush to reveal how cunningly virtue had been snatched.

A young philosopher, whose testimony may ruin the life and destroy the career of his opponent, won $49,500, which is valid sugar for a philosopher, and then was forced to "take a dive." Although he had cheated on every rung of the ladder, it was the cheat to lose that troubled him most. He heard his name mentioned in NBC commercials "day after day," with rhetorical speculations whether he would "crack the $100,000 mark."

Gloomily he recalled, "And I was sitting there saying, 'No, he won't; he's going to take a dive tonight.' " As diving time approached he became "very upset"; he urged the producer "to let me play an honest game." But no: dishonest it had been, and dishonest it was to be. "On top of that, when I took my planned nose dive, I was forced to go out on a question about a motion picture which I had seen only three days ago. This can be embarrassing." Dishonestly missing a question he honestly could answer now seemed more reprehensible than dishonestly answering questions he honestly couldn't. It was a stern moment for a philosopher who hadn't yet "cracked the $100,000 mark."

Some time after hitting the canvas, at a speak-bitter luncheon with the producer, he remarked that his successor was doing very well. "Oh no, he's playing it honestly now," said the producer. Honesty and success being natural enemies in the TV jungle, our philosopher hustled out to bet

$5,000 at two to one that his former rival would go down next time around. Sure enough! Instead of $49,500, $59,500 now nestled in the poke.

As it always happens, security gave rise to afterthought. He found himself assailed by general nervousness and pangs of guilt. He developed a "strong feeling that I had myself engaged in the fraud for too long." It took ten months in psychiatry, and a good chunk of the swag, to disentangle id from ego from libido. Then he talked.

Others who had won less than the philosopher, suffered fewer pangs, or none at all. A union organizer got "sick of the business" after collecting $15,000, but not sick enough to turn down the additional $9,500 still owing him and later paid. A restaurant manager, who dragged an impartial $15,200 from two shows—one on CBS and the other on NBC—asked if he thought taking a dive "fraudulent," blandly replied, "I am not convinced of it." A producer earnestly explained that he was "trying to put together an exciting and interesting show and I never did feel there was anything terribly wrong with it."

Some contended they appeared on the show, not as legitimate contestants, but as paid entertainers: "If the truth were known, there wouldn't be any entertainment." A high-spirited housewife, who made off with only $1,460, refused to beat about the moral bush. She admitted her fraudulent performance may have made it difficult for rivals to come up with anything as "spritely." For herself, "I feel perfectly blithe about it. . . . They were having a happy time, I was, everybody was." And—bless her wise little innocent heart!—no one had the slightest reason to disbelieve her.

A producer who lied to the grand jury did so because "I was panicked, I was terrorized, and I did it." In 1957 and 1958, NBC had held a series of meetings with producers and agents of a gyp-show, and thereafter had issued a statement that charges against the show were baseless, and the network had complete confidence in its integrity. In this dreary autumn of 1959, a vice-president of NBC delivered himself of the bizarre opinion that his corporate master had "no reason to suspect any rigging," adding that he and his rustic colleagues were "very badly deceived." On top of this, a vice-president of the sponsoring pharmaceutical company gravely announced that he "was shocked." He hastened to explain that "Our position was such that until such time as we had any facts to back up the newspaper stories, we would sit tight."

A sixteen-year-old schoolgirl, lured by an ad for folk singers, sworn to secrecy even from her family, and Fagined into the mysteries of cheating, found it not so easy to sit tight. When rumors circulated about the show, the child's mother asked for, and received, from her daughter—along with such tears as every parent can imagine—the bitter truth. Later, over a bugged telephone, the producer urged the child to perjure herself if investigated.

Questioned about her reaction to the fraud, the girl said, "I didn't even think about it. . . . The over-all excitement appealed to me, I saw it more as an opportunity to sing than anything else." The plan, she explained, was for her to tie twice with a contestant, and "finally beat him, and the interest would be enough that I could bring my sister on the show."

Said the mother, whose family the producers had violated, "Any parent would be outraged when their child had been deliberately led to be deceptive. . . . But when a youngster is asked to commit perjury or led in that direction, I think it's disgraceful." The girl, now eighteen, asked if she thought her conduct two years ago questionable, said: "Yes—I'm here now trying to help clean it up." Which is much more than the sponsors, the producers, the advertising agencies, and the network executives are doing or intend to do.

And so the sorry tale unfolded. As almost always happens in the climactic passion of Congressional investigation, the real fraud consisted in the exposure of fraud. No Congressman can hope for headlines if he dwells on the carcinoma factor in cigarettes, the worthlessness of snake oil as a cure for senility, the calculated falsity of practically all televised commercials, or the arrogant greed of men who have appropriated the free air and turned it into a witches' bazaar of howling peddlers hawking trash. A sixteen-year-old girl makes safer copy than the president of NBC and has fewer lawyers.

In only one aspect of the scandal can we take real satisfaction: all who participated in the fraud were certified, loyal Americans. The elaborate system of blacklisting, by which the networks deny use of publicly owned channels to those with whom they disagree, makes certain of that. Everybody connected with the shows had been cleared by the American Legion, the House Committee on Un-American Activities, the Senate Internal Security Committee, AWARE, ALERT, *Red Channels*, sponsors' check-ups, the agencies' private eyes, the networks' corps of dedicated snoops. And Heaven knows how many private nuts, crooks, and crackpots.

The people who dived, and the people who won, and all who arranged the cheat and sponsored it, and distributed it, had never been controversial; they had never publicly dissented from anything; they had never joined a *verboten* organization; they had never given money to unpopular causes. To the last child they were authenticated patriots, well-oathed and clean as the whistle that finally blew them up. Though tens of millions of dollars were earned by sponsors, broadcasters, and producers of the fraudulent shows, though the trust of a nation's children was ravished by them, at least the Republic could take comfort that it hadn't been gulled by a gang of subversives.

Once that crumb has been digested, the rest is nightmare. The eerie landscape of Madison Avenue, perceived through the private agony of gifted men who mirrored and now must atone for its knavery, reveals the future, which there has been projected for us—a future boldly rigged for the naked worship of things and self, animated by a materialism so primitive that it is incapable of developing either philosophic basis or moral objective: a future of true godlessness, of pure degeneracy, of corruption absolute.

It was not intended this way. In the early twenties, David Sarnoff prophesied of radio broadcasting: "As the picture will become plainer, there will emerge, in radio, musical foundations, operatic foundations, and lecture foundations, endowed or supported by great public-spirited Americans, who will see in this vast instrumentality of the air another means to become benefactors." What did we get? *Twenty-one, Tic Tac Dough, Dotto,* and *The $64,000 Question,* not to mention the assorted incitations to murder regularly purveyed in Western and detective dramas, and the general reduction of sex from love to lust.

In 1922, when it was learned that station WEAF, in New York, had sold air time to hawk real estate, Secretary of Commerce Herbert Hoover, whose department gave the boys their licenses, said: "I don't believe there is anything the people would take more offense at than the attempt to sell goods over radio broadcasting." (They did not take offense, and neither did Mr. Hoover. Instead, they surrendered their free air, as an absolute monopoly, to a gang of merchandisers whose ethical standards are defined in a single question: Will they believe it?)

By now it no longer matters whether we believe, only that we acquiesce. And we do acquiesce. We expect the news to be slanted; we expect the statesman to lie; we expect the politician to make deals; we expect the advertisement to be false; we expect the repairman to cheat us; we expect the fight to be fixed; we expect men to place self-interest above any conceivable social end.

And when our expectations are fulfilled—when the fraud is finally revealed—we are never surprised, and rarely angry. Publicly and before the children, we deplore it. Privately, we admire its audacity, and marvel that it went undetected for so long. We sharpen our wits on its details (but never its cause) and are wiser citizens for what it has taught us.

The unlucky young men of TV, whose downfall we shall applaud as all good Philistines must, haven't really harmed us. They haven't violated our innocence. We had no innocence. We never did believe.

That is the crime we have committed, not against ourselves alone, but against our children, our country, and even the broken victims of prostituted television. It is a crime that renders us morally unfit to judge those who have betrayed us. They have been judged enough already. They have suffered the anguish of exposure; their lives have been soiled; one or two of them may go to jail on peripheral charges of perjury; the networks and sponsors who financed their trickery and sold it and made millions from it have virtuously cast them into limbo.

Let's forget the whole sad mess. Let's close the book and wait for the next show, Congressional or quiz. Let's remember with John C. Doerfer, chairman of the Federal Communications Commission, that "sometimes we have to endure ill for the over-all public good. We can't have everything perfect."

Of course we can't. We don't even expect it. Yet there are occasions, late at night and alone, when I'm touched by the chill apprehension that we no longer live in America, we merely occupy her; that we no longer love her, we only sit in the bleachers and root.

13. Traitor to My Class

by Shelby Gordon

Mr. Gordon's perceptive look at television was based on more than a dozen years association with radio and television. Once a script editor for the CBS-TV science series, Conquest, *he was at various times a writer, producer, and director, and for three years was associate producer of the prize-winning CBS television series,* Adventure. *As one of the first Mass Media Fellows for the Fund for Adult Education during 1958–59, he spent the year studying at the University of Southern California. The article is reprinted from* Mass Media *for July, 1959, and is reprinted here by permission of The Ford Foundation. Mr. Gordon is deceased.*

I've had a vaguely disturbing sense of guilt lately. I try to tell myself it's just another baseless anxiety like my acrophobia, but the feeling persists. To begin with, another year has gone by and I'm still unnominated for an Emmy. With five candidates being named in each of forty-two categories over a period of something like eight years, the opportunities I've missed for distinguishing myself as a television writer are pretty considerable. It isn't that I don't try. I do, I do. I'm tireless. I'm careful. I can prove, through a series of mystical syllogisms, that I'm not a mediocre writer, but

I'm afraid I just don't have that tiny measure of greatness needed to qualify for television's highest award.

Still, I don't think that having anonymity thrust upon me is at the bottom of my disturbance. It's another phenomenon, quite the reverse, that occurs usually at cocktail parties, alumni meetings, and other similar gatherings where I find myself rubbing elbows with strangers, particularly intellectual strangers. Naturally, there's always going to be someone polite enough to ask me what I do.

I learned long ago the folly of telling them. Truth leads to complications, so I lie, but lying became a problem when I realized I was beginning to look for status with my fictions. It was all right as long as I said I was in awnings or ladies' purses, or even a middleweight fighter; but when I became, for the moment, an anthropologist or neurophysiologist or special assistant to the Director of the Office of Unemployment Stabilization, my wife accused me of cheap social climbing, and I abandoned the practice. She advises me to tell people I make ice cubes because everyone is already making ice cubes, and I must confess that it's been the perfect compromise in a bad situation: modest, provocative, and safe.

The trouble is that there's always someone present who knows what I really do, and it's just a matter of time until my secret is out. I'm confronted by a series of reactions ranging from curiosity through contempt to indignation. "Tell me," a nice lady says, "can't you do anything about those Westerns you fellows are always writing? I mean, I caught my kids looking at one the other day, and all that blood!" "So you're a teevee writer," says the portly gentleman with the bourbon neat. "If I couldn't knock out better stories than those corny whodunits you fellas are turning out, I'd quit writing if I was a writer. All you need is a new way to knock a guy's brains on the floor and you got a series." "Let me tell you, young man," says Princeton, '33, "you're corrupting the morals of our youth, trading on violence and brutality, and for what? Money! Can't you realize what you're doing is immoral? Can't you put an end to it?"

Well, no.

I have no quarrel with the premise. That there is a disproportionate number of Westerns and mysteries on television is a platitude. There's no argument, either, that they expose children to an enormous display of violence they could easily live without, and that a great portion of television entertainment is a questionable instrument for shaping and adjusting young personalities. The public believes it, the critics believe it, the government believes it, even broadcasters themselves believe it. I'll go along with the crowd. I suspect that Westerns and mysteries are at least partially responsible for some juvenile crime. I'm apprehensive about being employed by an industry that traumatizes young minds by an abiding presentation of brutality and deprives parents of the things they really want to see, like Shakespeare, the Philharmonic, and Dylan Thomas. Something ought to be done, of course, but what and by whom?

It seems fairly obvious that replacing horse operas and detective thrillers is no answer to improved program quality. When the big money quiz shows were discredited and dropped en masse, their replacements were no more representative of improved quality than the programs for which they were substituted. Many, in fact, were mysteries and Westerns. This is only surface evidence, but it should serve to illustrate what seems to be a basic point. The elimination of an entire generic form of entertainment is absolutely no guarantee that the holes will be automatically filled with material of a higher standard.

Perhaps this is an illustration, too, that almost all the criticism leveled against the brutal aspects of those programs, while valid, is pretty superficial, as platitudes often are. It seems strange to me that I've never heard any professional critic or educational broadcaster or educator, to mention the people with whom such conclusions ought to be most at home, utter what I think is an absolutely inescapable piece of logic: Westerns, crime shows, and violence itself are not the disease that troubles television. They're symptoms of a sickness that saps the strength, not only of television, but of our whole society.

Since I'm not an accredited philosopher, I don't want to speculate publicly on the nature of the disease, but I think I'll have little argument when I say that another symptom of it is a far-reaching conformity. Now, whatever else can be said about conformity, whether it's healthy for the social structure or unhealthy, it isn't good for entertainment. The irreducible, minimum requirement of drama, both fictional and nonfictional, is a conflict of forces. No matter how basic or how contrived they are, those forces are representations of human values or ideas, and the very presence of a conflict implies controversy.

It's been some years since controversy was a welcome commodity. It lost its appeal in television at about the same time it fell out of favor with the general public. Most people in television, after all, except for a greater susceptibility to gastric disturbances and coronaries, are fairly normal. Their attitudes, convictions, and fears are developed in common with their contemporaries. Their principal preoccupation is personal advancement, and they bend to the same economic pressures as other businessmen or the same social pressure as the country club set, the sewing circle, or the fellows down at the bowling alley.

Audience ratings aside, the pressures have been considerable. During the past ten or twelve years, a relentless parade of special interest groups, most of them probably not very large, has presented the networks with a series of demands for the expurgation of certain kinds of material, always with the threat of economic boycott; while virtually no similar pressure has come from the other direction to support the wisdom of controversy. It's true that subscription to popular notions doesn't in itself guarantee success in the continuing popularity contest between the networks, but the best evidence available about public opinion indicates an absolute and predictable relationship between the expression of unpopular ideas and inevitable unpopularity.

It shouldn't really be surprising that a profit-making television network, operating within the private enterprise system, reflects what it considers to be the public attitude. The trouble is that the great effort to offend no one in order not to jeopardize profits has created so many sacred cows, it's difficult to present any idea that isn't regarded as disputatious.

The result has been a systematic reduction of the power of television, and I think there is no other cause for the serious degeneration of program quality. Television drama fought a losing battle for years, and the holding action being maintained by some of the ambitious nighttime dramas is equally hopeless, I believe, because their repertory of ideas is so limited. It's tiresome when the same one triumphs all the time, and particularly dull when the opposing force is always the same. Even truth, or at least that which we believe to be truth, loses its edge under the circumstances, and the devil himself degenerates into a pitifully ineffectual bogeyman, too worn from repeated exposure to a sophisticated audience to have his proper effect, even on little children.

When ideas are unusable and the conflict of dramatic forces must be resolved in the triumph of clearly acceptable, publicly approved sentiments, it doesn't really pay to cloud the issue with expensive production and complex characterization. Why not reduce the conflict to its simplest possible form: lawlessness and law, the black hat and the white hat? I guess that's the reason there are so many Westerns and crime shows.

As for the violence, I suppose that the people who produce those shows, unlike the producers of more pretentious drama, are not deluded that they have any interesting point of view; and they're theatrically wise enough to realize that they must present their offerings as a purely emotional experience, devoid of any intellectual content. There are just about four ways, excluding the use of ideas, to get any emotion into a television drama: a fight, a chase, a song, and a love scene. Because all Westerns and mysteries are telling essentially the same story, and because not every actor can sing, competition is always in the direction of bloodier fights and more ingenious chases. We should be grateful for the unwritten law of the horse opera that cowboys don't kiss. If they did, American kids would probably become the most erotic in the world, and all through competition, all through the danger of ideas.

The wonder to me is that the damage hasn't been far greater. There is still an area of broadcasting called public affairs that gets a pretty good share of the available time, although I wish some of these affairs were made a little more public by programming the shows at better times during the week. The convention in broadcasting, however, is that people who think don't like picnics or Sunday drives with the family, and so Sunday is the time for virtually all the nonfiction television. The Columbia Broadcasting System is particularly generous with Sunday time for these programs. Many of them are universally well regarded. Still, I think that perhaps even some of the best of them suffer the same symptoms that trouble the dramatic shows. Historical programs are too often presented as a sterile recital of events without a sense of continuity with the present. Science, in most programs dealing with it, is treated as a world apart. It's only occasionally that either type of program has anything to say about the relationship of history or science with our present environment, and I think the reason might be the difficulty of doing so without rousing some new expression of public indignation.

I don't wish to imply that censorship is the official policy of the television industry. Although it occurs occasionally, the general practice is to let the producer make decisions without executive interference. The fact is that censorship usually is not necessary. There was a time, I think, when it was very severe. Congressional investigating committees, citizens groups, and even individuals who claimed the power of economic attrition badgered the networks into taking their stand and forming a policy. People were blacklisted as a result, and so were ideas.

Writers who took their function seriously enough to write plays examining the convictions that governed their lives or the behavior of society found generally that they couldn't sell their work, not because they were blacklisted, but because their themes were considered potentially offensive to some group or other. If their work was somehow sold, there was a good chance that it would appear with great, gaping holes where an idea was deleted or distorted to eliminate the possibility of controversy. In time, they found they couldn't make a living because their only stock in trade was ideas. They couldn't adapt themselves to the new realities, and left television. Some continued writing—plays, novels, even films. Others, it's sad to say, pumped gas or sold neckwear or went into factories.

Those who remained in television, of whom I'm an example, dealt with the situation by a self-imposed censorship. We expunged our ideas even before they were formed, knowing in advance they were unacceptable, and if it's true that our stories were pointless, at least they didn't have any great structural weaknesses, because every scene was inoffensive and stayed in the script. There were a few writers lucky enough to escape censorship altogether because they had nothing to express but platitudes anyway. A good many of them eventually became producers. Their great value is an infallible instinct for the lowest common denominator.

Today, there's no censorship problem to speak of. New writers who are attracted to television generally have nothing to say. The rest of us have forgotten how to say it.

I would say that's the great loss to television and to the public. That lonely fellow, the writer, who works always by himself and always in great agony, who grapples with ideas and tries to flesh them so that all of us can have a look at ourselves, is gone.

All that remains is the television writer. We deal with ideas, too, of course, but less idealistically. They're just elements of a puzzle to us, and if it means a sale, one's as good as another, even if it's diametrically opposed to the original. We're even willing to write material with no ideas at all, a difficult feat, but one which every one of us has mastered in the struggle to survive in our field.

We're businessmen. We fill a need. We supply a commodity, even Westerns. We shudder when our kids see them, just like anyone else, but we continue to write them just the same because they pay as well as any other kind of drama on television and because they're just as valid. Black hat, white hat; the law versus crime; there's not much more of an idea than that in the biggest productions on television.

I think we've found our proper niche. I'm absolutely convinced that no one writes down. No one deliberately adulterates his stuff to reach a level of intelligence or artistic acceptance lower than his total available capabilities. Beauty of style and depths of a penetration into human understanding of which a first-class writer is capable will be evident no matter what form the writing takes. Those other writers go out of television because they couldn't write down. If they were around today, they could do something about Westerns. You can't stifle an idea just by dressing it in Western costume. You have to cut the heart out of it, and that's what was done.

As for those of us who remained, we can't write better. I honestly think that if the public hadn't rescued us by supporting a program of conformity in the development of its most powerful medium of expression, most of us would be out of television today. As it is, we are where we belong. We're proud of our professionalism and grateful for the liberal rewards we're paid. We're worth every cent we get, too, and maybe more, because the public is getting just what it wants.

Perhaps that's not entirely true. It's probably more accurate to say the public is getting just what it deserves. If the economic threats of a handful of people can convince three networks of the folly of ideas, a hundred and seventy million people ought to be able to reverse the trend. The profit motive is very compelling, and there should be no difficulty in getting the networks and the advertisers to do something about Westerns, crime, and brutality on television if the public really has strong convictions about it; but as long as public morals continue to corrupt television, it will continue to corrupt the morals of our kids.

14. Here, We Would Suggest, Is a Program for the FCC*

from Consumer Reports

The following article appeared in Consumers Reports *for February, 1960. It is reprinted from that publication with permission by Consumers Union, publisher of* Consumers Reports, *who hold copyright.*

Early in December, the Federal Communications Commission held hearings in Washington, D.C., to which representatives of various consumer and professional groups were invited. The Commission asked for expressions of opinion on current TV programming and suggestions for improvements. CU [Consumers Union] was among those requested to testify. A summary of CU's proposals made at the hearing are presented below, and on two counts CU wishes to call special attention to the details of its program:

In the opinion of the best-informed members of the press who have been following the TV scandals, neither the FCC nor the Congress will take significant remedial action unless consumers—that is, TV-set owners—bring insistent and consistent pressure for reform. Congress is expected to turn soft on the issue because this is an election year, during which TV time for office-seekers will be sought, and also because a number of Congressmen are reported to have close financial ties to the broadcasting industry.

Although there was no previous consultation among the various groups and individuals invited to testify, many of the suggestions outlined in the CU testimony also were made by other witnesses. Thus, the most important revelation to come out of the FCC's recent hearings was evidence of a broad agreement, among groups interested in the public welfare, about what must be done to guard against the abuse of broadcasting privileges and against corruption in the governmental regulatory agency, the FCC.

A 12-STEP PROPOSAL

In its testimony, CU took a positive tone, spent no time reviewing the recent and multiplying exposures of malpractice on the part of both the broadcasting industry and the FCC, but put emphasis rather on what might be done to build into the FCC's operations a living concept of the

* In recent testimony before the FCC, CU proposed a practical plan to rid TV and radio of such ills as "payola" and commercial censorship. *Consumer Reports*, January, 1960.

public interest. The 12 steps outlined in CU's testimony would require no changes in legislation, only a change of heart, ethics, and point of view on the part of the Commission. They were:

(1) Set up, as advisory to the Commission, a Television and Radio Consumers Council with full power: (a) to review all FCC licensing decisions; (b) to request, if necessary, additional data on a licensee's performance; and (c) to publicize its findings.

Neither the FCC nor the Congress gives evidence of being aware of the importance of the TV- and radio-set owner in the economics of broadcasting. Nor are most consumers fully cognizant of the central role they play when they purchase receiving sets, or of their rights and obligations as set-owners to exercise a dominant influence over programming and advertising policies.

So far as investment goes, the consumer's stake in the radio and television industry outranks that of any other segment. Conservatively figured, consumers have paid out $10,000,000,000 for the 50,000,000 TV sets they own. Another $10,000,000,000 has been spent on repair and maintenance of these sets. The power that operates them an average of four to five hours a day also comes out of consumers' pockets. Furthermore, when they buy the goods advertised over the air, consumers pay all the costs of the programs as well as the operating costs and earnings on investment of the broadcasting stations.

For its 100,000,000 or more radio sets, the public has spent, roughly, another $2,000,000,000.

Thus, leaving out the cost of power to run the receivers and taking only $15,000,000,000 as a conservative estimate of what was spent for TV and radio advertising in the past six to ten years, we find that consumers have invested the impressive total of $37,000,000,000 in this industry.

(2) Make hearings mandatory in all license renewals. This point was emphasized a number of times at the hearings. To obtain its highly prized license to operate—the right to use the public domain: an air channel—a TV or radio station vies with other contestants by, among other things, promising programs of quality. Every three years, according to law, a broadcasting license must be renewed. Although the FCC has the power to hold a renewal hearing to determine how well the licensee has carried out his promises, it has been most lax on this score. In 1958, for example, out of the 1500 or so radio and TV renewals granted, only two renewal hearings were held (see *Consumer Reports*, January 1960).

(3) Hold all such renewal hearings, as well as new license hearings, in the locale of the broadcasting station, so that the community to be served may be heard.

Since community tastes and interests may differ, local set-owners are, obviously, the ones best situated not only to know what they want over the air, but what the actual program content has been. It makes little sense, therefore, to hold license hearings in Washington, D.C., where attendance is limited, practically speaking, to the industry's lobbyists, attorneys, and public relations personnel.

(4) Publicize the renewal hearings over the stations involved for a given number of days at fixed hours, and invite public participation in the proceedings.

A public hearing of which the public is unaware is hardly public.

(5) Require each broadcaster to maintain for public investigation the commitments he made with regard to programming and advertising when he was granted his license.

The promises made by the broadcaster as a condition to his receipt of a license are not, practically speaking, a matter of public record unless access to such commitments is made easily available to the particular consumers whose sets receive the particular licensee's broadcasts.

(6) Require each broadcaster to carry, at least once a week during prime time, a statement of the basis upon which he holds his exclusive privilege to the public domain. And as a part of this weekly announcement, licensees should be required to invite set-owner comment on the station's program content, advertising policy, etc. To receive this continuing public check on performance, the FCC should maintain in each broadcasting area a local post office box to which set-owners would be invited to mail their comments.

Advertisers and their agencies are now exercising a rigid censorship over program content. The fact that this is a private censorship exercised for private ends renders it doubly hazardous to the public welfare.

Only when avenues are opened up for *free* and *considered* response from set-owners and their organizations oriented to the public welfare can we design a means by which to escape the rigid commercial censorship now exercised over this all-important medium.

(7) Establish in each of the FCC's 24 district offices a consumer review staff to read and classify public responses, and then to further the material to the Consumer Advisory Council of the Commission. After review, this material should become a matter of record with the Commission and the broadcaster concerned. Both at FCC headquarters in Washington, D.C., and at the place of business of the station involved, a continuing account of these consumer responses should be maintained for public inspection.

The misuse of commercial program ratings to dictate program content long has bedeviled the efforts of the conscientious creative talent engaged in preparing TV and radio programs. These commercial checks on listener response are far from reliable. The results of one such poll will not infrequently contradict the findings of another. Furthermore, the method of collecting reactions via spot telephone calls is not only irritating to householders but cannot produce any *considered judgment* of program quality from listeners and viewers.

A function as important as the reflection of public reaction to program content should under no circumstances be left in the hands of the advertisers, whose first interest is not program content but competitive sales pressure. Further, because an accurate and reliable report of public reactions is vital to the conduct of broadcasting in the public interest, CU recommended that the first review of that response be done in regional offices in order to minimize the possibility of misinterpretation or misrepresentation of reactions. The comparisons and contrasts of response between and among the regional offices can serve as a partial defense against manipulation of the record. As a further spot-check, the Consumer Advisory Council should be empowered to conduct telephone polls periodically if the Council feels a counter-check is required.

(8) Require that all Federal Trade Commission and Food and Drug Administration citations against advertising carried over each station be on file and available to the public at both the FCC's Washington headquarters and at the licensee's place of business; and make these citations a part of any licensee's record submitted at renewal hearings.

A broadcasting licensee cannot, in the conduct of his business, be compared to most other private-enterprise ventures. The broadcaster's most valuable holding, his license, is given to him by the Congress via the Federal Communications Commission. He is, therefore, a trustee, a manager of public property.

Furthermore, this particular trusteeship over the public domain carries with it the unique opportunity to enter the home of every set-owner and to use the householder's private property, his TV or radio set, as a source of revenue for himself, the broadcaster. What a broadcaster sells to advertisers is not transmission of a program but its reception. It is the consumer's set which, basically, earns the broadcaster's advertising revenue.

Thus, the licensee has a special responsibility for the ethical standards of the advertising he transmits, since his license grants him a monopoly privilege to capitalize on the consumer investment in receiving sets.

(9) Require of all licensees, as a condition for obtaining either an original license or a renewal, a declaration of advertising policy which shall contain: (a) the standards by which the contents of commercials shall be evaluated by the licensee—standards to screen not only false and misleading statements but also bad taste; (b) a limit of both the number of commercials to be permitted in an hour and the total amount of time to be given over to commercials in each separate hour of broadcasting; and (c) a review of the means by which the broadcaster will screen from both advertising and programs all sources of commercial corruption such as "freebies" and "payola," etc.

A "freebie" is a pictorial boost for a nonsponsoring product, provided via a gratuitous display in either the program or commercial. And a "payola" is, as everybody knows, payment made to play particular records. These two examples do not exhaust the sources of commercial bribery in TV and radio, but they do serve to illustrate the nature of the problem.

Above and beyond such huckster pay-offs, however, TV and radio advertising suffers from almost a total lack of control at the local level because the Federal Trade Commission (the agency empowered to enforce the truth-in-advertising law) has power only over interstate commerce. When the offensive ad for a local product or service is broadcast over a station whose signal strength does not exceed state boundaries, the FTC has no jurisdiction. Of course, through its licensing power, the FCC could exercise control over local commercials; but it has failed to do so, with the result that TV and radio consistently carry much of the worst "bait and switch" advertising.

In addition to local frauds, TV commercials in particular, even those of national advertisers, frequently are both irritating and repulsive—the gamy statues in the body-odor ads, some of the words and postures used in soap and toothpaste ads, the blaring forth in the living room about hemorrhoids or constipation.

The fact that bad taste in advertising is not illegal does not mean that set-owners should be forced to endure it.

(10) Require the declarations of advertising policy to be posted for public inspection in each licensee's place of business and also to be available on file at the FCC's Washington headquarters. Include in renewal hearings a review of the effectiveness of a licensee's advertising control as well as his fulfillment of promises on program content.

The difficult problem of monitoring radio and TV commercials might be mitigated by volunteer viewers if advertising codes of ethics were made available to them.

(11) Prohibit the sale of any license without a full-scale re-hearing on the transfer of the privilege.

Under present FCC rules and regulations, a license granted to one party on a given set of promises made at an official public hearing may be transferred through private sale to another party. Frequently, such transfers earn great sums for the original holder, who obtains his license absolutely free but who sells it for thousands, and hundreds of thousands, of dollars. If full-scale re-hearings before permitting such a transfer were required—hearings where competitive bids could be entertained—the use of this transfer privilege as an avenue of corruption could be limited.

(12) Set up a graduated system of licensing fees based on two considerations: station signal-strength and advertising revenues.

The licenses which the FCC grants are, under present rules and regulations, absolutely free gifts to their holders. The high dollar-value of these gifts is obvious in the fantastic earnings of the networks and many of the stations. In most other industries operating under government license, the cost of regulation is borne not by the taxpayers, as it is now in the case of broadcasting, but by the licensees themselves, who pay a fee for their privileges.

The cost of putting the whole of the program recommended here into effect could be financed by such fees without hindering the development of TV and radio services.

THE BLUE BOOK

One additional suggestion made by several witnesses at the FCC hearings, but not included in CU's testimony due to lack of time allocated for presentation, dealt with the famed *Blue Book,* a statement of basic principle made by the FCC in 1946 to guide broadcasters in their determination of public interest in programming. Although no heed has been paid to those principles for many years now and the *Blue Book* is out of print, it never has been repudiated. Its immediate revival and re-issuance were urgently called for at the hearings and CU agrees that, in addition to opening up the FCC to public response, it also is important that both the broadcasters and the set-owners be made aware of what good broadcasting practice means.

One of the most respected newspaper commentators on the problems now besetting TV, Jack Gould of the *New York Times,* had the following to say about the Blue Book:

> A slim pamphlet called "The Blue Book," now out of print but once the target of violent abuse by the nation's broadcasters, may yet turn out to be something of a prize volume—thirteen years late. . . . The history of "The Blue Book" might well be used as a summary of today's controversy over television; all that remains to be seen is whether the ending is going to be different.

III. Freedom and Responsibility

All too often, critics frequently claim, broadcasters want freedom but shirk the responsibility that goes with it. The serious conflicts which sometimes arise because of the economic base or sponsorship approach of United States broadcasting deserve recognition and discussion.

How do high executives in various national systems see these responsibilities? How do the pressures and conflicts which arise affect writers and performers? How well understood is the role of the writer? How responsible is top management for compromises which newsmen make?

These generally unavailable selections are intended to generate discussion leading to clarification of the meanings of freedom, responsibility, and "in the public interest."

15. Who Forgot Radio?

by Edward P. Morgan

Mr. Morgan is a well-known Washington broadcasting commentator. His remarks were part of a speech delivered in the lecture series "The Press in Washington," at the American University in Washington, D.C., on March 30, 1965. The essay also appears as a part of a collection in the book, The Press in Washington, *published by American University. It is reprinted by permission of Mr. Morgan and the Washington Journalism Center, who hold copyright. Ray E. Hiebert, Director of the Center, edited the book.*

I'm not very hopeful but I'd like to live to see the day when the public, its ears aching from the drumbeat of banality, would demand a marketable Dick Tracy two-way wrist radio in order to answer back at the awful stuff on the air waves. Maybe people don't care, but as far as I'm concerned, the broadcasting industry has littered my street of dreams with garbage. I am, I confess, something of a romantic about the medium of radio. Long before I ever worked in it, I was bewitched by its magic. Sometime in the middle twenties my father brought home one of the early DeForest sets, a big black box with a console as complicated as the instrument panel of a jetliner, a combination tube and crystal job with "cat's whisker" tuning and a revolving diamond-shaped aerial in whose wires I, figuratively, became hopelessly entangled. My night sounds had been the sharp, haunting bark of coyotes but now the boundaries of my world suddenly dilated far beyond the sagebrush hills of Idaho, and through the hissing swish of static, like a bell pealing in a snowstorm, came the sweet, wavering voices of KHJ, Los Angeles, KDKA, Pittsburgh, and, one enchanted evening, Havana, Cuba.

Radio has not lost but technically refined its magic. To be sure, we now have the "tummy TV" (you've seen the ad for the miniature sets) and satellites can now bounce pictures instantaneously from London and Tokyo into your bedroom. But long before Telstar, radio was hauling in the noises of Istanbul and Hong Kong and reflecting them, in the full color of imagination, on the private picture screen of the mind. Today, you can take the pulse and temperature of Boston or Bangkok through an instrument no bigger than a diamond brooch, a tiny collection of transistors lost in the palm of your hand.

But turn it on and what do you really get? A medicine man's pitch to swallow a pill that will end backache by producing a mild diarrhetic action through the kidneys; a promise that some sugar-coated physic will regularize junior's bowels; a doom-cracking bulletin on the latest East-West crisis calculated to paralyze any adult intestinal tract; a report on the flow of traffic in the approaches to Main Street (it too is paralyzed), and then the loud rolling whang of the latest assault on folk music, to be followed in machine gun tempo by more commercials dinning, on ears already split, the world-stopping bargains in used cars, detergents, and cake mix plus the prestige and well being to be gained by switching to ragweed filter cigarettes. Now and then, as if he were extracting a pebble from the profitable shoe of commercialism, the announcer will toss out a public service message urging blood donations to the Red Cross, cash contributions to the Community Chest, and reverent attendance at the church of your choice.

Through the nineteen-twenties, the thirties, and most of the forties, radio reigned supreme as a medium of communications—though by no means supremely undeserving of criticism. There was Graham McNamee and the Rose Bowl games; FDR and his fireside chats; Bill Shirer reporting from Berlin the rise of the Third Reich and the beginning of Hitler's war; Ed Murrow from London bringing home to us nightly the agony and the heroism of the blitz; Elmer Davis twanging out his clean, cool five minutes of wisdom each evening from New York; Howard K. Smith with his incredibly sharp and informative Sunday half-hours from Europe; Eric Sevareid and his penetrating analyses from Washington and elsewhere. There was good drama along with the soap operas; *Information Please* along with the amateur hours; the *Town Hall of the Air* as well as dull harangues; not only vapid foolishness but dimensional documentaries and such wonderful recaptures of history as *You Can Hear It Now*. Fifteen-minute newscasts, if you can believe it, were not the exception but the rule, and Father Coughlin was not the only controversial commentator on the air.

Then the networks, national advertisers and, presumably, the public eloped with a brazen but seductive hussy called television and radio suddenly became an abandoned orphan. Like many a neglected juvenile it developed alarming symptoms of delinquency. It is against this tawdry background that the operation of radio coverage of the news from Washington must be examined.

In a puzzling but provocative recent book entitled *Understanding Media*, the director of the Centre for Culture and Technology at the University of Toronto, Professor Marshall McLuhan, writes "one of the many effects of television on radio has been to shift radio from an entertainment medium into a kind of nervous information system. News bulletins, time signals, traffic data, and, above all, weather reports now serve to enhance the native power of radio to involve people in one another." The frustrated broadcast journalist in Washington would like to involve people more perceptively with the news, to make radio, for example, an information nervous system rather than a nervous information system. Instead, with exceptions that only tend to prove the rule, he is forced to squeeze his report on the national scene into a breathless two-minute package that often is shrunk to the ridiculous capsule of 30 seconds.

In just 15 years the membership of the House and Senate Radio and TV galleries has more than doubled. In 1950 the total was 218 active and associate accreditations. (Associate members are based out of town, have gallery privileges when they come to Washington.) In 1964 the total was 491 accredited broadcast correspondents. Both the House and Senate galleries, ably presided over by Robert Menaugh and Robert Hough respectively, are proverbial beehives of activity. According to the most recent available count, they registered, for an 11-month period 1225 radio interviews; 188 radio panel shows; 1175 news "spots" from the Capitol, broadcast live, on tape, or by telephone "beeper." There were 877 TV interviews during the same interval. Members of Congress, both Senators and Representatives, have come increasingly to use the broadcast facilities provided for them in the bowels of the Capitol building. Some 335 Representatives, for example, regularly utilize the radio or TV studios (most of them use both) for weekly or bi-weekly recording and films cut for 5- or 15-minute programming on the stations in their constituencies. Congressmen frequently participate in special shows on such issues as poverty, medicare, immigration, et cetera. These specials average two a day.

Quantity-wise this all sounds very impressive. Quality-wise it is too often lacking in informative wisdom though there have been some excellent reporting and excellent programs emanating from Capitol Hill. CBS correspondent Roger Mudd's marathon daily progress report on the 1964 Civil Rights Bill was a unique if sometimes repetitive experiment. When he was New York's junior senator, Kenneth Keating's weekly interview and comment stint was a frequent must as a source of news, and colleagues—Republican and Democrats—considered their guest appearances with him almost as important (and often more comfortable) than the standard Sunday panel shows of the networks.

By and large, I suggest, radio reporting from Washington—and from almost everywhere else, for that matter—suffers most from what might be called instant spasms. Pour hot water over a dehydrated concentrate and you get instant coffee. Bring a processed cereal to a boil and you have instant oatmeal. Grind a news story down to a palmful of facts, pour on the audio for 40 seconds, and serve. The quality of this journalistic spasm does not compare with instant food and is likely to cause mental indigestion. I am not arguing that you cannot impart a lot of information in intervals of a minute or less. Sentences such as "I love you" or "drop dead" can transmit a world of meaning in less than two seconds. I am arguing that the frenetic framework into which most broadcast news is now compressed produces a dangerously superficial picture.

There are at least a couple of other abominations. One comes under the heading of "actualities." The other could be labeled the compulsive monster of microphone-itis. Radio is sound but to justify itself in a broadcast, sound must have meaning. That includes the sound of voices. A tape recorder is a marvelous electronic butterfly net in which to capture the noises of the news but too often the process is carried to lengths that are utterly absurd. A stunned victim's halting phrases may give an on-the-scene sense to the report of an accident. There may be a legitimate measure of disaster in the noise of a hurricane, a mark of authority to the warden's description of a prison break in his own words and accent. But too often these actualities, while audible, are unintelligible. I would rather get my facts on a story from a trained newsman than have the information impeded by the mumbling voice of a so-called eye-witness butchering the English language. How many times have you heard a forest lookout, a traffic cop, or a passerby drone out a completely unilluminating version of an event that an experienced broadcast reporter, once he had the facts, could have put more clearly in half the time?

I'm not talking about documentaries or interviews in which there is time to turn around with questions and answers, but about these quick and interruptive splices of strange voices to establish the mobility of datelines: Radio has a portability that TV may never be able to match. I have had the good fortune to cover Richard Nixon's vice-presidential trips to Africa and the Soviet Union, President Eisenhower's extraordinary journey to India, and President Kennedy's historic visit to Berlin. Thanks to patient engineers who applied their special talents to unusual circumstances and were able to put the right circuit jacks in the right holes on the right wavelengths I was able to maintain my nightly quarter-hour program with hardly a hitch. I was even able to get through from such unlikely places as Monrovia, Liberia; Tripoli, Libya; and Sverdlovsk in Siberia.

The so-called "actualities" could indeed embellish a newscast far more than they do if we Americans on the whole didn't have such abominable diction or enunciation and could speak a simple sentence. While we're on the subject, interviews and panel shows could be far more lively and "listenable" if the protagonists could articulate more lucidly their thoughts, if any. What is there about the American way of life, upbringing, and education that makes so many men speak in monotones and so many women squeak like shrill shrews? We could learn a thing or two from our British cousins in the art of speech and expression and I'm neither demanding nor expecting a nation of Winston Churchills or Laurence Oliviers when I say this. At the risk of sounding both pretentious and presumptuous, I long ago gave up the use of inserted interviews on my radio program because I found it easier to translate the gist of the interviews into my own scripts—thus minimizing persistent problems of sounds, length, and content of the interviewees' expressions.

This brings us to, in some respects, the more urgent matter of compulsivitis of the microphone. On stepping down from his lofty role as Secretary of State, Dean Acheson once told James Reston of the *New York Times,* in effect, that there was a basic incompatibility between the press and the government, that reporters were constantly trying to find out what officials were trying to conceal, with the implication being that the higher the involvement of the story at hand with national security, the more intense the incompatibility. There is a basic truth here but in an open society like ours there are two honest interpretations of it, one by a responsible press and one by responsible officials and out of the tension between them flows, erratically sometimes, a current of information vital to an informed public opinion. That is the theory at least. Sometimes it does not work. Paradoxically the addition of radio and television to the media of communications has often made it harder rather than easier to make this system work. The reason is the ubiquitous microphone and its companion tool the TV camera.

An important caucus breaks up at a political convention and some hapless official is waylaid in the corridor to say something, at gunpoint, in a manner of speaking, with the heavy artillery of radio and television zeroed in on him. Out comes banality or evasion. What else is the victim likely to say under such circumstances? The Secretary of State flies off to a crucial conference. Before his plane leaves he has to run the bristling gamut of the thrusted microphones. He says he is confident that the conference will produce a useful exchange of views. After his plane lands he goes through the same ritual all over again. He has high hopes that the conference will produce a useful exchange of views. After the conference is over a communique is issued and a spokesman assures the microphones that there *has* been a useful exchange of views.

This dubious travelogue is news? It is an exercise in the thinnest and most synthetic kind of journalism. The interviewee, a past master of the art of manipulation of the meaningless expression by now, has long since ceased to become the victim. The public is the dupe and broadcast journalism is an accessory before, during, and after the crime. When the tense three-hour confrontation between President Johnson and Alabama's Governor Wallace ended on that sunny but chilly Saturday afternoon of

March 13th, the President guided the governor through a writhing waiting-room full of reportorial flesh to a cluster of microphones outside the executive west wing of the White House not for a news conference but for a statement for the electronic gear, deployed and waiting. They had an exchange of views, the governor said, not with notable agreement but the President, he added, behaved, as he always does, like a gentleman and the governor hoped that his deportment had been gentlemanly too. Then he flew back to the state of Alabama, where there seems to be a conflict over the codes of what makes and who is entitled to be, not a gentleman necessarily, but merely a voting citizen. This kind of high jinks is substantive information?

Questions can be as ridiculous as answers—another liability of the actualities and ubiquitous microphone techniques. I caught one broadcast from Travis Air Force Base in California of interviews with some of the first service families to be evacuated from South Vietnam after President Johnson had expanded American counter-strikes against the Communists. Obviously desperate to get *something* on tape, one reporter asked a returning Army wife, "and do you feel your husband will be ok there, now that you've gone?" While we're attacking the clumsy query and the cliché I soberly suggest that we working broadcasters all sign a pledge of abstinence and never again ask a candidate's wife how it feels to have her husband nominated. I'd be willing to break the pledge, naturally, if there was reason to believe the lady would reply, "what a stupid question," or "this is the last straw. I've been trying to get Horace to quit politics for years. Now I'm leaving him."

These perhaps slightly ulcerated criticisms should not be taken as justifying the free and untrammeled transit of men and women who make the news to and from their appointments without any attempt by the press to find out the score but there is a distinction, or should be, between news and nonsense. The old-fashioned shoe-leather approach of the reporter patrolling his beat is still valid and for the most part is still the best way to dig out the facts. On the other hand, "live" coverage of certain events can be unbeatable, not just for immediacy but for dimension. If the principals and broadcasters can train themselves—restrain themselves, it might be better said—to concentrate on substance, not just the sensational, then Congressional committee hearings will be even more newsworthy and instructive to the public. It is possible that "live" coverage of the sessions of House and Senate themselves would be a benefit to the country and maybe even improve the quality of Congress, or at least its debate. Let us not go overboard with optimism on these prospects however, for if we are candid with ourselves, live radio and television coverage of national political conventions has not improved the function of those cumbersome events at all.

The thoughtful, conscientious broadcast reporter is probably the most frustrated member of the Washington press corps. He makes his rounds, as do his colleagues of the newspapers and magazines. He goes to hearings, background briefings, filibuster sieges, news conferences at the White House, the State Department, and the Pentagon. He goes endlessly to lunches and dinners with people in or behind the news. He seeks key officials out in their bureaucratic lairs for interviews. He reads mountains of material, from handouts to the *New York Times.* And when he has his story he has to squirt it through a tiny hole of time in a 5-minute or a 15-minute news roundup on the air.

The *CBS World News Roundup,* at eight o'clock every morning on radio, used to be one of the most vital sources of intelligence in Washington. The government, especially the executive branch, listened for news that might beat—or set in clearer perspective—the diplomatic dispatches from our embassies abroad. It wasn't nationally sponsored however and the local ad spots sold by affiliates on a "co-op" basis were not profitable. So at one point, when I happened to be director of news for CBS, the network brass decided to take that 15-minute journalistic gem off the air. I sounded the alarm in Washington and thanks in large part to James Hagerty, then White House news secretary for President Eisenhower, we got some high-level testimonials, some of which were almost threatening in their emphasis on the program's value. It stayed (and that is the only accomplishment I can think of worth remembering in my brief tenure as a junior broadcasting executive). The *World News Roundup* continues and so do counterparts on ABC and NBC but they are all so loaded with disconcerting commercials now that it is hard to separate the news from plugs for liver pills, laxatives, body deodorants, and cigarettes, filtered and unfiltered.

New York Herald Tribune columnist John Crosby once wrote that I had "one of the more enviable jobs around." He was talking about my five-nights-a-week assignment of 15 minutes of news and comment, sponsored by ABC radio and the AFL-CIO. I am well into my eleventh year at that same job and as far as I am concerned, Crosby's words are truer than ever. I wish, in a way, that they weren't so. That is to say, I wish there were more competition. Not just in the 15-minute time segment. The eclipse of the quarter hour by these 5-minute bursts that are called newscasts is bad enough. (They are nearer to three minutes when you subtract the commercial time.) But I am free to voice opinion and critical comment, even including criticism of broadcasting and organized labor. I wish that journalistic phenomenon were not so unique. If the trade union movement, with all its warts, dares encourage it, why can't General Motors, or the National Association of Manufacturers, or the American Medical Association, or the organized groups of the radical right sponsor broadcasts that report the news and in addition paint arrows of responsible criticism that point inward as well as outward? The sinews of our society would be stronger and Washington would be a livelier place for the electronic journalist to ply his trade, if they did.

Some people are shocked by the little-known fact that approximately seven identifiable right-wing extremist groups sponsor, by conservative estimate, some 7,000 radio and TV (mostly radio) broadcasts a week in 50 states. I am also shocked. But I am more shocked by the fact that so many broadcasting stations seem to welcome and encourage this sulphurous stuff, which might be called yellow journalism of the air (though I would not ban it if it does not violate the law), while at the same time powerful elements of the radio-TV industry and, it must be added, their advertisers, tremble with apoplectic opposition to the airing of progressive or moderately liberal points of view. These are condemned as "controversial."

I indicated earlier that everybody abandoned radio in the scramble to television. This is not quite true. There are almost as many radio sets in the country today as there are people and the number, like the population, is increasing. The Electronic Industries Association counted 179,476,000 radio sets in the United States at the end of 1964, including more than 51 million car radios. There is of course some proliferation within families, especially now that we have transistors bulging, literally, out of our purses and shirt pockets. The Bureau of the Census reported that in 1960 more than 30 million households owned one radio; nearly 18½ million households owning two or more. But any way you figure columns the audience potential is impressive. Indeed radio broadcasting remains a richly prosperous industry especially on an individual station basis. (Networks perennially have a hard time in radio profit-wise.)

In an angry and valuable speech at an Ohio State University seminar in Columbus just four years ago, Morris S. Novik, a radio consultant who helped Fiorello LaGuardia open and run New York City's station WNYC, sharply condemned commercial radio for its irresponsibility. "Radio today," Novik said, "is making more money, and has more listeners and more commercials and it also has less public service programming, less community action programs—and less standing in the community." Novik also revealed the depressing fact that while a relatively small number of independent and network-affiliated TV stations did not carry the Kennedy-Nixon presidential campaign debates in 1960, as many as 300 affiliates and more than 2,400 nonaffiliated radio stations failed to broadcast this unprecedented series.

Timidly, tenuously, radio stations are moving a little farther into the field of journalistic crusade and editorial comment. At first they dared do little more than back the United Givers Fund and condemn the toll of traffic deaths. There are some 26 radio stations in operation in the Greater Washington area. Some of them have boldly committed themselves to valuable public service. Station WAVA in Arlington, Virginia, for example, backs up an editorial campaign against smoking by refusing to accept cigarette ads. The editorials of WWDC are broadcast personally by the owner of the station, Ben Strose. The 50,000-watt voice of WTOP engages itself responsibly with community problems and WMAL has carried on such a sharp campaign against child abuse that corrective legislation has been introduced in Congress. I don't know of any station, however, in the nation's capital or elsewhere (and I would be happy to stand corrected) that has come near to the courageous aggressiveness of owner R. Peter Straus over his station WMCA in New York City. The station has produced breathtakingly factual documentaries about the scandal of housing in such slums as Harlem, has relentlessly criticized politicians, and demanded political action. Careful to honor the FCC's equal-time proviso, the station takes sides in election campaigns. WMCA and Straus through the courts and on the air led the fight to demand redistricting of the New York State legislature. The Supreme Court's "one-man, one-vote" decision included specifically an opinion supporting the suit that Straus had initiated.

Bernie Harrison, the careful radio-TV critic of the *Washington Evening Star*, says radio in and around the national capital "isn't yet doing the job it should be doing." In news, whether local, national, or international, too many stations are "headline hunters" and concentrate on the "rip and read" technique—the news is ripped off the AP or UPI tickers and read, not by experienced journalists, but by announcers. Harrison sees too much quantity saturation—weather forecasts, the time of day, police helicopter traffic reports, and the like—and not enough quality selection. Let nobody be fooled (and who is really?), the techniques of echo chambers, bell ringing, Morse code-signaling, and machine-gun delivery are no substitute for news on a newscast. These critiques, I'm afraid, apply as snugly to other communities around the United States as they do to Washington and they reflect a state of the industry which restricts and discourages more meaningful coverage and analysis of the news from this and other world capitals, not to mention everybody's home town.

Who forgot radio and its tremendous potential to inform and promote understanding? All of us, I'm afraid, in varying degrees. We are faced, it appears, with a growing dilemma. Instant communication, more or less, to virtually all parts of the planet is confronting us with an increasing volume of datelines and data from strange places, but the bigger the stream of information about everything the less we seem to know about anything. Here the dilemma mockingly tosses us on its horns: for the more spare time our still rising standard of living gives us to pause and reflect thoughtfully about the world and our problems in it, the less inclined we seem to be to put this leisure to good use.

Indubitably, we minions of the mass media, and especially our masters, the station and network owners, must assume much of the responsibility for our predicament. We do not purvey information, we merchandise it—especially in broadcasting. We do not provoke intelligent controversy, an exercise so vital to the health of an open society; we mesmerize, we sensationalize, we scandalize, and we temporize. Given a cataclysm, we can rouse ourselves to serious dedication, as the broadcasting industry did on the nation's black weekend of November, 1963. But as the *New York Herald Tribune*'s television critic, John Horn, has so aptly pointed out, in order to rise to such heights in journalistic excellence and public responsibility, broadcasting had to reject almost completely its daily code of operational ethics with its endless commercials and idiotic entertainment—to which it cozily returned almost immediately after John F. Kennedy was buried.

As for the newspapers, it isn't as if they didn't carry more than a whisper of meaningful information, it's that what informative dispatches they do publish are so thoroughly buried under the groaning poundage of supermarket, real estate, and used car ads that the reader is virtually exhausted after turning the pages before he can find what he is looking for—if he is looking.

Which brings up the startling question—if you'll pardon a few final variations from my assigned theme—are newspapers already dead, information-wise, and don't know it? After I, among others, mourned the demise of the Western edition of the *New York Times* in January, 1964, I received a communication from a high school history teacher near Los Angeles who said that while he was sad to see the edition die, he had become indifferent to newspapers, dull and ponderous as they were. He relies for his "instant information" on radio and television, backs it up with weekly magazines and his own general knowledge.

Another listener, who happens to be a teacher too, a professor of English at Los Angeles State College, had, perhaps, a more penetrating reaction: that our basic fault lies in education. Our schools are failing to produce the kind of citizens who would demand and support superior newspapers—and presumably, superior radio and television news and public affairs programs. Author of a textbook himself, the professor said his editor had informed him that "there seems to be a growing disposition among the school officials who buy textbooks to lump together analysis of the mass media with those really extraneous items which crept into the English curriculum during the past two decades and to dismiss them all as 'frills.'" "I do not think," he added, "we will produce an audience for the *New York Times* if we must confine our treatment of the press in high schools to an excerpt from Addison and Steele's 'Spectator Papers' tucked away in a literature anthology."

The most shattering observation of all on our intellectual grasp came from a long-time friend of mine from Pasadena, who also happens to be in the teaching profession. "The Greeks and Romans didn't make it with their civilizations," he said, "and after two thousand years I don't think we are any smarter than they were." More knowledgeable, yes, in science, in medicine, in some of the mysteries of the mind, but more intelligent and wise? No. And in this age of marvels in communication, have the forces of the information media—especially radio and television—done their best to change that negative to a positive? You, I hope, already know the answer and so, to your immense relief, I will not belabor the point further but now sit down.

Thank you.

16. The Conscience of the Programme Director

by Sir Hugh Greene

Sir Hugh Greene is Director-General of the British Broadcasting Corporation. His address was delivered in Rome, Italy, before the International Catholic Association for Radio and Television on February 9, 1965. It is reprinted here by special permission of Sir Hugh and the BBC.

I speak to you today with great humility on a subject to which all of you I feel sure have given much more thought than I have in the course of a busy journalistic life. We are all conditioned by our early life and in my approach to broadcasting I have always remained, incorrigibly, a journalist.

In preparing to speak to this international Roman Catholic association so soon after the Vatican Council's recent discussions on Ecumenism, which I followed with great interest in the press and in Patrick Smith's broadcasts from Rome, I was particularly struck by the parallels between the problems of Ecumenism and of Broadcasting.

The senior representative of my own country at those discussions, the Archbishop of Westminster, Cardinal-elect Heenan, described the purpose of Ecumenism in these terms:

> Its object is not the conversion either of non-Catholics or non-Christians. It sets out to break down barriers between religious denomi-

nations in order that each may come to know and better understand the other. Ecumenism is an essay not in polemics but in charity. The dialogue is not a battle of wits. Its intention is not for one side to score a victory, but for each side to emerge with deeper knowledge of the other.

Substituting for "Ecumenism" the word "Broadcasting," and enlarging the "religious" concept to which Cardinal-elect Heenan referred to include also the widest secular concepts, I believe the Cardinal-elect's definition provides some notable parallels.

CREATING UNDERSTANDING

Broadcasting's true objectives too (I believe) are not "conversions" but rather the "breaking down of barriers," so that those of differing views "may come to know and better understand" each other's attitudes. Like Ecumenism, Broadcasting's main purposes, I believe, are not "polemics" or "battles of wits"; not the "scoring of victories," but rather the emerging of each side in controversial matters "with a deeper knowledge of the other."

This does not mean that Broadcasting should try to avoid entirely "polemics," or "battles of wits"—or even some "scoring of victories." My own personal attitude is far too combative for me to think that. I believe that sometimes these things make for lively broadcasting—and without some liveliness there would soon be no broadcasting, or at least only broadcasting to a limited and intellectually moribund audience. But they are not its main purpose.

The main purpose of broadcasting, I suggest, is to make the microphone and the television screen available to the widest possible range of subjects and to the best exponents available of the differing views on any given subject—to let the debate decide—or not decide as the case may be—and in Cardinal-elect Heenan's words, "to emerge with a deeper knowledge."

The presentation of varying views does not mean that the BBC, for instance, merely seeks to foster an equivocal attitude towards all that it broadcasts; to attach a ubiquitous, unanswered, question-mark to everything it touches, in religion, culture, politics, or education. But it does mean, in my opinion, that the BBC and other broadcasters should encourage the examination of views and opinions in an atmosphere of healthy scepticism.

I say "healthy scepticism" because I have a very strong personal conviction that scepticism is a most healthy frame of mind in which to examine accepted attitudes and test views which may have been accepted too easily—or too long. Perhaps what is needed, ideally (though we cannot all, I certainly cannot, achieve the ideal), is what T. S. Eliot described as "an ability to combine the deepest scepticism with the profoundest faith."

It follows that in its search for truth—indeed, in whatever it undertakes—a broadcasting organization must recognize an obligation towards tolerance and towards the maximum liberty of expression.

As John Milton put it three hundred years ago, in one of the most famous essays in the English language against Censorship and in favour of freedom of expression: "Where there is much desire to learn, there of necessity will be much arguing, many opinions; for opinion in good men is but knowledge in the making."

SAFEGUARDS ON FREEDOM

Obligations towards tolerance and liberty of expression for serious thought are, of course, problems not only of the BBC, nor only of broadcasting organizations. Long before the discoveries of that great Italian, Guglielmo Marconi (to whom all broadcasters owe their professional origins), your own Church had some experience of these problems—going back to Galileo and beyond. Echoes of this were heard during the Vatican Council's recent discussions.

Indeed, in the presence in Rome of an organization with close on two thousand years' experience of judging what limits and safeguards need to be placed on total freedom of expression, you may well think it presumptuous of me to discuss the experience and views of an organization which goes back only a little over forty years—and in which I have had less than twenty-five years' personal experience. But in discussing the subject you have given me, "The Conscience of the Radio and Television Programme Director," I excuse myself with the reflection that all such programmes only go back those same forty (or to be precise forty-two and a half) years that the BBC goes back and that, therefore, no *broadcasting* organization in the world has a longer experience of the particular modern form which these problems of freedom of expression take.

First of all perhaps it would be helpful if I were to describe briefly the legal and constitutional limitations within which the BBC works.

Great Britain has no written constitution and the BBC too is governed much more by precedent and experience than by legal instruments. It is not a commercial company; therefore it is not subject to the legal limitations of Acts of Parliament regulating companies. It is not a Government department; therefore it is not answerable for its day-to-day operations, and particularly not for details of its programmes, to a Minister or Parliament.

In theory the Postmaster General (the Minister who in Britain is responsible for broadcasting matters in the very broad sense) and, behind him, the Government can, subject to certain safeguards, require the BBC to broadcast or refrain from broadcasting any particular matter. But, in practice, in forty-two and a half years the right of veto has never been exercised in respect of any single programme. This, you will see, is a very important precedent, in view of the significance and authority we notoriously give in Britain to such working precedents.

THE BBC'S CONDUCT IN PRACTICE

The BBC, in fact, operates under one of the least restricting legal instruments known in Britain, namely a Royal Charter, supported by a Licence to operate from the Postmaster General. These two instruments lay down a relatively small number of things which the BBC must *not* do. It must not carry advertisements or sponsored programmes. It must not express its own opinions about current affairs or matters of public policy. Almost the only positive thing which the Corporation *is* required to do, apart from the general requirement to provide a service of information, education, and entertainment, is to broadcast daily an impartial account of the proceedings of Parliament—and even that the BBC started to do on its own initiative, before it was made an obligation.

For the rest, the BBC is left to conduct its affairs to the broad satisfaction of the British people (and, in the last analysis, of Parliament) under the guidance and legal responsibility of a group of nine distinguished and independent individuals, known as Governors.

This Board of Governors, whose members are appointed for fixed periods of time, is free to guide the Corporation and its affairs and policies according to their best judgment, without detailed answerability to any outside body or Minister. The Governors in turn appoint the Director-General—at present myself—and then leave him, with his executives, to conduct the day-to-day affairs of the BBC and its programmes.

From all this you will gather that the BBC, its Governors, and Director-General are remarkably free of controls or restrictions from outside—from politicians, or written laws.

Of course, again like Britain itself, over the years a number of conventions have grown up which—almost with the force of law, but not quite—govern the BBC's conduct in practice. But I must emphasize again that, unlike, for example, theatres in Britain (which are subject to the censorship of a high official of the Royal Court) and unlike the cinema (which is subject to censorship by a self-established cinema industry board of censors and also to the rulings of local legal authorities known as Magistrates)—unlike all these, the BBC is subject only to its own controls and, naturally, to the laws of the country. These laws in Britain are especially severe—many people think too severe—in a field which is of especial interest to all broadcasters and newspapermen—the law of defamation.

But subject to these few restrictions the BBC is left alone to keep for itself the delicate balance between freedom and responsibility.

How do we in the BBC interpret and use this freedom? Straight away I should say we do not see this freedom as total licence. We have (and believe strongly in) editorial control. Producers of individual programmes are not simply allowed to do whatever they like. Lines must be drawn somewhere.

STIMULATING IDEAS

But, in an operation as diverse in its output as broadcasting, the only sure way of exercising control—here we come to one of my most strongly held personal convictions—is to proceed by persuasion and not by written directives; by encouraging the programme staff immediately responsible to apply their judgment to particular problems, within a framework of general guidance arising from the continuing discussion of individual programmes by themselves, by their seniors—and, when necessary, by the Board of Governors. In my view there is nothing to be achieved by coercion or censorship, whether from inside the Corporation or from outside—nothing, that is, except the frustration of creative people who can achieve far more by positive stimulation of their ideas in an atmosphere of freedom.

In stimulating these ideas we have to take account of several important factors, some of which are new to this age of broadcasting, some of which are as old as articulate man himself.

We have to resist attempts at censorship. As Professor Hoggart (one of our leading British writers on the themes of broadcasting and freedom) has noted recently, these attempts at censorship come not merely from what he describes as the old "Guardians" (senior clergy, writers of leading articles in newspapers, presidents of national voluntary organizations, and so on) who like to think of themselves as upholders of cultural standards although, in many cases they lack the qualities of intellect and imagination to justify that claim.

The attempts at censorship come nowadays also from groups—Hoggart calls them the "new Populists"—(one might call them the new Puritans)—which do not claim to be "Guardians" but claim to speak for "ordinary decent people" and to be "forced to take a stand against" what they arbitrarily call *unnecessary* dirt, *gratuitous* sex, *excessive* violence—and so on. These "new Populists" will attack whatever does not under-write a set of prior assumptions, assumptions which are anti-intellectual and unimaginative. Superficially this seems, and likes to think of itself as, a "grass-roots" movement. In practice it can threaten a dangerous form of censorship—censorship which works by causing artists and writers not to take risks, not to undertake those adventures of the spirit which must be at the heart of every truly new creative work.

Such a censorship to my mind is the more to be condemned when we remember that, historically, the greatest risks have attached to the maintenance of what is right and honourable and true. Truth forever on the scaffold, wrong forever on the throne. Honourable men who venture to be different, to move ahead of—or even against—the general trend of public feeling, with sincere conviction and with the intention of enlarging the understanding of our society and its problems, may well feel the scourge of public hostility many times over before their worth is recognized. I see it as the clear duty of a public-service broadcasting organization to stand firm against attempts to decry sincerity and vision, whether in the field of public affairs or in the less-easily judged world of the arts, including the dramatic art.

I believe that broadcasters have a duty not to be diverted by arguments in favour of what is, in fact, disguised censorship. I believe we have a duty to take account of the changes in society, to be ahead of public opinion, rather than always to wait upon it. I believe that great broadcasting organizations, with their immense powers of patronage for writers and artists, should not neglect to cultivate young writers who may, by many, be considered "too advanced," or "shocking."

Such allegations have been made throughout the ages. Many great writers have been condemned as subversive at some point in their careers. Long ago, for example, there was Euripides. Only a few decades ago Henrik Ibsen was regarded as too shocking for his plays to be staged in Britain. Indeed, I am informed he was at one time on the "Index" of your own Church.

THE DOGMAS OF TOMORROW

At least in the secular and scientific fields—today's heresies often prove to be tomorrow's dogmas. And, in the case of the potential Ibsens of today, we must not, by covert censorship, run the risk of stifling talents which may prove great before they are recognized.

I do not need to be reminded that broadcasting has access to every home, and to an audience of all ages and varying degrees of sophistication. We must rely, therefore, not only on our own disciplines, but on those which have to be exercised by, among others, parents. But programme plans must, to my mind, be made on the assumption that the audience is capable of reasonable behaviour, and of the exercise of intelligence—and of choice. No other basis will meet the needs of the situation. How *can* one consciously plan for the unreasonable or the unintelligent? It is impossible, or if not strictly speaking impossible, utterly disastrous.

Editorial discretion must concern itself with two aspects of the content of broadcasting—subjects and treatments. If the audience is to be considered as it really is—as a series of individual minds (each with its own claim to enlightenment, each of different capacity and interests) and not as that statistical abstraction the "*mass*" audience—then it would seem to me that no subject—no subject whatever—can be excluded from the range of broadcasting simply for being what it is. The questions which we must face are those of identifying the times and the circumstances in which we may expect to find the intended audience for a given programme.

Relevance is the key—relevance to the audience, and to the tide of opinion in society. Outrage is wrong. Shock may be good. Provocation can be healthy and, indeed, socially imperative. These are the issues to which the broadcaster must apply his conscience. But treatment of the subject, once chosen, demands the most careful assessment of the reasonable limits of tolerance in the audience, if there is any likelihood of these limits being tested by the manner of presentation of the material. As I have said, however, no subject is (for me) excluded simply for what it is.

The most recent Committee of Enquiry into Broadcasting in Britain described the responsibilities of broadcasting in these matters like this: "Broadcasting," it said,

> must pay particular attention to those parts of the range of worthwhile experience which lie beyond the most common; to those parts which some have explored here and there, but few everywhere. Finally, and of special importance: because the range of experience is not finite but constantly growing, and because the growing points are usually most significant, it is on these that broadcasting must focus a spotlight.

Does all this, I wonder, strike some of you as arrogant? What right have I and my colleagues in the BBC—even with the guidance of our Board of Governors—to decide where lines should be drawn? Why should we be more wise than outside censors? I don't suppose we always are more wise. But—here we come to another of my personal convictions and one which I think one can support from the experience of history—it is better to err on the side of freedom than of restriction.

INDEPENDENCE FROM PRESSURES

Attempts at both open and disguised forms of censorship are, of course, only one of the forms of pressure to which the BBC—like all other independent broadcasting organizations—is subject. In the case of the BBC, however, we are especially fortunate in our power to resist. The system of licence fees paid by our viewers and listeners is one which makes us financially independent not only of the Government but also of commercial pressures.

I hardly need to remind an international audience of examples of the kinds of pressure of which some Governments are capable. What may happen in various parts of the world where programmes are broadcast under the commercial "sponsorship" system is perhaps somewhat less familiar, at any rate in Europe. A former Chairman of the American Federal Communications Commission, Mr. Newton Minow, has described a classic example of such pressures. The United States gas industry (he has recalled) sponsored the presentation in a drama series of a play about the Nuremberg War Trials, under the title "Judgment at Nuremberg." Viewers noticed that a speech by actor Claude Rains about the killing by cyanide gas of thousands of concentration-camp prisoners was abruptly interrupted by a deletion of a few words. The editing was done by a television network engineer while the video-tape recording of the drama was actually on the air. The words eliminated were "gas chamber." The editing was done to accommodate the gas industry sponsor; and a broadcasting company executive later gave this explanation: ". . . We felt that a lot of people could not differentiate between the kind of gas you put in the death chambers and the kind you cook with. . . ."

One of American television's best known writers, Rod Serling, has also recounted the changes in a script that an advertising agency can force. Mr. Serling had based a one-hour drama on the lynching of a Negro boy in the Deep South. By the time the agency had finished with the story, the chief character was no longer a Negro boy but a former convict, and living not even in the South but in New England!

Nor can we on this side of the Atlantic afford to be smug about the dangers. Our own post-bag of correspondence at the BBC is full of examples of attempts to exercise pressure in favour of this interest or against that—usually by complaints or by thinly disguised threats to cause trouble by approaches to Members of Parliament. Without true independence, therefore, it is difficult for any broadcaster to maintain the highest standards of truth, accuracy, and impartiality. Conversely, of course, without a reputation for these things—truth, accuracy, and impartiality—it is difficult for any broadcasting organization to be recognized as truly independent and to be generally trusted.

Truth and accuracy are concepts which are not susceptible of legal definition. The Government in Britain is content (after forty years' experience of the BBC) to recognize that the BBC tries to honour these concepts and to treat "with due impartiality" all controversial subjects.

But although, in the day-to-day issues of public life, the BBC does try to attain the highest standards of impartiality, there are some respects in which it is not neutral, unbiased, or impartial. That is, where there are clashes for and against the basic moral values—truthfulness, justice, freedom, compassion, tolerance.

Nor do I believe that we should be impartial about certain things like racialism, or extreme forms of political belief. Being too good "democrats" in these matters could open the way to the destruction of democracy itself.

I believe a healthy democracy does not evade decisions about what it can never allow if it is to survive.

The actions and aspirations of those who proclaim some political and social ideas are so clearly damaging to society, to peace and good order, even in their immediate effects, that to put at their disposal the enormous power of broadcasting would be to conspire with them against society. Here it is extremely difficult to know where to draw the line. The answer must vary from case to case, from country to country.

Finally, I come to a field which is of special concern to UNDA, as a Roman Catholic organization seeking to establish a philosophy of broadcasting. I speak of religious broadcasting. I would suggest that in radio and television it is for the Church, all Churches, not to adopt a merely negative or critical attitude towards these new forms of communication. They should recognize that the techniques of communication in the modern age are something which the Churches must acquire, and show a readiness to experiment in the forms such communication should take.

I think that coming from the BBC I can speak with some authority here because, as far as I am aware, our range of religious broadcasting is unrivalled anywhere in the world, because, rather than in spite of, the increasingly secular nature of our society.

The field of religious broadcasting is one in which, probably more than any other, broadcasters have to honour and respect the feelings of their audience—if only because it is a field in which beliefs and feelings are more deeply (and often more easily) aroused.

This does not mean that in our broadcasting on religious matters we avoid the difficult or controversial questions. The BBC long ago added religious controversy to the other forms of controversy which it is prepared to broadcast. It regards it as its duty, too, to broadcast views of *un*belief, as well as of differing beliefs. We do not, for example, think it wrong to allow broadcasting opportunities to Humanists. We believe, in fact, that it is our duty to help to remove blinkers from believers—and unbelievers too—who may be inclined to wear them.

One of Britain's leading present-day Anglican theologians, Dr. Vidler of Cambridge, has declared: "Real Christianity can be healthily recovered only by having to contend on equal terms with competing faiths and no-faiths, and by being forced to rediscover its own bottom."

Or, as John Milton, to quote him again, put it much earlier: "Though all the winds of doctrine were let loose to play upon the earth, so Truth be in the field we do injuriously, by licensing and prohibiting, to misdoubt her strength."

In all these matters of belief and conscience I believe that the fundamental rule is—*not* to avoid everything that is difficult or controversial—but, on the other hand, to be sure that we honour other people's consciences, even where we, as programme producers, do not (or even cannot) share their beliefs.

In the last analysis the decisions of all broadcasting organizations about what to broadcast (and in what form) come back—as UNDA has truly observed in giving me the title for this lecture—to the "Conscience of the Programme Director."

THE BASIC STANDARDS

In their output of television and radio programmes, producers and their programme directors have to put into effect—by daily, hourly, or even instantaneous decision—their own judgments of what it is proper to put forward to an audience which may run into millions.

As the last Chairman of the BBC's Governors has pointed out, these judgments must spring (since they have to be exercised at such speed) from some ingrained code, which in its turn is derived from some basic standards. These standards, in practice, are composed (I believe) of a number of different elements:

1. The personal attitudes of producers, programme chiefs, and directors—which depend upon their own beliefs and upbringing.
2. A general code of practice established in the BBC, though not made rigid, by experience; and
3. The proper sensitivity of production staff to the world around them, so that they are concerned with a relationship to the audience which cannot exist if the language in which they are talking, and the assumptions they are making, seem to be too remote from the language and assumptions of the audience and of the times in which they are communicating.

These things, I believe, add up to "the Conscience of the Programme Director." And I think I have made clear the sort of atmosphere—an atmosphere of freedom and tolerance—in which I am convinced that that conscience should be exercised.

17. The Screen Writer and Freedom

by Michael Blankfort

Mr. Blankfort is a free-lance writer who lives in Los Angeles. He is the well-known author of many successful screen plays and novels. His remarks are reprinted from Speeches of the Seventh Annual Freedom of Information Conference, *held at the University of Missouri School of Journalism, November 16–17, 1964. The address is reprinted here with Mr. Blankfort's permission.*

INTRODUCTION BY Dean Earl F. English, School of Journalism, University of Missouri

A William Faulkner or F. Scott Fitzgerald aside, it is not often that the movie-going public has any awareness of the motion picture script writer whose lot has been summarized by Edmund Hartmann, formerly national chairman of the Writers Guild of America, in this way:

> The writer has never held the throne. The writer is too many steps removed from the moment of decision before the cameras. The script has always been a tool used by the struggling mammoths. Movie-making is essentially the telling of a story. But the simple function of the story-teller has been obscured by a complex machinery of manufacture. Each step of the assembly line has become a madhouse of distortion to press the story in the image of the producer, the director, the star. The movie script too often has been a chunk of meat thrown to the lions. The producer tears off a chunk. The director claws what he can get. The star roars and snarls for the largest bleeding bits.

Stepping out of this carnage and on to the campus today is the author of many screen plays, among them: *The Caine Mutiny, Broken Arrow, My Six Convicts, Halls of Montezuma,* and *The Juggler.* A play, *See How They Run,* was his first full-length film made expressly for television. He is the author of eight published novels and the ninth, *Behold the Fire,* will be published next spring.

Michael Blankfort was born in New York City and educated, as he phrases it, "in the great public schools of that city." The education begun so happily was continued at the University of Pennsylvania where he received a Bachelor of Arts degree and at Bowdoin College and Princeton University where he received a Master of Arts degree in psychology.

He has been an instructor at Princeton and New York Universities, a writer and director of plays for the Theatre Union in New York. He is a member of the Executive Council, Writers Guild of America West, The Authors League, and the Dramatists Guild.

It is my pleasure to present to you now, Mr. Michael Blankfort.

My first experience with censorship took place when I was about 14 years old. I can recall that about that time I was troubled not only by intimations of sex but also by the first rebellion against religious dogma. I had begun to question the God of my fathers—as well as my father, and that took greater courage. In those years I was a buyer of five-cent blue books published by Haldeman-Julius of Girard, Kansas, a powerful educational instrument in the 1920's now forgotten, unfortunately, by the social historians. Some of my favorites were "What Every Young Man Should Know," "Poems of Passion," by those pornographers, Horace and Ovid, and the "Love Sonnets of Shakespeare," all sent in plain wrappers. To support my burgeoning atheism, I sent my five cents for a copy of *The Philosophy of Spinoza,* who I heard had also broken with the God of Isaac, Abraham, and Jacob, and had been excommunicated for his daring. I thought he would give me all the arguments I needed.

Now, I make no claims for precocity; I am sure I did not understand the God-intoxication of Spinoza. Yet I must have found something in that heretical tract that was embarrassing my parents, for one day I overheard my mother talking about me to our rabbi. At the end of the conversation on the telephone, she said, "All right, I'll look and see what books he's been reading." The outcome was inevitable. The little blue book of Spinoza and *French Love Stories* by De Maupassant, both of which I had hidden behind *The Boy Scout Manual,* disappeared. It became clear to me that sex and the questioning of religious dogma were forbidden games. I learned my first lesson in censorship—and its worthlessness, for since that day, I confess, both these subjects have had an over-riding interest for me.

As I contemplate the present condition of the writer, chiefly in television and to a far lesser degree in films, I have the uncomfortable feeling that behind the policy committees of the major broadcasting networks, the ad agencies, the censorship efforts of the various state boards, and even the self-censorship of the industry, my mother is still at work, and, I add, probably with the same results.

So that there will be no question or misunderstanding about where I stand on the subject, "Freedom and Film," I want to say emphatically that I oppose censorship in all its forms, outside or within the industry, unintelligent censorship or intelligent censorship, if there is such a thing. For, in truth, there is no harmless censorship, anymore than there is a harmless cancer.

My motives for this stand derive from the principle that in a democracy, taste, like a citizen's franchise to vote, should be free of constraints. In addition, there is the matter of self-interest. For the sake of the freedom I need as a writer to pursue my visions of human behavior no matter where they lead me, I must protest all limitations to freedom. I cannot accede to the idea that any civic or industrial body has the right to impose its taste and standards of morality on the expressed thoughts of any other writer, lest it take advantage of its power and apply them to my work.

Now, if there are other equally responsible citizens who feel that there is a danger inherent in complete freedom of expression, they will do what they can to restrict it. And it is the writer's task to oppose them. The fact that screen writers work under the Production Code of our industry as a condition of their employment has never meant that we accept it. On the contrary, writers in Hollywood have been fighting against industrial self-censorship since Theda Bara first bared her shoulders.

Parenthetically, I want to point to a curious misapprehension which fills the mind of many of our censors. It is the idea that the opposite of a state of innocence is a state of sin or depravity. It seems to me that the opposite of innocence is quite different. It is knowledge, not sin. And if one argues that knowledge leads inevitably to immoral acts, then I think we are wasting our time coming here or supporting universities. When Adam and Eve ate the seductive apple, they learned that the loss of innocence, and I am being nontheological here, opened their hearts and minds not to the enjoyment of sin but to the hard knowing of life. In this sense, the whole force of censorship works to delay if not distort the facing of truths and maturing of society.

Before going on to a further discussion of the Motion Picture Association Code, I want to reflect for a moment on the fate of some of my colleagues, the writers for television, who are not represented at this conference limited to feature film. The hard knowing of life is experienced these days with great agony by the authors of teleplays. (Perhaps it would be more accurate to call their work "telepathic plays," since there seems often to be a lack of any sensible communication between what the writer has written and what is seen on the glass box.) As for control of his material and his vulnerability to constraints, not so much from the producers, but from the networks and ad agencies, he is far worse off than the screenwriter ever was, for he has twice as many bosses and even a greater number of pressure groups and censors who set up taboos with a whim as strong as steel. I recommend a new book which covers some aspects of the television writer's travail. *Only You, Dick Daring* is by Merle Miller and Evan Rhodes. It is a kind of *Uncle Tom's Cabin* of the writer in the clutches of the Simon Legrees who run the air-conditioned plantations of television.

Now, let's turn our attention to feature-length films, products of Hollywood. I am pleased to affirm in front of some of my colleagues that the old phrase, "Without the writer we are nothing," so often mouthed in the past and so often discarded in practice, is now coming to have some real meaning. There are reasons for this. Stars, free from the old long-term contracts with studios, have now become more powerful, their own bosses and bosses of production, and are no longer willing to commit themselves to films that haven't yet been written. Directors demand the opportunity of working with writers, not like in the bad old days when writers were frequently discouraged from discussing their intentions with the man in charge of communicating them to the actors. Finally, and this is perhaps the most important of the new elements in Hollywood, there is the break-up of the old studio oligarchy. More and more writers have become producers and directors, with greater control over their material. This has had an enormous impact on all writers who take their work seriously, giving them an importance and influence they rarely had before. So I bring you, in a sense, good news about Hollywood, that too-often-maligned place.

More has to be said, however, about our Production Code. In the old days, the industry's move to censor itself resulted from the attacks of pressure groups on the morals of the citizens of Hollywood and its films. These caused the studio executives some small amount of spiritual distress and a great amount of economic uncertainty. The question was how to head these censorship Apaches off at the pass. And so the Code was created on the theory that if any scalping was to be done, the producers knew better how to do it.

Writers, as I mentioned earlier, write scripts with the provisions of the Code in mind. Sometimes, we hope that the good scouts of the Code, honorable and talented men like Geoffrey Shurlock, when they read our manuscripts will miss those lines and situations which imply a contravention of the Code. But, with sadness I say, they invariably see through us. True, they try to help us keep what we want by suggesting some alternative approach. But in the end, sin has to be punished. And this stricture creates a spurious morality which has little, if any, connection with the contradictory complex morality of real life.

Today there are signs of great changes. One almost feels that the long, long war against industrial self-censorship has been won. We have entered a period of new freedom. The list of films which couldn't have received the seal of the Production Code five years ago is a long one. I mention only *Irma La Douce, Tom Jones, The Carpetbaggers,* and *The Night of the Iguana.* The *Los Angeles Times* recently reported a most unusual event. Dialogue was edited out of a film *after* it had received the Production Code seal. I'm shaken almost to unreason by the thought that in these matters Geoffrey Shurlock, the administrator of the Code, is more radical than Billy Wilder.

It would be pleasant to say that this new dispensation came through the splendid efforts of our freedom fighters—the Guilds, the anti-censorship committees, etc.—but it wouldn't be true. While I don't want to deprecate the victories won at this or that civil liberties barricade by the Motion Picture Producers Association and other groups, whatever triumph evolved has come, as Gordon Stulberg told us yesterday, through the ineradicable need of business to "make a buck."

As is well known to you, our film industry suffered great losses until recently as a result of television and foreign competition. It had no other choice than to make pictures bigger and freer, for the buck remains the *sine qua non*. Not so long ago, an organization of exhibitors pleaded with a group of producers to cut down on sex in films. It must not be understood by this, by the way, that the theater-owners are Galahads of purity. They are, in fact, the most easily frightened men in America. No, these gentlemen were looking for a way to appease their local censorship boards by pointing to someone else's guilt. When the producers asked, in turn, whether the theater owners would stop showing films which had not received the seal of the Production Code, these hinterland defenders of our morals accused the producers of trying to force them into early bankruptcy. In hard cash, morals are always the other fellow's affair. If I may rewrite the gospel: "Profit is not without honor, especially in one's own business."

Though we are freer than ever before, let's not hang up our swords. Civil liberties and civil rights as subject matter are still beyond the reach of Hollywood films. Not because of prejudices or principles, but simply because of the conviction that such films won't make enough money to return their investment. I was told recently by a distinguished writer-director that he was forced to cut out of his finished film, a Western by the way, one of its most meaningful and moving scenes because it showed the hero, a self-confessed quadroon, compassionately touching the cheek of a white woman he loved. The producers feared that the scene would hurt them at the box office.

Basically, the effect of this new freedom on screen writers is very good. It permits us to explore more deeply the characters we write about and our own as well. We have more ammunition with which to attack restrictions on subject matter and even dialogue. And while we are not all rushing back to our Krafft-Ebing, we are hopeful of avoiding those script compromises of the past which sometimes engendered more smirk than truth, more leer than light. Yet can there ever be too much of freedom? To ask the question is to revoke the whole concept of free expression, for if there are limits, there is no longer freedom. But what will come of it in Hollywood? To answer this is not too difficult. In my opinion, Hollywood will enjoy its new vistas for a little while longer and then will lose what it has gained. Lest I be mistaken for an unrecognized prophet, I hasten to add that I am a disciple of that theory of cultural history which some have called cyclical but which a philosopher friend of mine, Professor Abraham Kaplan of the University of Michigan, described to me as the "ladies' skirts" theory. As far as styles are concerned, skirts can go up just so far, then they have to start going down again. Then they start going up. I anticipate the time when our hard-won freedom as film-makers will create a counter force of demand for censorship. And that this demand can, in time, have an effect. Then new forms of old regulations will settle on us again—on the studios as well as the creators—and we will need all the help we can get from people like yourselves. But eventually, just as dresses have to start going up after they have become too long for comfort we will have a newer freedom. And if you forgive my cynicism, it will come as before, just when business at the box office is threatened.

Now, I want to talk briefly of another kind of freedom, perhaps one that is not precisely in the scope of this conference, but one I feel deeply about, and one that is very little known. I suggest we leave the outer space of society, so to speak, where censorship boards fly around in ellipses of their own, and probe a little into the precious inner space, the heart and mind of the writer who is, by common agreement, the first cause, the creator without whom there is nothing, not even *Beverly Hillbillies*. In that mind and heart, in that inner space, are waged great battles for freedom. In the lonely, heart-breaking struggles to find creative truths, the writer is both hero and victim, for in films, once he has accepted an assignment, he is no longer master of himself and his work. He can fight for days for his conception of theme, but the ultimate decision lies with someone else—producer, director, star; and to the degree that he fights and wins, to that degree the work of the screen writer is his own. One of the mildest men I know, and also one of the most talented of film writers, is called "the killer" by some producers. He fights for his script with a zeal and resourcefulness that mark a great quarterback. The fact that, at the moment, he is Hollywood's most successful screen writer, with the right to select director and actors, may not be the result solely of his talent. The moral is clear.

As Dean Earl English mentioned earlier, don't forget that most of what you see on the screen has first been written. Don't fall into the trap of that woman I heard say, "The picture was bad, except for Jack Lemmon's wonderful wit." Mr. Lemmon happens to be a witty man off the screen, but his lines in a film are written by someone else.

It is one of the great secrets of the movie-going world that writers write directions to the director, detailed descriptions for mood and setting, suggestions for music, sound, and silence—all this in addition to dialogue, the structure of plot and theme, the relationship of characters and so on.

Despite all the publicity to the contrary, the script is the prime and on-going impulse that makes the film. And it is created and shaped in what I have called the inner space.

To understand the terrain of these battles of the creator's inner space, we must remind ourselves that the film, like the theater, is a collective art, and not one in which there is a singular and private connection between the creator and his work, as in the novel and the plastic arts. Standing between the screen writer and his work there is a living wall of other tastes and personalities, so massive and involved that it is often impossible to separate the contributions of the writer from the producer, director, and actor. It is because of this that the screenwriter writes with an oppressive and sometimes unconscious tension which has little to do with the normal attrition of his labor.

There are some elements in this phenomenon which recall the old movie scene in which the heroine is tied to the railroad tracks by the villain. From the distance comes the sound of the onrushing train. We see the dark smoke curling above the far horizon, and we ask ourselves, "Will she free herself before the train reaches her? Will she be freed by the hero in time?" And as we watch, two factors operate in our minds. Having been conditioned in our American culture not by the tradition of tragedy but by melodrama, we know with absolute certainty that the heroine will be saved in time, yet we are equally convinced that she won't be saved in time, for otherwise there would be no suspense, and we would laugh instead of holding our breaths or biting our nails. As audience, we are held in constant tension between knowledge and belief, tradition and imagination.

The film writer, too, as any writer in a collective art, works in the same kind of tension. He knows with certainty that his script will not be done as he has written it, yet in order to do his best work, he must believe or imagine that it will. He must give his all, while knowing that no matter how good his all may be, it will not and cannot satisfy the tastes of the others who have equal, if not greater, power over his material. But what may be healthy in an audience is schizoid in the writer. Whether he is or isn't aware of this division in him, his work is affected and his freedom to be himself is curtailed. Who can know how seriously this distorts his work? His capacity to pull out from the recesses of his experience all that he has learned about life is diminished. Although he is able to function under this tension—and it is extraordinary how well he does manage—there are areas of psychic inhibition beyond which he may not be able to move. This is the worst kind of loss of freedom, for it is self-inflicted, a misery beyond calculation.

It may be that to avoid this subtle and dangerous self-censorship many talented young men are not coming to Hollywood at all but are writing, producing, and directing their own films in New York, Philadelphia, Chicago, and elsewhere. They are in that still-innocent state where they are free to make films as they see them, and also to make mistakes which are all their own.

I am not offering panaceas for the situation I've described. I'm not sure any exist but I think it is worth including in a conference on "Freedom and Film," for the inner constraints of screen and television writers, so often forgotten, are as important as the constraints imposed by conventional censorship. When there is a considerable loss of creative energy within that precious inner space, we are all diminished.

18. Through the Regulatory Looking Glass—Darkly

by Robert W. Sarnoff

Mr. Sarnoff's address was delivered before the National Broadcasting Company's Affiliates Meeting in New York City on March 18, 1965. It is reprinted here through special permission granted by the network.

It is a great pleasure to see all of you again. I hope that this annual meeeting is giving you—as it does me—assurance of a dramatic expansion of NBC's leadership in the season ahead. Each of these previous gatherings has had its own distinguishing feature. But this is the first time, in my memory, that one has been marked by a parade up Fifth Avenue.

It is less than ten months since our last full-scale get-together. In a way, this shortening of the span between conventions is symbolic of the pace of our industry. It is not the advancing age of the family that makes time seem to move faster, but rather the quickening pace of developments and accomplishments, in broadcasting generally and at NBC specifically.

The relatively brief period since we last met has been particularly demanding but immensely gratifying. This morning's presentation clearly documents NBC's success in establishing leadership in entertainment, achieving front position in sports and extending its pre-eminence in news. And this leadership has been attained without sacrificing the diversity of programming that is the hallmark of our network service.

Over-all, television today is attracting larger audiences, for longer periods, than ever before. Color is accelerating very rapidly, and with NBC opening the era of full-color network service, that pace will quicken and bring striking advantages to the public and to those carrying our schedule. Advertising revenues are at new highs and the number of sponsors using the medium is at a record level.

Yet despite its growth and achievement, its commitment to flexibility and innovation in response to developing needs, broadcasting still finds that in one of the most important areas of its capability, the more things change, the more they stay the same.

This results from an official attitude that refuses to recognize that broadcasting, as today's foremost instrument of journalism, is entitled to the same journalistic freedom as other media of information.

OLD ENEMIES

Exhibit A continues to be the equal-time restriction on proper coverage of political campaigns. The record of outstanding coverage in 1960, when the equal-time shackles were loosened, and the lack of any meaningful development of campaign issues in 1964, when the shackles were replaced, offer convincing evidence that the books should be rid permanently of this worthless device.

Another regulatory cloud still hovering over broadcasting's journalistic function is the official meddling fostered by the so-called "fairness doctrine." Broadcasters' opposition to a doctrine that establishes a federal agency as the arbiter of journalistic fairness is certainly not directed against fairness, nor to how the FCC applies its judgments in passing on fairness. It is against the impropriety of a federal agency operating in this area at all, drawing a government curtain between the public and a free press. This is a power over the press that is expressly denied to government by the Constitution. Its very existence is a threat to the proper exercise of the journalistic function the public has come to expect of broadcasting.

In still another area—although outside the realm of direct regulation—broadcasters are foreclosed from providing a full measure of service, and the citizen is deprived of a clear view of the process of his government. This situation arises from the variety of laws and regulations that prohibit television access to legislative and judicial proceedings. We are still seeking total access, at federal, state, and local levels, on the premise that as the nation's primary medium of communication, television should be allowed to go wherever the public is admitted, just as the print media are permitted to do.

These limitations upon television's service—in coverage of politics, and controversy and access to public proceedings—are, of course, old enemies.

But our problems with regulatory interference are by no means limited to our function as journalists, where, at least, they have their shaky precedent in custom.

REGULATORY RAMBLING

On the contrary, the champions of broadcast regulation appear constantly to discover new worlds to conquer, with or without legal sanction and often in the absence of demonstrated public need. Indeed, we seem to be confronted by a sort of regulatory rambling that seeks to shape the medium by theory and abstraction, far removed from the realities of practice and experience.

This chronic urge to regulate has now propelled the FCC deeply into broadcasters' programming and business practices, in the form of proposed rules to regulate the television networks' organization of their program schedules.

The Commission has under consideration a staff recommendation that rules be promulgated requiring, in effect, that 50 per cent of each network's evening entertainment programming be supplied and controlled by advertisers. The proposals contain various other restrictions. In acquiring programs, networks would be limited solely to the right to broadcast, without any other rights or financial interests, even though they had financed production of the programs. Network companies would also be completely barred from domestic syndication, and in foreign distribution, they would be confined to their own productions.

Apparently, the staff has kept in mind that the law gives the Commission no direct regulatory power over the networks—a power it has unsuccessfully sought by legislation. But noting that each of the network companies also holds five VHF licenses, the staff has engaged in a new game of words. Calling networks "television network licensees," it seeks to create jurisdiction by phrase-mongering.

This assertion of jurisdiction-by-device gives rise to some sobering thoughts. If the Commission can extend its authority into the program and business operations of networks merely by coining a phrase, what is to prevent its phrasing its way into direct regulation of a "newspaper licensee," or a "magazine publisher licensee," or a "producer licensee"? In the latter case, as a parallel to the presently proposed rule, it might then direct production companies that are station owners to limit program sales to networks to 50 per cent of their output, with the remaining 50 per cent available only to other customers.

The proposal devised by the staff constitutes a sterling example of ruling with the authority of office, rather than the authority of experience. For it undertakes to repeal the fundamental patterns that have developed to meet a unique requirement: the organization of a tremendously costly national program service offered on a continuing basis, every night of the year, which matches the interests of enormous audiences and attracts the necessary advertising support. By a designed disruption of these patterns, the proposed rules would shatter the structure of network television which, unlike the rules, was not tailored to theory or preconceptions. This structure evolved gradually, through the interaction of advertisers, networks, and the public, cultivated by experience and economic necessity.

OUTLANDISH RESULTS

During television's growth, the pressure of rising costs and the complexities of programming resulted in a shift away from the full-sponsorship pattern of radio. Fewer and fewer television advertisers were willing to enter the program business, or able to assume the multi-million-dollar risks involved in the development and production of even a single program series. It was left to the networks to shoulder this responsibility and to organize the program schedules, so that advertisers could employ their budgets with maximum efficiency through partial sponsorships or minute participations.

It is this structure, developed over the years to meet the needs of the audience and the business, that the FCC staff proposals would destroy, with resulting chaos in network programming and advertising.

It is hard to believe that the FCC, once it has discovered the outlandish results these proposals promise, would move to adopt them. Nevertheless, let's look at some of the probable effects of this incredible 50–50 rule.

First, it would undoubtedly force a substantial reduction in network service. There is absolutely no assurance that there are enough advertisers to supply half the evening schedule. In fact, all the evidence is to the contrary. Currently, less than 10 per cent of the network schedules are filled by programs supplied by advertisers. Even if this percentage should double or triple under the forced draft of the proposed rules, limiting network-supplied programming to the volume supplied by advertisers would automatically curtail the total amount of programming available to affiliates and the public.

Secondly, there would be a lessening of the networks' responsibility for their total schedules. Presently, although advertising goes in and out of the schedule, the program structure is maintained by the network as a continuing service to the audience and the affiliated stations. With advertisers controlling a large portion of the schedule, programming would stay or go depending on the sponsor.

It is no solace to suggest that if the rules produced these abortions in network service, they could be changed or waived to meet the realities of a given season. Disrupted schedules cannot be repaired or extemporized overnight. But even worse, the networks would be hard put to know how to begin organizing their schedules for any season—a task of the most intense creative activity that requires more than a year's advance planning.

ENORMOUS ECONOMIC LEVER

Of course, the proposed rules would hand some advertisers an enormous economic lever, and this apparently is what the staff intends. Networks would be so dependent upon advertiser-supplied programs to fill half their schedules that advertisers could dictate the assignment of time periods and terms of sale. This, in turn, would prompt rate-cutting, with resulting loss in station compensation, and would reduce the networks' financial ability to produce their present volume of news, informational, and cultural programming.

The rules, nevertheless, would be an albatross to many advertisers, forcing the larger ones into the costs and risks of the program business as a price for using television, while penalizing in particular the smaller ones. The latter can now use network television flexibly and efficiently, buying commercial participations in a schedule offered by the networks, gearing the amount and duration of their advertising to their budget and marketing requirements. Under the rules, these opportunities would be drastically reduced. To overcome this handicap, smaller advertisers would have to approach major sponsors, soliciting opportunities in the latter's programs, with advertisers themselves becoming time and program brokers.

Finally, the rules would work against program diversity and against the interest of special and minority groups. A network controlling its own schedule can provide for such programs, and serve as a balance wheel against undue emphasis on a few program types. Dividing responsibility for the schedule between networks and advertisers would weaken the networks' ability to fill this vital role.

The effect on the presentation of news documentaries would be even more severe in the case of a network which, like NBC, adheres to the policy of producing such programs itself. Since we have no intention of abandoning the news responsibility our policy reflects, the proposed rules would confine documentaries to half of our schedule. Consequently, there would be a substantial reduction in the number of these programs.

A premise of the proposed rules—that advertisers in large numbers are unsuccessfully storming the networks' gates with fine programs of special interest—is wholly contrary to the facts. A few are interested in such programs, and we have found places in our schedule for acceptable programs they have offered. But, in each network's schedule there are now programs of this character still seeking sponsorship. For example, NBC's award-winning *Profiles in Courage* was not developed by an advertiser. It was committed to the schedule prior to any sponsorship, and despite its recognized quality, it has fallen far short of attracting full advertiser support. If advertisers are not drawn to this type of existing programming, what basis is there for believing many will supply it themselves?

It seems to me that the proposed rules would not only create great confusion in the industry but might well succeed in giving chaos a good name. Someone has suggested it is the first step in a long-term scheme to bring back Mah-Jongg.

What has brought us to the threshold of this new and bizarre regulation?

MISCONCEPTIONS AND ASSUMPTIONS

For an answer we must go back to the FCC staff study of network programming, which began in 1955 and proceeded at a leisurely pace until 1962, when the record was closed. That record brims with the subjective views of numbers of witnesses on the quality of programming, expressions of personal taste and special interest, nostalgia for the early, rudimentary days of television, and discussion of what might or might not be meaningful drama. But nowhere does it offer a shred of support for the assumption that advertisers would undertake to supply the nearly $500 million worth of programming that constitutes half the television networks' evening schedules.

One can only assume that having been exposed—even fleetingly—to the advertising fraternity, the staff has adopted one of the techniques attributed to Madison Avenue—running it up the flagpole to see whether anyone salutes. In the varied comment generated by publication of the staff's recommendations, some have saluted. But the salutes are for a gaudy flag of misconception and false assumption that, like Alice's Wonderland, grows curiouser and curiouser under scrutiny.

For example, it is a too widely held misconception that the networks now own 90 per cent of the programming in their schedules, and that by occupying the field of program ownership they have frozen out other creative sources. The fact, of course, is that the networks produce and own only a tiny fraction of the entertainment programs they present. Almost all the programming comes from a variety of independent production sources, the very same sources from which advertisers themselves obtain programs. In the current season, NBC itself produces only two series in its evening entertainment schedule, while 17 independent production companies provide the rest.

The idea that networks lock out meritorious programs in which they cannot obtain financial interests or syndication rights flouts the economic facts of life of network programming. Surely there is justification for a network to share in the profits, if any, of a series whose production costs it underwrites. But no network can afford to subordinate program merit to financial participation. The ramifications of a successful, long-lasting program in the schedule far outweigh any potential revenue from subsequent use of the program. It's axiomatic that success must come first—profit hopefully later.

In addition, those saluting the staff proposals seem to have no valid explanation of why network companies should be barred from syndication, which may be an advantage to other competing syndicators but serves no public purpose at all; or why the reasoning that justifies foreign syndication by networks of their own productions does not also apply to domestic syndication of these same programs.

The proposed rules have also revived the catchy old canard that "three men" in New York decide what the nation shall see. Anyone experienced with the complications of network program development and scheduling knows what a silly over-simplification that is, even though it is true that the networks exercise final control over their own schedules. Actually the judgments of literally hundreds of creative people at networks, independent production organizations, talent agencies, and advertising agencies ultimately contribute to the whole programming process.

DEVELOPING THE SCHEDULE

On this score, NBC's own experience in developing its 1965–1966 schedule is enlightening. In preparing for the coming season, our program specialists considered nearly 450 different program ideas from various outside sources. Of these, more than 200 were the subject of follow-up discussion, with the creative people from the network and independent producers working together. Many of the ideas fell by the wayside, but well over 100 went on to treatments or scripts, in many cases financed by NBC. From these, 26 of the most promising were developed as pilot films, with NBC putting up several million dollars to help bring the projects along. And from these pilots, together with committed series, 15 new programs were chosen for the fall schedule.

The competition for selection is intense, and it is natural for a producer to be disappointed when his program idea, script, or pilot fails to make the grade. However, because the number of programs that can be accommodated is limited, not every program or production company can find its way into the schedule. But it is fantasy to argue that this competitive process, through which network schedules are built, freezes out independent production and reflects only the ideas of three network chieftains, whoever they may be.

It is true that, since no one else has responsibility for the total schedule, the final choice of programs must be made by the network itself. The Commission on many past occasions has insisted that the public interest demands that networks maintain this role, and has assailed advertiser influence on network programming. But many paradoxes appear through the regulatory looking glass. Even the principal author of the proposed rules recently insisted that they were not intended to reduce a network's responsibility for accepting or rejecting programs for its schedule. Yet, it's clear that the rules are designed to create a mechanism whose pressure would force the networks into less responsibility. In other words, the staff ventriloquist speaks in two voices—in one, admonishing us to be masters of our own house; in the other, demanding that we sublet part of it!

SUBSTANTIAL PROBLEMS

The recommendations seem to stem from an expectation that their adoption would bring to the air programs of better quality—that is, programs the proponents prefer or, in any event, with which they would like to be identified.

But as a prominent American, for whom I have great respect but with whom I sometimes disagree, recently said: "I see no evidence that the FCC's influence is likely to be a particularly good one in the field of programming." On the same occasion he also pungently observed: "Those hero types who criticize the vast wasteland of programming—when the chips are down, these guys don't care what the public wants. It's what they think is good for the public."

I am sure that Commissioner Lee Loevinger won't mind my having quoted him.

I hope that the Commission, rather than bogging down in attempts to repair its staff's mechanical monster, will, instead, consider carefully whether any proposal is warranted; whether the problem assumed is real or artificial; what purposes are sought; and whether they are valid for the Commission to seek.

Surely the fact that the staff has labored so long and now brings forth this morsel is not adequate reason for plunging ahead with "let's-do-something" exuberance. The Commission must concern itself continuously with broadcasting. But there are currently substantial problems pressing upon it that, in my opinion, should have higher priority than the questionable tampering with a system of television programming that, whatever its faults, is still the best in the world. Among them are the rapid growth of a world-wide system of satellite communication and the future role of the present complex of international communications; the unresolved problems of UHF, CATV, and pay-TV; the demand for more effective use of the frequency spectrum and coordination of government and civilian needs; the proper development of educational television; and the expansion of personal communications.

In the face of these demanding matters, the Commission—to justify undertaking to revamp the network program structure—should be required to demonstrate the public need, and define its power and ability to meet it.

NEITHER NEED NOR AUTHORITY

I submit that there is neither the need nor the authority for the type of regulation proposed. It is founded on a series of fallacies:

The first is that the networks' responsibility for their schedules forecloses diversity of program sources. The fact is, the bulk of network programming comes from a broad variety of independent creative sources.

The second is that with direct advertiser control or influence, program quality would benefit as advertiser selection replaced network selection in the program process. The FCC staff that authored the proposed rules argued to the contrary in its report issued barely two years ago, when it said:

". . . The factors which go into the judgment of advertisers and their agents . . . are bottomed in an effort to attract an audience which will consist largely of potential customers for the product. . . ." Under advertiser influence, the report said, "diversity remains within prescribed limits and will not include many program types and formats which are not of proven appeal to the audience."

This observation is underscored by recent example. One of television's largest advertisers—among the few who bring in programs for scheduling—underwrote nine pilot films for potential series next season—all family comedies. This is no reflection on the quality of the programs or their place in television, but it emphasizes advertisers' basic quest for programming with the broadest audience appeal—programs of the type the supporters of the rules criticize for "sameness."

The third fallacy assumes some direct relationship between quality of programming and the number of production organizations supplying the programs, as if the programs for our current schedule, selected after the most careful winnowing process, would be improved if they came from 25 different producers instead of 17.

Underlying this drive for regulation, there seems to be an effort to change network programming by operating on its structure. Both the programming and the structure derive from the environmental fact that network television is a mass medium, supported by advertisers. As such, through a variety of entertainment forms, it strives for programming of broad appeal to attract maximum audiences, without excluding programs for more specialized interests.

CHALLENGE TO PROGRESS

This is the fundamental character of our American system of television. The staff, and perhaps some members of the Commission, may not like the results. They may not all enjoy the fare that attracts the largest audiences. They may feel the medium should cater more to small audiences. They may have no confidence in the tastes of the majority and may want to see television devoted to refining those tastes.

But these personal judgments are not relevant. The Commission was not given the mission of reforming broadcasting to serve the interests of the few rather than the many; and if it undertakes to tinker with the basic mechanisms of the medium, it will create more problems than solutions, more questions than answers.

It must be kept in mind—and indeed impressed upon our advocates and critics alike—that television's great moments in emerging as the nation's foremost medium of communication were not prompted by the wielding of a regulatory baton or the application of abstract formulas. They came from the unhampered initiative and honest efforts of responsible broadcasters. The medium has continued to develop strongly, progressively engaging the interest of more people for more of their time. It seems fair to assume that this is because of, not despite, present program policies and practices.

The challenge—not only to the networks and our affiliates, but to the vast multiple society we serve—is whether the medium will be allowed to continue its natural development through the broadcaster's conscientious response to the demonstrated interests of his audience, or whether it will be forced to fit into a governmental mold of untested premise and unproven theory.

I am confident that this challenge can be met successfully, but only if all of us remain alert and continue to serve the public properly and in good conscience.

19. Some Aspects of Freedom of Information

by Albert Namurois

Mr. Namurois is the legal adviser of the Belgian National Institute of Radio-Broadcasting. His essay appeared in European Broadcasting Union Review, Part B, *for January, 1960. It is reprinted here by permission of the author, who acknowledges his debt for material to the administrative offices of the EBU and the United Nations.*

1. Nowadays people are wont to refer to freedom of information as if this were some new-fangled notion, whereas in fact it is a facet of one of the oldest freedoms proceeding from natural law, the freedom to express one's opinions.

The origin of this freedom must be sought in the principle of individual freedom, which may be defined as the right to be unmolested and unconstrained by the powers that be or by one's fellow men in the very private domain of the human conscience. Clearly enough, the exercise of this right cannot be allowed to create a breach of the peace or to encroach upon the rights of others.

Individual freedom has a bearing on every form of human activity, and as it is part of the essential nature of man as much as the conscience is, the majority of those who framed our constitutions have confined themselves to recognising the existence of this natural right in its various forms, prominent among which is the free expression of opinion.

The fact is that an essential component of individual freedom is freedom of thought, which in law does not consist merely in the act of thinking—for no coercive measures have yet been devised to prevent a man from espousing the opinions which appear most plausible to him—but also in the expression of his ideas and opinions, whatever they may be, without let or hindrance, and the communication of those ideas and opinions by word of mouth or otherwise. To believe in one's heart that a given proposition is true is to believe it true for all, whence the desire to impart it to one's fellows; for the man of conviction the possession of the truth and the wish to hand it on are inseparable. Freedom of thought therefore implies the faculty of expressing one's opinions, and freedom of information is only one specific and particular aspect of freedom of opinion.

But like the other concomitants of individual freedom, this freedom of speech would have been incomplete and illusory if man had not had freely available the means by which he could convey his opinions and win support for them. This corollary of the right of free speech is the right to employ the various media of communication to air opinions in public;

more specifically it is the freedom of the press which every law-abiding State has granted to its citizens.

In a State with a representative form of government where the repositories of power derive their authority from the nation, the press is a means whereby the country can keep in touch with what the government is doing, and whereby the government can keep its finger on the pulse of public opinion.

More than this, the press will act as a powerful regulating influence through the spotlight it throws on the acts of public authorities, and will thus be a safeguard for the other liberties of the subject (Damoiseaux: *Les Institutions nationales de la Belgique*). Nevertheless the press cannot play its role as effectively as it might if it is hedged about with preventive measures such as censorship or the posting of caution money.

2. While the general movement towards political independence that asserted itself particularly from the late eighteenth century onwards could count among its achievements the granting of freedom of speech, it must not be forgotten that the seed had been sown long before. Aristotle had recognised the existence of the will and free agency, which was termed by the Scholastics *libertas a necessitate,* the power of the will to decide for itself without being under the constraint of any outward or even inward force, and hence the foundation and the *raison d'étre* of civil and political freedoms. The postulate of freedom of opinion was taught by St. Augustine and St. Thomas Aquinas; it was taken up again by Locke, who held that man had the natural right to safeguard and develop his freedom.

3. One of the first provisions recognising the right of free speech is to be found in the Declaration of the Rights of Man and of the Citizen of 1789: "The unrestricted communication of opinions being one of the most precious rights of man, every citizen may speak, write, and publish freely, provided he be responsible for the abuse of this liberty in the cases determined by law."

With the passage of time the majority of States have recognised the right of free speech in their constitutions, along with other freedoms. Certain constitutions achieve this end by a general statement of rights; in many cases certain freedoms are "guaranteed," as in the Belgian Constitution of 7 February 1831 (article 14), the Iraqi Constitution of 21 March 1925 (article 12), the Lebanese Constitution of 23 May 1926 (article 13), the Jordanian Constitution of December 1951 (article 15), and others; elsewhere the freedoms may be declared "inviolable," as they are in the Mexican Constitution of 5 February 1917 (article 7), the Constitution of the State of Arkansas (USA) of 1874 (article II), or the Constitution of the State of Maryland (USA) of 1867 (article 40); or again it may be provided that certain categories of persons may adopt a certain course of conduct, or that they have the right to do so. Other States have defined certain liberties by way of prohibition of any interference with the free exercise thereof, examples of this being the Constitution of the Argentine Nation of 11 March 1949 (article 23), the Constitution of the Philippines of 1935 (article III), the First Amendment (1791) to the United States Constitution, the Constitution of Denmark of 1953 (article 77), the Belgian Constitution of 1831 (article 18), the Constitution of El Salvador of 7 September 1950 (article 158), and the Brazilian Constitution of 18 September 1946 (article 141), while other constitutions contain a prohibition on restrictions abridging these freedoms, this being the case, *inter alia*, of the Belgian Constitution (article 14), the Danish Constitution (article 77), the Netherlands Constitution (article 7), the Syrian Constitution (articles 14 and 15), and the Jordanian Constitution (article 15).

In countries of the British Commonwealth—the United Kingdom, Australia, Canada, New Zealand, and the Union of South Africa—the right to freedom of expression and other freedoms is not laid down in constitutional instruments. In these countries the right to freedom of expression and opinion is legally defined largely by reference to limitations placed on its exercise.

4. It has been said above that the right to freedom of expression will not be genuine and complete unless a man is free to avail himself of the means by which he can propagate his opinions. One cannot but notice, however, that constitutions generally make reference only to the press in the strict sense, i.e., to the dissemination of ideas and news through the written word, and ignore modern media of communication and broadcasting in particular ("The broadcasting service may include transmissions of sounds, or transmissions by television, facsimile or other means"—Buenos Aires Convention). There are, however, some clauses which guarantee freedom of expression by any means whatsoever. Article 29 of the Constitution of Uruguay of 26 October 1951 provides: "The expression of opinion on any subject by word of mouth, private writing, publication in the Press, or by any other method of dissemination is entirely free . . ."; article 10 of the Political Constitution of Chile of 18 September 1925 guarantees to inhabitants "freedom to express their opinions without previous censorship, by word or in writing, through the medium of the Press or in any other form . . . ," while the Political Constitution of Ecuador of 31 December 1946 guarantees "freedom of expression of thought, and of speech, through the Press or other means of utterance or diffusion. . . ."

The reader will find other examples below of provisions in which the proclamation of the right of freedom of opinion and expression is attended by certain provisos. Article 6 of the Constitution of Bolivia of 23 November 1945, as amended on 20 September and 26 November 1947, provides *inter alia* that "Every person has the following fundamental rights . . . To express freely his ideas and opinions by any means of diffusion"; article 59 of the Constitution of Honduras of 28 March 1936 asserts the right of every person freely to "express his opinions orally or in writing by means of the Press or by any other procedure"; article 22 of the Constitution of Libya of 7 October 1951 states: "Freedom of thought shall be guaranteed. Everyone shall have the right to express his opinion and to publish it by all means and methods"; and the Panamanian Constitution of 1 March 1946 declares in article 38 that "Every person may freely express his thoughts, by word, in writing, or in any other medium, without subjection to previous censorship." Further instances are to be found in article 63 of the Peruvian Constitution of 9 April 1953, which provides: "Everyone has the right freely to express his ideas and his opinions by means of printing or by any other method of diffusion," and in paragraph 1 of article 14 of the Constitution of Syria, according to which "The State shall guarantee freedom of opinion, and all Syrians shall be entitled to express their views freely in writing, speeches, graphically or by any other means of expression."

Article 30 of the Constitution of Eritrea adopted on 10 July 1952 provides that: "Everyone resident in Eritrea shall have the right to express his opinion through any medium whatever (press, speech, etc.), and to learn the opinions expressed by others." Article 21 of the Constitution of the Italian Republic of 27 December 1947 provides in part that: "All men have the right to express their personal views freely in word, in writing or by any other means of communication."

Article 33 of the Constitutional Law of the Republic of Cuba of 4 April 1952 provides that: "Every person shall be free, without previous censorship, to express his thoughts in speech or in writing or by any other graphic or oral mode of expression, and may use for that purpose any available means of dissemination." Article 19 of the Constitution of Haiti of 25 November 1950 states, *inter alia*, that: "Every person has the right to express his opinions on any subject and by any means within his power."

Some constitutions have made specific reference to radio and/or films. Paragraph 6 of article 40 of the Constitution of Ireland of 29 December 1937 includes the following provisions:

> 6. (1) The State guarantees liberty for the exercise of the following rights, subject to public order and morality:
>
> I. The rights of the citizens to express freely their convictions and opinions.
>
> The education of public opinion being, however, a matter of such grave import to the common good, the State shall endeavour to ensure that organs of public opinion, such as the radio, the press, the cinema, while preserving their rightful liberty of expression, including criticism of Government policy, shall not be used to undermine public order or morality or the authority of the State.

> The publication or utterance of blasphemous, seditious, or indecent matter is an offense which shall be punishable in accordance with law.

The Basic Law for the Federal Republic of Germany of 23 May 1949 provides in paragraph 1 of its article 5 that:

> Everyone has the right freely to express and to disseminate his opinion through speech, writing and pictures and, without hindrance, to instruct himself from generally accessible sources. Freedom of the Press and freedom of radio and motion pictures reporting are guaranteed. There is no censorship.

The special category into which films, radio, and television are sometimes placed is reflected in the Constitution of the Czechoslovak Republic of 9 May 1948, section 22 of the Detailed Provisions of which reads as follows:

> 22. (1) The right to produce, distribute, publicly exhibit, as well as to import and export motion pictures shall be reserved to the State.
> (2) Broadcasting and television shall be the exclusive right of the State.
> (3) The exercise of these rights shall be regulated by Acts, which shall also lay down exceptions.

Section 14 of the Constitution of Greece of 22 December 1951 contains, among other things, the following stipulation:

> The provisions for the protection of the Press contained in this Article do not apply to the cinema, to public spectacles, to gramophones, to broadcasting or to other means of transmission of speech or of performance.

To sum up, many constitutions have only guaranteed the exercise of the right to freedom of expression through the medium of the printed press, while a few others have countenanced its exercise by any means, or at least by means of radio and motion pictures, and yet others have excluded the two last-mentioned media from any guarantee. Moreover, a number of states have passed special legislation establishing *ad hoc* provisions applying to broadcasting and motion pictures, which are not in keeping with the concept of freedom. This state of affairs as it affects broadcasting commands general acceptance on account of the limitations imposed by the circumstances: the spectrum cannot be split up indefinitely and the number of frequencies allotted to each country is limited. For all that, the broadcasting organisations authorised by law should be in a position to accept, and should in fact accept, suitable contributions from individuals and associations, whatever their nature.

5. It has been established above that there is generally some guarantee or recognition in a state's domestic law of the individual's right to freedom of expression and his possibility of exercising it through the medium of the press, but that in most cases access to the other media such as broadcasting or motion pictures is denied him unless he complies with various prior requirements.

Consideration must now be given to what international law has to say on this subject. There are two conventions which contain peremptory provisions in this behalf, namely the Universal Declaration of Human Rights approved by the United Nations General Assembly on 10 December 1948 and the Convention for the Protection of Human Rights and Fundamental Freedoms prepared under the auspices of the Council of Europe, signed at Rome on 4 November 1950 and since ratified by all the signatory countries (Belgium, Denmark, France, Federal Republic of Germany, Greece, Iceland, Ireland, Italy, Luxembourg, Netherlands, Norway, Sweden, Turkey, and the United Kingdom).

The former instrument declares in article 19 that:

> Everyone has the right to freedom of opinion and expression; this right includes freedom to hold opinions without interference and to seek, receive and impart information and ideas through any media and regardless of frontiers.

Article 10 (1) of the second instrument states that:

> Everyone has the right to freedom of expression. This right shall include freedom to hold opinions and to receive and impart information and ideas without interference by public authority and regardless of frontiers. This Article shall not prevent States from requiring the licensing of broadcasting, television or cinema enterprises.

In these two texts which do not differ as to substance one finds two concepts which throw light on the constitutional arrangements in the several states. The first is that freedom of expression does not consist in the right to express, utter, or convey *ideas* alone, but also that of imparting *information,* and it also comprises the right to seek and receive information. The second point of interest in these provisions is that they contemplate the exercise of the right to freedom of expression by any means whatsoever, though it should be noted that the Convention of the Council of Europe makes it a matter for national legislation to require the licensing of broadcasting or cinema enterprises. It must be admitted that as regards the majority of countries this is merely a confirmation of the position that already prevails.

The Convention for the Protection of Human Rights is accompanied by a Protocol, article 1 of which guarantees private property and does not permit any derogation from this principle except in the public interest and subject to the conditions provided for by law. Thus, whenever there is a clash between the right of freedom of expression and private property—a theme to which reference will be made later—the problem is as thorny as ever, unless resolved by express provisions of national law.

6. It was said above that the United Nations Convention and the Council of Europe Convention enshrine the right to seek and receive information as well as to impart it. This enumeration, in the view of the present writer, is redundant, since everyone will concede that the possessor of a right to impart information has an implied right to seek and receive it. This, indeed, may be held to be a principle of natural law which will apply so long as the rights of others are not encroached upon.

At the beginning of his *Metaphysics* Aristotle asserts that every man is naturally athirst for knowledge. Indeed, curiosity is an irresistible inclination which admittedly may fasten upon all sorts of subjects, but which, in one way or another, affects us all. This inclination is not the product of habit or education; it is innate and has its roots in human nature itself. Habits may be formed and broken; education may vary according to the time and the place; but intellectual curiosity is a universal and invariable trait which manifests itself in the child even before he has had time to acquire habits or receive an education. It has its *fons et origo* in the intelligence of man.

It can therefore be taken that the freedom to seek and receive information, as much as the right to convey it, is firmly founded in natural law and may represent a valuable argument for such media of expression as broadcasting, in demanding access to places where an event is taking place, for instance. But while it is no new idea, it would have been all the better if it had been spelled out in some provision of law.

7. In the field of copyright, which may be involved when freedom of expression is exercised, the Berne Convention contains a more affirmative provision in its article 10 *bis* than that in article 10 of the Council of Europe Convention, although it too only refers the problem to the national legislation. Under article 10 *bis* of the Berne Convention—

> it shall be a matter for legislation in Countries of the Union to determine the conditions under which recording, reproduction, and public communication of short extracts from literary and artistic works may be made for the purpose of reporting current events by means of photography, cinematography or by radio-diffusion.

This article has already brought forth sundry provisions of domestic law such as section 33 *bis* of the Swiss Federal Copyright Act, section 6 (3) of the United Kingdom Copyright Act, and section 21 *bis* of the Belgian statute on copyright.

In Germany there is a special enactment on filmed newsreels which is still considered to be in force and which permits the motion-picture reproduction of copyright literary and artistic works for the purpose of reporting the events of the day.

The various bills now in process of preparation in countries of the Berne Union suggest that this exception to the exclusive rights of the author will soon enjoy more general recognition, so that from this point of view at least the right to information will no longer be impaired.

8. After this attempt to determine the foundations and discover the sources of the right to freedom of expression, of which the right to information is one aspect, it is logical to consider whether the media of information, particularly broadcasting and the cinema, are in a position to fulfill the mission assigned to them by virtue of the existing law, as set forth above.

This inquiry will be confined to the question of access to the event, and will not dwell on other aspects of information such as the circulation of news.

9. At the outset it may be well to determine the nature of this mission, in order to be able to assess the cogency of the claims advanced by the information media.

It will be common ground that the cinema and, to an even greater extent, the broadcasting services perform a service to the community in general, which is why governments often regard them as public services. This is true in any event of the broadcasting services constituted as public services through the devices of concession, licensing, and government recognition and approval (e.g., the British Broadcasting Corporation and the Société Suisse de Radiodiffusion), through establishment as an endowed trust in the nature of a public service (*fondation sous forme d'établissement public*—the case of the Institut National Belge de Radiodiffusion) or as a State-owned enterprise, or by their organisation as agencies under direct government control.

Broadcasting performs a service to the community at large not only when it purveys instruction or entertainment but also when it gives the news. By inventing ever more efficient methods of transmitting information, modern technology has created a daily need for news which is permeating into every stratum of the population. News has become a social necessity, and one which must be catered for by the media of information.

It must be borne in mind that information is not the only role of broadcasting and the cinema, but it is only in the pursuit of this activity that they would appear to be able to invoke the right to information. It is self-evident that this right will not serve them when their intention is to amuse or educate their audiences.

It remains for us to determine the subject-matter of the information mission assigned to the broadcasting services.

It is contended by some that the word "information" in itself suggests that what is meant is spectacles on which the public has a legitimate claim to be informed; that this will seldom be the case with concerts or stage shows; but that national ceremonies, traditional pageantry, sporting events, or other news events will as a rule fall into the class of affairs that are "fair game" for public curiosity, and hence proper subjects for "information."

In the writer's view this is a somewhat arbitrary distinction. It has been shown above that every man is by nature a seeker after knowledge, and that his innate curiosity springs from his natural intelligence and constitutes an irresistible urge that may fasten on the oddest things. And that is why it is suggested that the right to information cannot be limited in scope to only a few of the world's doings.

But because the role of presenting the news differs from that of purveying education or entertainment, and because the right to information can be asserted in respect of the first-mentioned activity only, these facts will of themselves impose a limitation of sorts, which has to do with the length at which an event can be reported on the strength of this right. While in the printed press any report, however lengthy, will still qualify as news, this will hardly be the case with films and with sound and visual broadcasting. One has to draw the line between the "newsreel," on the one hand, and the documentary—or even the showing of the event in full—on the other. Televising or filming the whole (or all that matters) of a sporting event cannot be called information, a concept that will only warrant a very brief and "potted" version giving just the highlights of the event.

10. Having thus defined the nature and the limits of the information function of broadcasting and the cinema, it now remains to be seen whether these media are given the proper statutory powers to carry out this function. After all, it is not enough for national constitutions or international conventions to recognise or guarantee the right to information; the proper steps must be taken to see that the exercise of this right by the information media does not clash with other rights—copyright, the right of property in entertainments, the right of property in enclosed premises, and the individual's right in his own likeness (the *jus imaginis*).

11. In the field of copyright, it has been seen above that the Berne Convention affords a partial solution to the problem by permitting national legislation to provide that recording, reproduction, or public communication of short extracts from works may be made without a licence from the author for the purpose of reporting current events by means of photography, cinematography, or by radio-diffusion. As already stated, a number of national copyright statutes have availed themselves of this power, but it must be pointed out that it is somewhat awkward of application to television and motion pictures, where it is difficult not to reproduce and show works of figurative art in their entirety if the author's "*droit moral*" is

to be respected. For this reason, some acts already countenance the reproduction of works of this nature in their entirety, and one can only hope that the next Berne Convention revision conference will have this question on its agenda.

12. Turning now to the possible clash that may arise between the right to information and the right of property in the event, it is interesting to quote parts of the decree handed down on 15th November 1955 by the Praetor of Rome (*EBU Bulletin*, No. 40, p. 815) and upheld by the Milan Civil Court (*EBU Review*, No. 48B, p. 30). The judge took the view that the promoter of a sporting event cannot be regarded as the "proprietor of the event." The event is undoubtedly a public spectacle. There is, however, a fundamental difference between a sporting spectacle and a theatrical or motion picture spectacle: the latter provide the spectator, who from the business angle may be said to be the consumer, with a form of merchandise which although not material may nevertheless be seen as something protected by law, namely the work performed in the theatre or shown in the cinema. This work is the fruit of the spirit and the creative intelligence of man, which is more than can be said of a sporting event which, even if it is the product of the human intellect, is at best a glorified trial of strength and skills of various kinds. Moreover, "shows" in the proper sense of the word are susceptible of repetition, whereas sporting events are not, and are consumed in time and space as "irreproducible" events.

And it is because of this very "irreproducibility," together with the distances and the travelling involved, that the bulk of the public, although legitimately interested in the event and desirous of watching it, is necessarily and irrevocably prevented from being there. But this public must not be kept from participating, if only indirectly, on account of the interest aroused by sporting events which, as social phenomena, must be regarded as belonging to history and, as spectacles, as belonging to the public as a whole. It cannot therefore be right by any standards to refuse admission to information media such as television, sound broadcasting, and the cinema newsreel firms. These are, in brief, the grounds adduced by the Italian judge.[1]

In this connection attention may be drawn to a passage from the judgment delivered by the Berlin *Kammergericht* in 1952:

> The mission of the Press is to inform the public on all the events and happenings that are of interest to it. In conformity with this public mission, section 5 of the Bonn Basic Law, read in conjunction with articles 1 and 87 of the Berlin Constitution, guarantees the freedom of

[1] A more recent judgment, reported below in this issue of the *Review*, indicates a development in Italian judicial practice that is unfavourable to the right to information. This, in the writer's view, makes it all the more urgent to give effect to the conclusions in this article.

> the Press and freedom of information. Like the Press, the cinema newsreels have the task of apprising the public of all the important happenings in politics and the arts. The information they provide is only complementary to that of the Press, and consequently they too must be given freedom of information. This view is borne out by the Act of 30th April 1936 to facilitate newsreel reporting. Under this Act newsreel firms are allowed to show and perform works in copyright for the purposes of reporting current events; thus even copyright has to yield in the interests of unrestricted information through the medium of the cinema. When the defendants, in the intention of catering to the public's interest in the boxing match in question, took a sequence of some 30 to 50 metres of film with a tele-lens from an adjoining ground, they did nothing contrary to public policy and established usages, in view of the principle of freedom of information.

The Italian judge expressly denies the organiser any proprietary rights in his sporting event; the German courts do so by implication, in recognising films taken from adjoining ground to be lawful. It can be safely said, however, that in no case will the sporting event in itself be deemed to be a work in which copyright subsists.

In support of this statement mention may be made of a fairly recent Canadian judgment (Exchequer Court of Canada, 21st May 1954) and also the position adopted by the Copyright Committee in the United Kingdom which during the deliberations that preceded the new Act flatly rejected the claim by sports interests for a copyright. There is also the opinion of the Italian judge in the case reported above that a theatrical or cinematographic work is the fruit of the spirit and the creative intelligence of man, whereas a sporting event, though a product of the human intellect, is at best a glorified trial of strength and skills of various kinds. It is certain that those legal systems in which the creation of a work of the mind is the basis for copyright protection will fall in with the opinion of the Italian judge.

When reading these court decisions, one should not lose sight of the fact that the Basic Law of the Federal Republic of Germany guarantees freedom of reporting by means of broadcasting and cinematography, while the Italian Constitution guarantees freedom of expression, and hence grants the right to information not only via the spoken and the written word, but also via any other medium of communication.

The fact must therefore be faced that as regards the organisers of concerts and other similar shows the problem is no nearer solution, and that even in the case of sporting events there is some doubt as to the line that the courts would take in countries other than Germany or Italy, even assuming that the judicial decisions reported above are upheld.

13. In the exercise of the right to information, the most serious clash

that is likely to arise is the clash with the right of ownership in enclosed premises, and hence with the principle of inviolability of domicile. In the absence of legislation to the contrary, it would seem that the owner of private enclosed premises in which an event is being held can refuse admission to the employees of the information media and their gear.

This raises the question of what is meant by "enclosed premises." It is submitted that a restrictive interpretation is called for in this connection, and it is not enough, for instance, that the circuit on which the event is taking place should be closed off for reasons of safety or public order for it to constitute "enclosed premises," enjoying inviolability of domicile; domicile cannot be such a haphazard and short-lived thing as a stretch of the public highway closed to traffic for a short period of time. The writer thus adopts the contentions of the Praetor of Rome in the case reported above.

It may be noted that in Italy the Copyright Act of 22nd April 1941 empowers the broadcasting corporation to make outside broadcasts (within certain limits) from theatres, concert halls, and other public places, the owners, impresarios, and others concerned in the show being required to admit the technical installations needed for such relays.

14. A few words must be said of the right of the individual to his likeness, the *jus imaginis*. While recognising this personal right in principle, it is important where the right to information is concerned to set limits to it so as to prevent vexatious claims that may involve costly lawsuits.

The Italian Copyright Act may again be cited in this connection. Section 97 of that Act states, *inter alia*, that the consent of the person portrayed is not necessary when reproduction of the picture is warranted by fame, public office, or for scientific, educational, or cultural reasons, or again when reproduction is connected with happenings, events, or ceremonies of public interest or taking place in public. Similar provisions are contained in section 62 of the German Academy's draft bill.

In the case quoted above the German court held that as sporting events are part of the news in which the public may properly take an interest, the boxers had to be regarded as "figures of contemporary history" within the meaning of the Act, and could therefore not claim a right to their own likeness, particularly as they had consented to the defendants' making and distributing throughout the world a film of the fight.

Reference may also be made to a judgment of the Vienna Civil Court of 28th October 1957 (*EBU Review*, No. 50B, p. 26). In the case in point, a public event (a car crash), at which the plaintiff, of his own free will, had been an onlooker, had been the subject of a visual *reportage*. The court took the view that freedom of information as it has developed in recent years, whether in relation to the press, the cinema, sound radio, or television, authorises such visual representations provided they do not interfere with the individual's private life. No one present at an accident that happens in public can object to his picture appearing in the course of a news item concerning the accident and thus becoming visible to viewers.

15. One can only conclude that although the information media may in principle avail themselves of the right to freedom of information, there are many obstacles that hinder their access to information, and there are only a few scattered provisions of positive law in this field.

There can be no doubt that it is primarily a matter for national legislative bodies to recognise that the information media perform a mission that is in the public interest, and to alleviate the greatest difficulties, which mainly proceed from the right of property.

Where literary or artistic works are concerned, the countries of the Berne Union can fall back on a provision in the Convention that enables the requirements of the news services to be met to some extent; this provision is Article 10 *bis*, to which reference has already been made, and which was added at the Brussels Conference with a view to enhancing freedom of information. It has been seen that several countries have already legislated on this basis and have even gone beyond the limits it lays down.

In the case of sporting events, which involve the right of property in enclosed premises and possibly in certain countries the right of property in events, the problem is a difficult one to solve.

Some domestic legislation on the subject already exists in draft form. One such draft contemplates two different cases; the first is where the event takes place on the public highway, in a public place, or a private enclosed space that is open to the public free of charge, and the second is where the event is held in enclosed premises open to the public on payment of an admission charge. In the first case the promoters and others cannot restrain the live or deferred televising or sound broadcasting of the event, nor the making of reports by the newspapers or the newsreels; moreover, they cannot demand a fee for the presence of employees of the news concerns or the installation of their equipment. The same provisions apply also in the second case, except that only excerpts may be given in broadcasts and in newsreels, and that these reports must be deferred.

A second draft bill would give sound and visual broadcasting services the right to record and broadcast, for the sole purposes of radio news bulletins and television newsreels, current events taking place in public places or in places open to the public, the owners, impresarios, and all persons taking part in the spectacle being required to permit the exercise of this right.

16. It is somewhat surprising to note that in the field of copyright the international legislator has been prepared to sacrifice to some extent the author's interests for the sake of news, whereas he has shown himself to be much more solicitous of the interests of organisers of sporting events.

The fact is that although the Convention for the Protection of Human Rights and Fundamental Freedoms establishes the freedom to seek and receive information, Article 1 of the Protocol to this Convention stipulates that no one shall be deprived of his possessions except in the public interest and subject to the conditions provided for by law and by the general principles of international law. In other words, the effect of this Convention is that national legislative bodies can only limit the property rights of promotors of sporting events by resorting to expropriation on the grounds of the public interest. This procedure presupposes lengthy hearings and the payment of compensation beforehand.

The procedure is reasonable enough in the event of the sporting event being broadcast or televised in its entirety: for instance, a Spanish decree of 4th June 1959 authorises the television service in certain circumstances to televise contests of an international nature and such national contests as are, in its opinion, of widespread interest, but in conformity with the legislation on compulsory expropriation, the television service must make good the damage and loss occasioned by the broadcast, such loss to be assessed in each case at a flat rate, proportionate to the total value of unsold tickets (*EBU Review*, No. 57B, September 1959).

But this is not the case in the matter under review, for as has been said above the right to information should normally operate only for an outline report of the event, whether the report be in print or on film.

Under these conditions the restrictions placed on the right to information by the Convention for the Protection of Human Rights are rather difficult to swallow.

17. After examining the majority of the existing legal provisions on the subject both in national constitutions and statutes and in international conventions, one is forced to the conclusion that these provisions are inadequate to enable the information media in general and broadcasting in particular to fulfill the social role assigned to them.

While they can invoke the principle of the right to information, it must be admitted that most of the time they lack statutory powers, especially the possibility of gaining free access to the event. The right of property and the principle of inviolability of domicile—rights which figure in national constitutions on an equal footing with that of freedom of information—are the main potential stumbling blocks to the latter, in the absence of legislation to the contrary. Until such time as special legislation grants priority to the right to information when it conflicts with these other rights, it is submitted that the right to information is doomed to remain inoperative, at any rate whenever the event takes place in private enclosed premises.

To this end the law should, in the writer's opinion, forbid the organisers of sporting or other events to obstruct the recording, reproduction, and communication to the public of extracts from such events to illustrate reports on current affairs.

The authors have to put up with a rule of this kind in respect to their intellectual works, and it is hard to see why the promoters of events should not have to accept it likewise.

At a time when the media for the communication of knowledge and ideas are increasingly unconfined by national frontiers, at a time when international organisations are seeking ways of facilitating international exchanges, it is highly desirable that there should be an international—or at least a European—instrument on the subject if the right to information is really to be accorded and to be effective. The real danger is that in the absence of an international standard to give a lead to national parliaments, the latter will either do nothing or will legislate along different lines, which will not make international exchanges any easier.

IV. Educational Television and Radio

THE ROAD to educational broadcasting has been a rocky one. Documented elsewhere and in *Congressional Records* are the interests which have clashed, and the declarations of networks and others opposing the reservation of frequencies for U.S. education.

But the problem is broader than this. It is the problem of finances, which Mr. Burns, then President of RCA, raised in his address, "The Challenge of Quality in Education." It is the problem of the "image of education" created by our commercial mass media, as Dr. George Gerbner traces it in "Teachers and Schools in the Mass Media." And it is the question of just how much of America's resources are to be devoted to education—especially if this reduces resources available to commerce.

These questions deserve continuing review. The selections included here, but a few of many available, are intended to generate discussion and review of the issues involved for both education and the mass media.

20. A Look at the Wagner-Hatfield Amendment

by R. Franklin Smith

The author is an Assistant Professor in the Speech Department at Western Michigan University. The article appeared in the National Association of Educational Broadcasters Journal *for March–April, 1964, and is reprinted here by permission of the author.*

With the first small appropriation to ETV by Congress last spring, with hopes for more to come, and with the development of UHF and implementation of the all-channel receiver bill in July, 1964, it is hoped that ETV's growth rate will be greatly accelerated.

Educational broadcasting has come a long way since the twenties and thirties when educators made futile and disorganized efforts to gain a place in the sun in the burgeoning new medium dominated by private industry. Several attempts were made during these early years to obtain reserved channels for education. Success was finally achieved in the late thirties with FM reservations and, of course, in 1952, with TV reservations.

One of these early attempts to obtain reserved channels for education occurred during the debate on the Communications Act of 1934. In fact, the bulk of the discussion on the Act in the Senate, as recorded in the *Congressional Record,* was taken up with a proposed amendment to reserve one-fourth of all radio channels for education and other nonprofit groups.

Before the issue of educational reservations ever came to the Senate floor, the Rev. Fr. John B. Harney, representing the Missionary Society of St. Paul the Apostle, licensee of a New York station, had submitted an amendment to the Communications bill to the Senate Committee on Interstate Commerce. His amendment called for the reservation of 25 per cent of all frequencies for nonprofit stations. The Committee rejected the proposal; however, Senator Robert F. Wagner (D-NY), late father of the present mayor of New York, and Senator Henry D. Hatfield (R-WVa) sponsored a similar amendment and tried to get it adopted during the Senate debate.

On May 15, 1934, after committee amendments were disposed of, Wagner opened what proved to be the most explosive issue in the Senate on the bill. He introduced a three-part amendment which called for the reservation of one-fourth of all radio broadcasting facilities for nonprofit stations. The amendment was designed, ". . . to eliminate monopoly and to insure equality of opportunity and consideration for educational, religious, agricultural, labor, cooperative and similar nonprofit making associations. . . ."

Part I of the amendment called for the rescinding of all broadcast licenses within ninety days; Part II would have required the new FCC to re-allocate all frequencies, power, and hours of operation of all stations within the same period. In short, a major overhaul of broadcasting channels was to be undertaken in a relatively short time. Part III of the amendment read:

> The Commission shall reserve and allocate only to educational, religious, agricultural, labor, cooperative, and similar nonprofit making associations one fourth of all the radio broadcasting facilities within its jurisdiction. The facilities reserved . . . shall be equally as desirable as those assigned to profit-making persons, firms or corporations. In the distribution of radio facilities to the associations referred to in this section, the Commission shall reserve for and allocate to such associations such radio broadcasting facilities as will reasonably make possible the operation of such stations on a self-sustaining basis, and to that end the licensee may sell such part of the allotted time as will make the station self-supporting.

It was this last phrase that provided the central target for the opponents of the amendment. The first question in the debate, however, concerned the re-allocation of radio facilities. It was pointed out that not only was ninety days insufficient time to re-allocate facilities, but that Congress did not have the right to cancel licenses in less than six months, except for specific violations. Wagner quickly agreed to change the amendment to allow six months instead of ninety days for the re-allocation of facilities. Supposedly the first issue in the debate was settled.

The opening assault on the "self-supporting" provision of the amendment was led by Senator Clarence Dill (D-Wash), the Senate sponsor of the Communications bill. He claimed:

> . . . this amendment does not propose at all what the Senator proposed in the amendment. He proposed that the time allotted should be used by educational stations, presumably for educational purposes; but subsection (g) of this amendment provides that the so-called "religious,

educational, or agricultural nonprofit stations," are to sell time in the commercial field sufficient to pay for the maintenance of the stations.

Dill returned to this argument often during the debate. He claimed that the "nonprofit" stations would be in competition with the existing commercial stations. In stating that two-thirds of existing commercial stations could only make enough money to cover maintenance costs, Dill implied that there was no actual distinction between these existing stations and "nonprofit" ones which would sell time only for maintenance.

Senator James Couzens (R-Mich) noted that there was nothing in the amendment which required the stations on reserved frequencies to broadcast any religious or educational program at all. The entire time allotted to the station could be used for commercial purposes; presumably such programs would not be of an educational nature.

Senator Wagner would not yield. He insisted that stations on reserved channels be allowed to sell time. He insisted that there was a difference between a "self-sustaining" station and one "making a profit." He was willing to exclude wages and salaries paid to employees of "nonprofit" stations, and he said that he was willing to safeguard the matter in any other way "except that I think it is fair that the station should be permitted to do sufficient business to make it self-sustaining."

Dill again had the floor to assert that religious and cultural organizations simply did not have the money necessary to operate stations, even if part of their broadcast time were to be sold.

Finally, it was time to hear from Senator Hatfield. The West Virginia Republican, a doctor in private life, began his speech, the longest single speech of the debate, with a discussion of education in the United States. He showed how the American people had supported education, and claimed that radio, too, was a part of our educational experience. "Education was a state function until our schools developed extension courses and radio became an interstate rather than a state or local matter," Hatfield asserted.

The Senator relied on ridicule in attacking commercial stations for their lack of educational programming:

> The graciousness of these commercial stations may well be gauged by the time allotted such organizations as the American Farm Bureau Federation, the National Grange, the 4-H clubs and several other farm groups who collectively receive one hour each Saturday noon from the National Broadcasting Company, or the American School of the Air, promoted by the Columbia Broadcasting System, on weekdays about noon.

In comparing our use of frequencies with other countries, Hatfield said that the United States was the only important country in the world placing control of radio in the hands of those who seek private profit. He concluded:

> Radio broadcasting reaches many millions of our people daily. The educators and others in our country who are seeking to build a higher standard for all Americans are denied opportunities which they should have. To my mind, these worthy organizations should be accorded the facility which they can so effectively use for the common good. . . .

Dill spoke again in rebuttal. From 1927 to 1932, only 81 applications for educational radio stations had been received by the Federal Radio Commission, Dill pointed out. Of these, 32 were granted in full, 27 in part, 10 were denied, and 10 dismissed. Dill asked his colleague, "Has the Senator stopped to consider the fact that it is financially impossible for these institutions actually to build and operate these stations without becoming commercial and advertising stations?"

Dill indicated that he was in sympathy with the sponsors of the amendment. He was "extremely anxious" that some plan be worked out for a larger use of radio for educational and religious purposes. But, in summary form, Dill emphasized his three primary objections to the amendment. (1) First of all it would be impossible to re-allocate all radio channels in the country even within six months. (2) His second, and "strongest objection" was that these stations were not to be "what we understand as education and religious stations merely, but they are to be stations that are to sell time on the air to advertisers who will make use of the stations for advertising purposes." (3) Even if the stations could borrow money, they would still have to sell between 60 and 75 per cent of their time to maintain stations and pay back the money needed to build new stations. He said:

> I cannot believe . . . that there is any hope of their using 25 per cent of our radio facilities effectively, even if we gave them the right under this bill. They have not the money and there is nowhere they can secure the money except they go into the commercial field and themselves become commercial stations.

With the assistance of Democratic Senator M. M. Logan of Kentucky, Dill reduced to absurdity the meaning of the terms, "similar nonprofit," as they applied to the licensees of proposed new stations.

> If we should provide that 25 per cent of time shall be allocated to nonprofit organizations, someone would have to determine—Congress or somebody else—how much of the 25 per cent should go to education, how much of it to religion and how much of it to agriculture, how much of it to labor, how much of it to fraternal organizations, and so forth.

When we enter this field, we must determine how much to give to the Catholics probably and how much to the Protestants and how much to the Jews.

LOGAN: And to the Hindus.

DILL: Yes, and probably the infidels would want some time.

LOGAN: Yes, there is a national association of atheists. They perhaps would want their part of the time.

The vote was finally taken. The amendment lost, twenty-three Senators voting yea: forty-two voting nay.

Why did the amendment fail? One can speculate on the reasons on the basis of the recorded argument in the debate, personal, emotional, or unknown factors excluded.

In the first place, the problem of re-allocating all radio facilities would have been enormous. Even though Wagner quickly conceded the need for six months instead of three months to re-allocate a fourth of the spectrum for educational broadcasting, the physical task of the undertaking probably would have been so great that it could not have been done in that time. Furthermore, explosive opposition more than likely would have come from the established commercial interests. Broadcasting, having begun chaotically some fourteen years earlier, by this time had come to live with a comparatively orderly pattern of frequencies; certainly there would have been great resistance to upsetting technical arrangements so long in the making.

The sponsors of the amendment revealed a stubborn reluctance to alter the amendment's provision allowing the educational stations to sell time only to meet expenses. It was this provision upon which the opponents of the amendment, led by Senator Dill, concentrated their fire relentlessly throughout the debate. They argued that this provision would obliterate any real distinction between commercial and noncommercial stations. Many commercial stations, they argued, were doing little more than meeting expenses. Moreover, they argued that many of the programs probably would not be of an educational nature; such programs were difficult to sell, and if time were to be sold to meet expenses, that time would be filled with programs of more general popular appeal.

The real concern of the opponents of the measure was the effect of the financial problem on the nature of the programming of educational and cultural stations, rather than the sale of time per se.

Had Wagner and Hatfield been willing to delete this provision from the amendment, it might very well have secured enough support for passage. Although they tried to alter the proposal to satisfy their opponents, they remained firm in their conviction that the only hope for the survival of the educational station was through the selling of time. Yet, it was understandable why they were adamant. In his history of the National Association of Educational Broadcasters, Harold Hill says, "During this depression period, most of them began commercial operation."

Not only were educational stations tempted to sell time because of the depression situation; they were also being encouraged to sell time by the commercial broadcasters. Hill says:

> Businessmen all over the country attempted to get educational stations to sell time. In view of the very precarious financial position of many educational stations and the difficulty of staying on the air during the depression, several educational broadcasters began to sell time, some leased their facilities to commercial interests, and some sold out to commercial stations.

Because the sponsors of the amendment failed to consider other means of financial support, their case was weakened. It may well be that no such consideration was given because the few educational broadcasters at the time were too preoccupied with their own problems to explore effectively and make known to their representatives other alternatives. It was during the same month in which the amendment was being debated, that the Association of College and University Broadcasting Stations had to cancel its convention because "so few members attended that no officers were elected, and it was decided that a meeting should be held in the fall."

Finally the wording of the amendment created the additional problem of determining exactly who would be eligible to operate the nonprofit station.

Although nonprofit organizations never obtained reserved AM channels, 1938 found FM channels and 1952 found television channels allocated for the exclusive use of educational stations.

Certainly it is not the purpose of this paper to imply that the debate over the Wagner-Hatfield Amendment emphasized all of the issues involved in channel reservations for nonprofit groups, or was the sole cause of future new approaches toward the development of educational broadcasting. What is evident, however, is that the issues that were raised in 1934 were either not relevant or were resolved when FM and TV channel reservations were authorized. It would seem, then, that the issues emanating from the Wagner-Hatfield debate might have contributed, in some unmeasurable degree, to a more thorough awareness on the part of educators and government officials of the key problems involved in the reservation of spectrum space for nonprivate use.

In the first place, in the case of both FM and television, the question of upsetting established allocation tables was not pertinent. Proposals and pressures for reserved channels for education were being considered *before* either FM or television had become a fully developed system. Admittedly, television was with us in 1948, but, thanks to the "freeze," the educators were given time to present their case before the system was established.

Second, it was clearly recognized in both FM and television reservations that reserved channels could not be used for commercial purposes if the character of the program service was to be different from the private service.

The FCC stipulates in its rules:

> Each station shall furnish a nonprofit and noncommercial broadcast service. No sponsored or commercial program shall be transmitted nor shall commercial arrangements of any character be made.

In issuing its *Sixth Report and Order,* the FCC denied the request of two universities to operate on a partially commercial basis on a reserved channel. The FCC said:

> A grant of the requests . . . would tend to vitiate the differences between the commercial operation and the noncommercial educational operation . . . in our view achievement of the objective for which special educational reservations have been established, i.e., the establishment of a genuinely educational type of service—would not be furthered by permitting educational institutions to operate in substantially the same manner as commercial applicants. . . .

The Commission also quoted from its previously issued *Third Notice.* It said that the need for noncommercial educational television stations was based upon the important contributions which these stations could make in educating the people at all levels. Recognizing the need for a noncommercial operation to insure a distinct type of program service, the FCC said:

> The need for such stations was justified upon the high quality type of programming which would be available on such stations—programming of an entirely different character from that available on most commercial stations.

How, then, as Senator Dill persistently queried, were educational stations to be financed? In 1948 the FCC authorized the operation of relatively inexpensive low-power (10-watts) noncommercial stations. This action was fairly successful as an impetus to educators to seek institutional funds to invest in these smaller stations.

But to develop a system of educational broadcasting, it was obvious to leaders in the movement that they would have to seek substantial funds from outside their parent organizations. With the formation of the Joint Council for Educational Television, and with contributions from the Fund for Adult Education and other foundations, educational television was given early assistance. Financing for educational broadcasting has come also from taxes, business and industry, and from audience contributions. One can suppose that the general economic upturn from the depression days of the thirties helped make this possible.

Finally, in May, 1962, the federal government authorized its first direct contribution for educational stations "to assist (through matching grants) in the construction of educational television broadcasting facilities."

Certainly, this kind of financial assistance would not have been forthcoming had not the educational broadcasters become better organized to promote educational station development. A year after the Communications Act—minus the Wagner-Hatfield Amendment—was adopted, broadcasters were reorganized into the present NAEB. Harold Hill writes:

> If one were to pick a point at which the fortunes of the educational broadcasters seemed to improve—not any sudden change but the beginnings of better things to come—he would probably have to choose the year 1936. . . . The 1935 convention (NAEB) had seen encouragement in the matter of exchange programs, ideas and information, and the position of educational broadcasting in general was more stable.

When television came on the scene, the educators were ready. During the hearings on television allocations in 1951, "educational broadcasters were well organized and in a position to present their case effectively." Walter Emery writes in *Broadcasting and Government*:

> More than 70 witnesses appeared before the Commission and urged that TV channels be reserved for the exclusive use of education. More than 800 colleges, universities, state boards of education, school systems, and public service agencies submitted written statements urging the Commission to make the reservations.

Finally, the FCC specifically defined those agencies eligible to hold a license on a reserved channel. Avoiding the obvious confusion in the Wagner-Hatfield Amendment by allowing religious, agricultural, labor, cooperative, and "similar nonprofit making associations" to apply for licenses, the FCC limited eligibility to "nonprofit educational" organizations, provided the channels are used for the furtherance of the organization's work.

The Wagner-Hatfield Amendment was significant as one of the early serious efforts to allocate channels to education. It was an important link in the chain of events leading to television channel allocations. Certainly the debate over the measure was beneficial to all concerned with educational broadcasting by drawing attention to several major issues—allocations, financial support, the need for educational broadcasters to better organize, licensee eligibility requirements—and the need for their resolution.

21. The Challenge of Quality in Education

by John L. Burns

At the time of writing this address, Mr. Burns was President of the Radio Corporation of America. The remarks included here were delivered before the National Association of Educational Broadcasters in Detroit, October 28, 1959. Reprinting is by permission of Mr. Burns.

INTRODUCTION

Those of us who work in commercial television and radio have a special appreciation of the serious problems confronting the educational stations, and a tremendous esteem for the superb job you have done.

Through your broadcasts to the classroom and the community, you have built yourselves into a major educational force.

You have helped turn the four walls of the schoolroom into picture windows that look out upon a limitless new world of knowledge. Taking full advantage of the matchless ability of radio and TV to reach into the home, you have created programs that contribute immeasurably to the enrichment of adult education.

I heard recently about a professor who complimented his adult nighttime television students on getting far better grades than his daytime classroom students. Next day, the professor got a note from one of the adult television students saying: "Don't be too critical of the daytime students. After all, they take their report cards home to understanding parents. We have to face our wife and kids!"

You in educational broadcasting are pioneering in an arca of vital importance to our national welfare—an area that today is alive with excitement and experimentation.

Nevertheless, because of the critical problems faced by education in general, there is an overriding need for still greater pioneering effort. In this, the educational broadcasters have a beckoning opportunity for leadership. For it is my steadfast conviction that *television offers a practical and immediate means for the greatest forward stride in education since the invention of the printing press and the textbook.*

THE PROBLEM IN EDUCATION

The dimensions of the current problem in education are clearly evident even to the layman.

They are reflected almost daily in grim headlines about the explosive increase in school population, the serious shortages of teachers and facilities, and the steady rise in the cost of education. They are revealed, too, in sobering statistics which show that one out of three high school graduates, who finishes in the top quarter of his class, does not go on to college because of a lack of funds. And in other statistics which show that fully 50 per cent of all college graduates over the next ten years could well be used in the teaching ranks.

Against this background, the inadequacy of our teaching methods is becoming painfully apparent. In schools, generally, the chief reliance is still on the teacher, the textbook, and the blackboard, as it has been for generations, to the exclusion of far-reaching innovations that have been developed in recent years.

And this brings me to my main thesis. As a former science teacher whose interest in the educational process has deepened with the years, I believe there is still another educational problem of even greater dimensions than those so far mentioned, a problem inextricably linked with all the others.

The real challenge, I am convinced, lies in the need for a massive upgrading in the quality of education.

By this, I mean:

1. A higher level of instruction and methods.
2. A higher level of instructors.
3. A greater attention to the individual student.
4. The fullest development of the human potential.

The challenge is all the more compelling in view of the quickening tempo of technology, the growing complexity of our social organization, and the mounting menace of Communism.

To meet the challenge will require new ideas, new approaches, and a new willingness to experiment and innovate. Many of the methods and tools for bringing about an immediate upgrading of quality are at hand. But we have made only the barest start on learning how to use them. In this area, I believe that the experience of American business and industry suggests a constructive approach.

IMPORTANCE OF NEW METHODS

Let me say, at the outset, I am fully aware that anyone who attempts to apply to education certain concepts from business and industry is automatically suspect.

He is as wide open to criticism as the alleged teacher here in the Detroit area who placed an ad in one of the local newspapers. The ad read: "If you are not satisfied with your child's progress in school, why not have *he* or *she* tutored by an experienced teacher."

Critics are quick to claim that the fellow who tries to translate education into dollars-and-cents terms really wants to cheapen it. They take the position that while automation in office and factory may be beneficial, automation in the classroom simply will not work.

They seem to forget that the biggest forward surge in the history of education was brought about by automation—in the form of the printing press. Then, for the first time, outstandingly gifted teachers were able to set down their ideas in books that spread their wisdom and influence far beyond the narrow reach of their personal contacts. It is well to remember that the book is one of the most important products of automation in the history of man.

The fact is that education has now become our biggest business. It has an annual budget of $20 billion, and more employees and a larger "plant" than either the steel or auto industry.

In industry, over the past half century, output per man-hour has shown a steady increase. This increase has been due basically to our skill in working out constantly better methods, and providing more and better tools to back up each worker.

Experience suggests that we can multiply the effectiveness of the good teacher with improved methods and appropriate tools, just as we have multiplied the effectiveness of the factory worker, the office worker—everybody right up to and including top management. In fact, just as the printing press did.

There are now available a wide variety of new tools and techniques for improving the quality of our educational system. Foremost among them, of course, is television.

Closed-circuit installations are extending the influence of talented teachers from one to several classrooms.

The noncommercial educational stations are carrying their lessons and lectures into the home, as well as the classroom, to a potential listening audience 50 per cent larger than the nation's total school enrollment.

The commercial stations and networks are offering an ever increasing fare of educational and cultural programs to entire regions, and, in the case of NBC's *Continental Classroom,* to the whole nation. Incidentally, *Continental Classroom*'s course in Modern Chemistry points up sharply the great value of color television in education. Those early risers among you, who have seen Dr. John Baxter's laboratory demonstrations, can appreciate color TV's amazing capacity for enlivening educational presentations.

I envisage the day when all the nation's schools will be linked in one comprehensive educational television network. Such a network, far from imposing uniformity on local curricula, could help to provide richness and variety.

As interest in educational television grows, TV-set manufacturers are working hard to improve the style and utility of their products for the classroom. For example, RCA is displaying here at your convention for the first time, a new television receiver designed specifically for classroom use. This model, which will be available after the first of the year, features greater picture brightness, higher audio levels, precision tuning, ability to be locked, greater mobility, and a special stand that can raise the set to a height of six feet.

This new model was designed on the basis of suggestions offered by educators and educational broadcasters themselves. And I want to take this opportunity to thank the education fraternity for their help and cooperation in this project.

Great as its potential is, however, television is not the only new tool available to improve educational quality.

The school of tomorrow will have electronic teaching machines that will free the instructor from routine tasks and give him more time for personal counseling. The teacher's desk will be equipped with a tiny electronic scanning device linked with the library and the records office so that references can be checked quickly. A small-size electronic computer will correct many types of examinations, process student records, survey performance, and determine areas of difficulty.

Magnetic tape, which is even now finding use in the schools as well as in your own educational stations, will extend its usefulness in the years ahead. In fact, schools might well start on a program of automation today with a relatively inexpensive magnetic tape sound system, and build up from there.

Now under development at the RCA Laboratories is a magnetic tape player capable of reproducing pictures as well as sound. It works through a standard television receiver.

This video-tape player will eventually be a natural complement to a school television system. It will permit the flexible use of a library of pre-recorded programs.

By comparison with present professional video-tape recorders used in television stations, the new apparatus will be low in cost. It may be set up to supply a pre-recorded program to one classroom, a group of classrooms, an entire school, or a whole school system.

NEW TOOLS, HIGHER QUALITY

When these exciting new educational tools are in general use throughout our schools and colleges, their effect on quality in education can be tremendous. Just consider the advantages of closed-circuit television alone.

1. *It can help bring about a higher level of instruction.* It can extend the great influence of the best teachers far beyond the confines of their own classrooms, and give them a dramatic medium for projecting their ideas. As one student in Hagerstown, Maryland, put it: "In class, the teacher talks to *us*. On television, she talks to *me*."

Closed-circuit television enables schools to call upon men of specialized talents for occasional lectures. By drawing on a central video-tape library, the closed-circuit system could present lectures by the greatest minds of our times. By linking up with commercial and educational stations, it could take students to the missile range at Cape Canaveral or inside the nuclear submarine *Nautilus,* to the halls of Congress or the Chamber of the Supreme Court.

2. *Closed-circuit television can help raise the level of instructors.* The big need is to relieve the teacher of all repetitive tasks through automation and give him more time for working with individual students.

Closed-circuit TV can greatly ease the burden on teachers by giving each one a chance to do the thing he is best suited for. One teacher may be best suited for lecturing to a group of several thousand students simultaneously. Another may be able to do an outstanding job of conducting a follow-up classroom session with a handful of students. Through the more effective use of teaching talent, television can make higher salaries a reality throughout the teaching profession.

With many schools and colleges participating in a program and sharing the cost, the salary of a particularly gifted television teacher might well be in the six-figure realm of the highest paid businessmen or other professionals. The entire salary scale could be raised, at lower cost per student. With higher salaries would come increased stature for the teacher in his own community. These factors would, in turn, keep able men and women in the teaching ranks, and attract new teachers of the highest caliber.

Since our nationwide requirements for teachers can never be met by present methods, television certainly poses no problem of unemployment for teachers.

3. *Closed-circuit TV can make it possible for the teacher to give greater attention to the individual student.* Once the classroom instructor has been freed from many of the chores he now performs, he will be able to devote far more time to personal counseling. Indeed, this opportunity for individualized instruction is one of the great advantages of television, magnetic tape, and other electronic aids to education.

They permit the teacher to reach the student on a highly personal basis. They enable each student, in effect, to set his own pace. The fast-learner in a particular subject is challenged to work up to his full capacity. The average-learner is encouraged to develop what gift he possesses. The slow-learner is encouraged to develop what gift he possesses. The slow-learner is assured of the kind of attention that will prevent his falling hopelessly behind.

A student might be in the third grade in spelling, the second grade in arithmetic, and the fourth grade in history. We do not have equal abilities.

Every pupil has potential talent of some kind, and his value in our democratic society lies in the extent to which he is helped to use his special talent for the common good.

Former President James A. Garfield once said: "A pine bench, with

Mark Hopkins at one end of it and me at the other, is a good enough college for me!"

This one-to-one teacher-student ratio has become impossible under modern classroom conditions. But with television and other electronic devices, teachers can have both the time and the means to reach the pupil once more on a person-to-person basis.

RESEARCH AND DEVELOPMENT

There is no doubt whatever about the capacity of educational television and other electronic devices to help us realize our democratic ideal of the fullest development of the human potential.

However, despite the fact that there are some 300 experiments, which we know about, now going on in closed-circuit TV, few minds have begun to apply themselves as yet to the all-important questions of the best uses of these new tools and the implications they will have for teaching methods, facilities, and curricula. Few have weighed the question of how to bring about a satisfactory combining of the teaching, broadcasting, and graphic arts.

To answer these questions there is urgent need for a full-scale program of research and development in education, comparable to the R & D programs now under way in defense and industry.

Today, industrial research is a $12 billion operation—the "industry of discovery," as one economist calls it. In my own field of electronics, there is an axiom to the effect that "either you get into research, or you get out of business." Many industrial corporations set aside 5 to 10 per cent of their annual budget for research.

By contrast, the total amount spent annually on educational research and development comes to about one-tenth of one per cent of the over-all education budget. Some of our industrial corporations are spending more money on R & D all by themselves than is being spent in all branches of education combined.

It is true, of course, that a few organizations are carrying on notable experimental work in the educational field.

The Ford Foundation and its allied groups are doing a remarkable job in exploring the frontiers of teaching. Typical of their imaginative approach is the recently announced airborne educational television project in which programs will be relayed by plane to a six-state area of the Midwest.

Another example of worthwhile educational experimentation is to be found in my own home community of Greenwich, Connecticut. Recently, Greenwich joined forces with the neighboring areas of Westport, New Canaan, and Darien to initiate a joint research and development project. The four participants are seeking a grant from one of the foundations to get the program off the ground. Once it is under way, they plan to allot a certain percentage of their annual budgets to finance it.

Still another example of experimentation is the Center for Instructional Television which, I am happy to report, RCA has helped to establish at New York University this fall. The main purpose of this Center is to conduct research and development on the most effective ways of using TV as a teaching tool, especially in uniting the teaching, broadcasting, and graphic arts.

All of these projects are serving a useful function. But they are "islands of experimentation" at a time when we need not islands but entire continents. We need, above all, a broad blueprint.

We need scientific answers to questions which are fundamental to the whole purpose and process of learning—questions relating to the best use of facilities, and the most satisfactory curricula.

We need a broad plan of approach to bring into being all the good things made possible by modern technology—especially closed-circuit TV. The present experiments, I think, are lacking in two things.

One is scope. We must learn to think in terms of state-wide and region-wide programming instead of strictly local operations.

Another lack is a real union of the teaching, broadcasting, and graphic arts. We must provide means whereby students may move through the educational process in accordance with their ability, and we must provide greater individual counseling so sorely needed in this technological age.

To achieve these aims, it is my proposal that we undertake a full-scale research and development effort in education, on both the local and the national levels.

The program would concern itself with both short-range and long-range problems and objectives.

The short-range aspect would deal with immediate steps that could be taken to broaden our educational advance.

The long-range aspect would deal with our educational needs a decade or two hence, when the children born in the "baby boom" of the 1940's will begin to have children of their own and school enrollments will soar still higher.

To finance such a program of research and development, I propose that each state and locality set aside no less than 2 per cent of its education budget each year.

On the basis of an aggregate education budget of $20 billion, this would amount to some $400 million. This may seem tiny in comparison with the sums spent by industry and defense. But it is enormous in contrast to the $20 million—about one-tenth of one per cent of our over-all school budget —now being spent on educational research.

Can we afford such a research and development program in education? I would suggest that a far more reasonable question would be: Can we afford to go on any longer without it?

Walter Lippmann put the matter clearly and concisely when he said:

> We must measure our educational efforts as we do our military efforts. That is to say, we must not measure by what would be easy and convenient to do, but by what it is necessary to do in order that the nation may survive and flourish. We have learned that we are "rich enough to defend ourselves whatever the cost." We must now learn that we are quite rich enough to educate ourselves as we need to be educated.

Over the long run, I am confident that research in education would pay for itself many times over—just as research in industry has done.

Let me give you an example of how this might work out in practice. Over the past decade, the annual per-pupil cost of running our public schools has increased by more than 70 per cent. Under conventional methods of instruction, the prospects are that this rise will continue because the cost of instruction goes up almost in direct proportion to the number of students.

However, even the preliminary research that has been done on the economics of classroom television, suggests strongly that once a break-even point is reached—at about 200 to 250 students—the per pupil costs decline sharply.

A study conducted by the Southern Regional Education Board—covering some 300 colleges and universities in sixteen states—showed that the cost of televised instruction would run about $2.80 per student semester hour, compared with the present cost of $12 to $18 for conventional instruction.

It is my feeling that additional research could lead to even greater economies in this area. But, above all, the quality of education could be greatly enhanced.

In a research and development program such as I have proposed, you members of the educational broadcasting fraternity could play a vital role. For you have in your hands the most potent means of communication between educators and the public. You can bring to your own communities a vivid awareness of the problem of quality and the means of solving it.

SUMMARY AND CONCLUSION

The quality of education can be greatly upgraded by modern technology, while costs can be lowered and the increase in students can be taken care of at the same time.

The advent of new electronic tools, especially closed-circuit TV and its future derivatives, makes possible an advance in methods of education comparable to that made possible by the invention of the printing press in the fifteenth century.

Indeed, it is an advance which can be even greater than the development of the atomic reactor. The atomic reaction is an explosion of a material kind. The education explosion, which can be made possible by our new electronic tools, is an explosion in thinking—affecting men's minds.

The prime needs to accomplish this advance are: (1) broader scope in the concept and the application of educational electronics; and (2) a combining of the three arts—teaching, broadcasting, and graphics.

Present experiments must be drawn upon to lay out a broad business and research program to get adoption of plans for areas large enough to permit fundamental advances.

The need for greatly enlarged expenditures for research in education itself is stressed by the increasingly large number of basic questions that are urgently demanding answers.

In view of all these factors and the present critical situation in education, I therefore propose that:

1. Plans for much broader use of educational television be drawn up for, say, a state or group of states.
2. A tape library of courses by distinguished teachers be created to support such a system. This scope is needed in order to demonstrate the great benefits of educational TV in visual illustration of a magnitude too great for a single classroom to undertake.
3. The three arts—teaching, broadcasting, and graphics—be united in the creation of the tapes.
4. A library of follow-up books to supplement the tapes be developed.
5. An R & D fund be set up by all educational units so that advances may go on apace.
6. Other tools of educational automation be brought into existence as quickly as possible.

The task calls for bold experimentation and imaginative new approaches across the whole broad range of education, supported by a very considerable original outlay of money to put an educational television system into being. The funding and implementation of these programs must come from a close collaboration between the localities and the states.

As the authoritative Rockefeller Report on Education pointed out recently: "Such innovations as the teacher aide and television should not be thought of as stopgap measures to surmount the immediate teacher shortage, but as the beginnings of a long overdue revolution in teaching techniques."

Major advances in education have a way of occurring in cycles of thirty years or so. For example, the 1830's saw the acceptance of community responsibility for public education and the beginning of systematic teacher-training programs.

The 1860's witnessed the extension of public responsibility to the high schools and the development of land-grant colleges.

In the 1890's there came the great widening of the college curricula and the establishment of the kindergarten.

The 1920's and 1930's saw broad acceptance of John Dewey's philosophy of learning through experimentation and practice.

Today, the astonishing advances of technology give promise of effecting reforms fully as significant and far-reaching as any of these, and of raising the quality of education to new heights. An advance of the magnitude of the printing press and educational television comes only once in several centuries.

You, in educational broadcasting, have a priceless new medium for the communication of knowledge and for upgrading educational levels in many areas. This should make the challenge of quality all the more meaningful to you.

"There is a tide in the affairs of men," wrote Shakespeare, "which, taken at the flood, leads on to fortune."

Such a tide is running now in education. It is up to us to take advantage of it—for the welfare of our children and for the future of our country.

22. National Jukebox or Educational Resource?

by Harry J. Skornia

Professor Skornia is a well-known critic of the broadcasting industry. He has written extensively and articulately on the subject. He is a full-time Professor of Radio and Television in the College of Journalism and Communications at the University of Illinois, and served in 1967 as Senior Specialist at the Center for Cultural and Technical Interchange between East and West at the University of Hawaii. This article appeared in Audiovisual Instruction *for April, 1964. It is reprinted here by permission of the author.*

On December 17, 1959, I testified before the Federal Communications Commission in hearings on the problems of radio and television network broadcasting. When I implied at that time that the FCC had been something less than generous to education, I was challenged. In the somewhat heated discussion that followed, I indicated that I did not believe that education was only one-tenth as important to America as advertising. I thought it was more important. I would not be satisfied, I said, until education had at least as many and as fine facilities for broadcasting as commerce had. I even had the audacity to suggest that half of the Commission—or at least a certain number of the Commissioners—should be educators. This stand has gotten me into difficulty before and since; yet I reiterate my position. I feel that a nation that claims to be civilized and pretends to respect education will not find this position unreasonable.

I find completely contrary to logic and justice the necessity for a public educational group to buy, for over $6 million, a public frequency for a New York City educational station. The original allocation of a single frequency to New York City—and that in UHF—makes no more sense to me than it would to restrict New York schools to six acres of land for buildings.

With regard to education's rights in the electromagnetic spectrum, I am impressed with the past record of neither the FCC, the Congress, the trade press, nor industry. Many of the failures can, of course, be attributed to past policies and personnel. Present and future signs are more hopeful. Let us therefore examine with optimism a few philosophical and practical considerations which will affect the positions we can logically take on the frequency proposals of the FCC and the National Association of Educational Broadcasters. Which proposal, if either, comes nearest to meeting our needs will depend to considerable extent on how each contributes to the solution of at least two problems. One I shall call the problem of *interaction;* the other I shall refer to as the need for *integration* of all the nation's radio and television services.

To clarify the problem of interaction, let me use as an analogy, alternating and direct current. The earliest type of electricity to find use in the United States was direct current—current which flowed in only one direction. The story of the patent and industrial battles between the DC (Edison) interests and the AC (Westinghouse) interests need not delay us here. Suffice it to say that DC soon proved to be ill-adapted to economical national distribution along long lines and cables in the United States. Even though huge investments had been made in DC, electricity was only half used until national conversion to AC was achieved. A new era in electrical power distribution was opened up with that decision.

I fear that in ETV and ITV we have been concentrating on one-way flow—to the student. We have been concentrating on *teaching* rather than back-and-forth flow, dialogue, or *learning.* It is this interaction that we now need to pay more attention to. As you know, there are rating services that use push buttons to permit the viewer to register instantaneously his approval, disapproval, or other reactions to programs. This may provide a clue to the kind of interplay that might be possible in educational television.

The use of electromagnetic energy could, of course, have been two-way from the first. It would have cost more; it would have reduced the frequencies available for the one-way flow from what we now call broadcaster to recipient; it would have been a different system; but it would have been possible. What we now call viewer or listener would then have been "participant." When faced with the alternatives of mass use and small-group or individual use of radio and television, mass use was chosen. These media were then defined as mass media. And generally we don't argue with this decision or even think that the other uses were equally possible. Yet we should, for education can never be a mass operation. Whatever the tools are, in education they need to be used individually.

The 1957 Canadian National Commission Study of Broadcasting (the Fowler Commission) challenges the mass concept. Listeners and viewers must be treated as individuals, not as a mass, we are told. The report goes on: "Packaging individuals for easy handling is a totalitarian device. It is this variety in the individual that gives our society its character and civilized life its richness."

The various British studies (Beveridge, Pilkington, etc.) agree. So do French, Italian, and German studies. Even the Soviet system, in an effort to keep flow going in both directions, emphasizes such programs as "Letters from Listeners" and "Letter and Answer."

Whereas in the United States we usually find broadcast officials and staff *speaking* to recipients who question their product, those of many other democratic systems stress *listening to* these voices of protest and inquiry. European stations have huge listener mail departments, for European viewers and listeners do feed back very much.

Communication in any true sense facilitates interaction, not only between broadcaster and recipient but among recipients. Communication facilitates interaction among individuals, between individuals and their leaders (not merely *vice versa*, which is totalitarian), and among the different institutions of society itself.

For centuries education was dialogue. Abelard, Rousseau, and Mark Hopkins represent this tradition. Then came the book, ending the regime of dialogue and scholastic philosophy. But now, after 400 years of book domination, which has caused some of us to confuse education and information, the electronic media make dialogue again possible.

Communications scholars like Colin Cherry and Marshall McLuhan recognize the need for dialogue. McLuhan states flatly: "AC electric circuits are basic dialogue forms."

"With the coming of electronic tools," he tells us, "the Western World has encountered the form of the dialogue in learning and teaching once more. . . ."

Teaching machines, with or without radio or TV, also illustrate dialogue. Although teaching machines have some mechanical features, as Dr. McLuhan notes, their basic feature is dialogue. Radar is also dialogue.

Jerome S. Bruner of Harvard has warned that the textbook, if read only, can produce "chronic somnambulism." The same is true of listening to and viewing TV. "The educator's job," Dr. Bruner tells us, "is to change the student's role from passive to active. Programed instruction, for example, ends the passive plight of the reader. It requires a dialogue between the printed word and the beholder. . . ."

Is such interaction impossible in the applications of TV and radio? Everyone seems to know that two-way flow is essential, but no one in TV seems to be doing much about it. We keep asking only for more and bigger pumps to increase the one-way flow we already have.

One way to promote feedback, or at least to reduce the totalitarianism of heedless, monolithic, one-way flow, is to insist on more "smallness" as opposed to "bigness." It is well known that new ideas, or educational content, are better accepted and understood if, as early as possible, the recipient is brought into a discussion of them, given an opportunity to question freely, and permitted to make suggestions. The best education seems to involve a partnership between learner and teacher in easy communication, with neither individual deprived of the means to be heard.

Feedback and interaction are not only democratic. They are essential to democracy. They enable leaders and teachers to tap the strengths and ideas of the whole populace. Feedback makes such leaders and teachers stronger than those in a dictatorship in which the channels run in one direction only.

I have a haunting fear that as we concentrate on carving up the spectrum for flow in one direction only, we are condemning the broadcast media to anachronistic, totalitarian uses which posterity will someday denounce. I think new approaches and techniques need to be devised. The public can't devise them; it is not properly organized for such action. Industry won't devise them; present one-way uses are too profitable.

Can we suggest solutions? I don't think we can until we recognize the problem and seek to identify its many aspects so that communications specialists and technicians who might contribute to its solution may join us in at least thinking about it. I do not believe that the new instruments we are discussing will be used fully until the problems of interaction, feedback, and flow from student or viewer to teacher or broadcaster are finally attacked and solved. Some devices exist in raw, undeveloped form to suggest the direction that certain studies of this problem may take. More attention to it is needed, however, if ITV and ETV both are not to indoctrinate instead of educate.

But the problem is indivisible. It goes beyond education and the educational uses of television and radio. It will not be solved by allowing occasional questions to be telephoned to the teacher. Dialogue and democratic dynamics involve interaction to a far greater degree; they call for an AC-TV system in this sense.

I referred earlier to the need for an *integrated* broadcast system for America. By this I mean a system that includes educational broadcasts of all levels and variants, commercial broadcasts and other possible variants such as toll-television, and a publicly owned and operated political and public-affairs network comparable in the print media to the Government Printing Office. But these variants must all be controlled as a single unified system whose parts support rather than fight each other. The FCC grew out of commerce. We might ask whether an agency so conceived can coordinate, or be expanded or changed to coordinate, an integrated system for granting equality if not priority to education, or whether some drastic new concept involving a new regulatory home may not be necessary.

I do not believe that a sum of disunited parts, growing Topsy-like with no policy coordination, can add up—in either education or commerce—to the balanced communications structure this country needs. To replace this Topsy-like growth, some planning involving far more than education is necessary.

We are at a point where the very term *broadcasting* begins to limit our thinking. We need to think in terms of *narrow*-casting as well as *broad*casting. Are not narrow uses of the electronic media, which education sometimes requires, as valid an employment of the electronic spectrum as shotgun uses? Is there any law that says that all casting must be broad?

Different kinds of education, of course, require different kinds of uses and techniques—from the widest to the narrowest. I foresee, then, an integrated system, perhaps no longer called "broadcasting." Such a system will be equally efficient in both directions—from people to leaders and students to teachers, as well as in the DC direction they are now restricted to.

I believe the essentially segregated system we now have, in which the educational system promotes certain values (thrift, peaceful solutions, rationality) while the commercial networks encourage their opposites (spending, violence as an acceptable solution to problems, deceptive advertising), is bad. I think that to separate in-school television from educational broadcasting is bad. I think that to say that certain levels of school shall have closed-circuit while others have open-circuit, is ill-considered. Are elementary schools any less entitled to frequency space than 50-kilowatt radio stations that offer little more than the phonograph records anybody can buy? I believe that any agonizing appraisal or reappraisal that includes education should include commercial broadcasting as well. Radio and television are unique. Their uses for all purposes are of sufficient social significance and power to require the utmost of care and planning, whether the label be education or commercial.

In the tug-of-war for educational frequencies—in which internecine warfare between and among the levels and kinds of education is forced by alleged shortages—facsimile, radio, lasers, and a score of other resources and developments are forgotten.

For the kind of coordination required to integrate these fragmented services, national leadership is necessary. In education generally, federal efforts have been necessary to break down the segregation of the races. It is unlikely that many of the "separate but equal" solutions—which are of course not equal at all—will solve the educational problems for which they are offered. Surely there are common goals towards which all the responsible minds in both education and broadcasting can march—goals consistent with the aims of public education and the public interest, not just the interests of the nation's advertisers and those who serve them.

There will of course be mutterings about socialism if commercial broadcasting is restricted to its proper role and resources. Many people still think, as Calvin Coolidge is quoted as saying, that the principal business of the United States is business. But if we are to find the proper role of educational broadcasting—and the proper balance and function of its parts—we must break new ground and challenge old practices. This will require coolness of nerve, sureness of hand, and the utmost of cooperation among all segments of U.S. education: public, private, lower, and higher. The ideas needed are bigger than any one person can develop. But together we can, I hope, generate them, and put each of the possible services in its proper perspective.

The fabric of both communications and education in America will be more varied, and stronger, if we have public and private ownership at least moving in the same direction—towards goals which all Americans recognize as in the nation's interest. Of the four functions generally recognized for broadcasting—to inform, to enlighten and elevate, to entertain, and to sell goods—education is disbarred from the latter. All of the others must be shared. Much of adult education, for example, is political, economic, and public-affairs in nature. These are not concerns of the schools alone, but of the entire nation. The retraining required following automation illustrates this point. It is actually more a problem of industry than of education. School-based facilities and services are no more obligated to fulfill functions of this nature than are the commercial ones. That is what integration means. To integrate American education, only, would hardly do the job.

If we are to have an integrated broadcasting system in which education finds its proper role, and which is communicative rather than merely propagandistic, in one direction, some changes in policy, law, and in the FCC itself must come. Most of the faults of the FCC are not, of course, of its own invention. Congress keeps the FCC as a watchdog over wire and wireless communications, but persists in doing its own barking—and sometimes its own biting. Rarely, however, does Congress bark unless prodded by large and influential firms. Whether pressure group politics of this nature is democracy or its opposite is another important question we need to ask.

Certainly, something better than a hasty decision is necessary. The frequencies of which we speak now are scarce natural resources. They should not be hoarded; but neither should they be committed with undue haste. These are assets of such scarcity that any rush to occupy them for selfish uses must be resisted firmly and finally. Americans are traditionally impatient and incapable of laying out long-range plans. Here, however, so that the growth of tomorrow is not given short shrift, it is essential for the FCC and other agencies to make plans covering a number of years. Areas in the United States that are not yet even settled must be provided for. Someone must speak for the yet unborn. Services which are as yet new, or not yet established in the curricula, need to be kept in mind for the day when we discover their requirements for spectrum space and facilities. The great planners have always reserved resources for the unborn—whether individuals or institutions. We must therefore ask ourselves: Are the plans we make today adequate for tomorrow? For a hundred years from now? For an age of lasers, space satellites, and a federated world? These considerations are more important than those of adequacy for present needs.

The diversity of services needed is all the more imposing when set against the present twin goals of the commercial services: advertising and money-making. At the elementary level, pre-school, in- and out-of-school, credit, and noncredit teaching must be provided for. At the secondary level, in- and out-of-school course work, in- and out-of-school enrichment, and extra-curricular needs must be met. The fact that schedule and bell problems have delayed high school usage of television should not prevent adequate reservation for the future. The needs in ten years—as university rejections press back upon high school standards and floods of students surge at the doors—will be desperate. High schools then will need these new media most urgently.

At the higher educational level, communication media and classroom space have become interchangeable parts of the same problem, whether for direct instruction, formal, or informal uses.

At the adult education level, the need is perhaps greatest of all—and all levels of schools, not just higher education, share responsibilities here. Many of the adults to be educated are of elementary school, or high school level, and can in no sense be considered graduates. Commercial broadcasting, also, must help. There is no reason why the commercial media should be permitted to do the easy, the profitable, and the popular, while educational media are left with all of the opposites. These are not merely tasks: they are inescapable responsibilities. They must be shared in an integrated approach to communication in which private and public interests reinforce each other.

Education's needs, when contrasted with those cited for commerce, are dramatic. Responsible educators in many areas have said they need six, ten, twelve, or fifteen more frequencies. Is commerce to have all it wishes, in the face of this shortage? Alabama needs 36 educational channels to serve a state desperate to lift itself educationally and industrially. Illinois has three VHF and five UHF frequencies for education. Chicago alone needs six or more on the basis of present needs, to say nothing of the future.

Ohio, with 61 four-year colleges, has nine frequencies—all UHF. Pennsylvania, with 121 colleges and universities, has five reservations. When Penn State, as late as 1961, applied for a VHF frequency, it was turned down. Rochester suffered the same fate. Los Angeles alone has over 20 four-year colleges and about 15 junior colleges, some very specialized and as large as average universities. Should they be expected to share one or even half a dozen frequencies? Certainly not until or unless commercial operators are also required to share allocations in the same area. Surely education is more than one-fifth or one-seventh as important as advertising, to the nation and to the world.

There are 40,000 school districts in this country. How do the needs of these units rate as compared with the interests of the owners and operators of the top 50 big commercial stations and networks?

Westerns, detective programs, family comedy, soap opera, sports, news—how many different kinds of programs does commercial television provide? Don't most stations use old films as "filler" at least half of the time? The average 12-grade school, by contrast, has to teach at least 100 different subjects and grade levels. Should it be expected to accomplish this task with mass uses of media and one or two frequencies? If a store or other business wants a piece of land, and educational interests need the same piece of land, who gets it? What is the answer when the frequency spectrum instead of land is involved?

When Sputnik was launched, educators were not alone in saying that education was the most important thing we could be promoting. Politicians, businessmen, and professional men joined in. So did broadcasters. But so far, the actions of the latter have often belied their words. Their opposition to educational frequencies would sometimes lead one to believe that education represents an enemy power.

There is a multiplicity of tools available now to meet the needs of education. There are "regular" VHF and UHF services, such as those tested and proved in the Salt Lake City area. There is what Adler describes as closed-circuit broadcast—the instructional television fixed frequency service. (And I believe tribute should be paid to the firms who are developing these many new and alternative services—low-power and low-cost.) There is airborne: MPATI, which as a condition of its license, promised to develop and test a three-megacycle, split-frequency approach. This development, which if it works, will double the availabilities, is awaited with interest. It is a fine type of experiment for a foundation to sponsor. There is wired closed-circuit. Here again I raise the question of why it should be restricted to education. Many radio and TV stations that serve only to deliver records and films—that, in essence, operate as common carriers—could well be limited to such facilities to the same extent that education is.

There is facsimile, used in Japan and elsewhere to advantage, but here, like radio, largely forgotten in our preoccupation with television. There is slow-scan facsimile, experimented with in England, which uses little spectrum space. There are lasers, although it is to be regretted that most experimentation in the laser-maser field—controlled by telephone interests—seems relatively inaccessible to education. There are community antenna approaches that show great promise, if education is not squeezed out. There are toll-TV approaches and there are satellite and other extraterrestrial possibilities. But most of these are long-range resources. The needs of education won't wait for them. Our children and our schools need help now.

Although there is a pressing need for deliberate and careful study of low-cost, low-power, non-mass approaches, toll and satellite approaches deserve brief additional comment. I am not, as an educator, interested in commercial toll-TV, although I think that two kinds of commercial services—one direct box-office service and one indirect box-office service via advertising—would be a good thing. It would make more direct feedback possible. I am more interested in the ways in which variants of this approach (special keys for doctors, etc.) could make the use of scrambled materials feasible on a more economical scale than closed-circuit is now. Emergency health instruction to public health officers and posts, directions to civil defense authorities, and many educational uses including even pay-as-you-go correspondence courses—these illustrate my interest in toll or subscription TV.

With reference to satellites, I would remind you of the degree to which we have become one world. To see satellites apparently destined for business instead of education is to me a tragedy. What support did Professor Oliver Reiser of the University of Pittsburgh get when he proposed a UN-UNESCO satellite devoted to international, educational, scientific, cultural, and goodwill purposes? His proposal for a "Project Prometheus," made at the Accra Conference in June 1962 was scarcely mentioned in the U.S. press—and certainly not on the network newscasts. As far as I know, reference to it in the press was limited to one letter to the editor of *Saturday Review*. Neither the executive branch, Congress, nor the FCC supported this plea for specific educational uses of satellites. Yet to my knowledge, no more exciting challenge to this new tool has been advanced. Perhaps we as educators should press more vigorously for such projects, if we are not to be squeezed out of this area, too, by commerce.

In accordance with the Pilkington Report and the government's new policy, Britain, at great expense, is converting to 625-line picture standards for television—the standards used generally in Europe and in the Soviet Union. Should the United States be considering the technical and other compatibilities of its services with those of the rest of the world? What stake has education in such matters?

Returning to land-based facilities, which it appears will be all that education can count on to meet its more immediate problems, several ownership patterns emerge. These include community stations; state or city-system stations; individual university, college, or school stations; stations operated essentially by instructional centers (illustrated by the Louisville Free Public Library FM services to the schools of Louisville); municipal stations like WNYC and WNYC-TV which might form the nuclei for an eventual public-service, public-affairs, noncommercial network separate from educational stations; joint educational-commercial operations like the VHF-TV station of Michigan State University; partially commercial stations like that of Iowa State University; and various other ownership patterns.

Conspicuously absent from the spectrum are joint commercial stations, arrangements whereby several commercial interests pool and share the same facilities, as schools and colleges must do. Before further commercial allocations are decided, the validity of joint arrangements for commerce as well as for education might well be examined. A national study to decide who gets what—a study cutting across commercial and educational lines—is sorely needed. In view of the shortage of frequencies should any station, educational or commercial, be licensed to broadcast mostly rock-and-roll films? Is a national jukebox or second-run movie-house a valid communications use of the spectrum?

When the courts ruled that broadcasting was not a common carrier, Mr. Sarnoff had not yet said that "of course, broadcasting is only a pipeline"; stations had not yet become essentially retail outlets for records, films, and other canned goods produced by outsiders, or by record and film companies owned by the same corporations—RCA, CBS, and others—who already control stations and programs, either by ownership, affiliation, or syndication. Now that stations have begun to operate as common carriers, are revisions in the allocations and rules not due for consideration? Or are there some choices here, depending upon the kind of operation we define as broadcasting and as common carriers?

Under the pressure of quantity-oriented industry standards, we need to reconsider whether "bigger" is always "better"; whether TV is better than radio for certain purposes—or poorer; whether color will be an advantage or a distraction for certain types of TV programs; and many other such questions that pose a choice. It has been found in many instances that the concentration possible via one sense permits more penetration and learning than an appeal to several senses in which the senses become blurred and distractions begin to cancel out the supposed advantage of multi-sensory appeals. The positive superiority of radio, facsimile, and other neglected media for certain purposes in many subjects is notable. Since what we ask is not for ourselves, nor to deprive commerce or any other entity of its just due, we must be sure that in all of our requests, our motives are honest, based on real facts and real needs, not on a mere sense of competition.

Perhaps a word about the real or imagined dangers of federal influence and centralization is in order. Senator Goldwater was quoted as saying on January 22 that he was less opposed to federal aid to universities and colleges than to high schools and elementary schools because there is less danger of influencing older students than younger ones. I personally believe emphasis on decentralization for education can be carried too far, at all levels.

Programs involving the arts, or expensive talent or production, should certainly be centrally produced. Programs involving ideas, social and human relations, controversial public affairs, discussions, and regional and local problems should, I think, bear the local stamp. But the problem is not one of "either/or." In most of the studies of viewer attitudes towards ETV and ITV, the objections seem traceable to the local type of production, to inadequate quality, not to the educational nature of the programs as such. To condemn any level of TV, whether open- or closed-circuit, to local production is in many cases to condemn it to amateurism. We do need to tap the grass-roots and, as commercial broadcasting has generally failed to do, keep local education and participation alive. But the system we call for, whether for elementary, secondary, or higher education, should also provide for significant enrichment and reinforcement by the best centrally produced programs that the collective efforts of American education and broadcasting can come up with. Some of these, most certainly, should be live.

Whatever changes we press for will be recognized, I trust, as born out of altruistic, professional motives. None of us has anything to gain personally or financially from one decision or another, made now or later. At the Institute for Education by Radio and Television in Columbus last spring, Professor Ned Rosenheim of the University of Chicago said:

> The service of truth can be a lonely, infinitely demanding, even an ugly task. Yet the service of truth is the only task for us . . . teachers and broadcasters . . . entrusted with this magnificent power of communicating wisdom. It is our job. It is our way of life. It is our way simply because, if humanity is to survive, there is no other way.

As another educator, I endorse Dr. Rosenheim's statement.

Many efforts in education are too timidly engaged in. Educators are often visualized as milquetoasts, requesting what they ask for with less than required firmness. When I say I believe education deserves more and better frequencies and facilities than private broadcasting, I am not saying it merely for effect, or as a trial balloon. I mean it. I invite you, whatever you do or ask for—not for yourself but for unborn generations and our nation's survival—to ask for it with firmness and confidence, the confidence that comes from knowing that there are no selfish motives involved in your requests. Then let all of us reiterate and stand by our requests, firmly, vigorously, skillfully—and if necessary thunderously. Such a position will help, not hinder, the FCC and other agencies interested in education and America's position in the world generally.

The day when education could *dare* to speak with a small voice is past. Let us be sure greatness, not pettiness, characterizes our planning and our requests for the tools we need to do the task expected of us.

23. Smaller Than Life: Teachers and Schools in the Mass Media

by George Gerbner

Dr. Gerbner heads the Annenberg School of Communications at the University of Pennsylvania. He divides his time between research and teaching in the social aspects of mass communications. *This article is a revision of a talk delivered at the Illinois Association of Secondary School Principals Conference at Urbana, Illinois, in October, 1962. It appears here by permission of the author.*

Two cultural offsprings of the industrial revolution grew up side by side. These siblings are the system of formal universal public education—the schools—and the system of informal universal public education—the mass media.

The mass media system is the direct descendant of technology, mass production, and mass markets. It was ideally suited to the demands of industrial culture, to its need for rapid, standardized reproduction and distribution of commodities to heterogeneous, anonymous, mass audiences, too large to interact face-to-face.

Formal education had to fight all the way. It is not so easily mechanized, not so cheaply organized, not so readily standardized, and not so handily merchandised.

The Founding Fathers tried to protect the integrity of both systems of popular culture from the main threat *they* knew: strong, unrepresentative government. The press, by constitutional commission, and education, by constitutional omission, escaped centralized public development and control. But, although exempt from the laws of the Republic, the mass media were subject to the laws of industrial development. These laws required organization, concentration, mechanization, and control—if not public, then private. By comparison, public schools remained the last major folk institution of advanced industrial society.

While the political, economic, and cultural centers of gravity have steadily shifted from the local to the national scene, major responsibility for educational development remained delegated to relatively weak political subdivisions. While television was made available free and equal to all across the nation, and while broadcasting tripled its revenues, our massive national effort in higher education—the G.I. bill—came to an end. Each year, a hundred thousand college-caliber youth could not go to college, for financial reasons. Higher education, we found, never got 40 per cent of the top one-fourth of our high school graduates.

The mass media appeared increasingly to take over democratic national responsibilities for illuminating the realities of today and settling the agenda for tomorrow.

How did they fulfill that responsibility? As well as could be expected; sometimes even better. Being free from public control but lacking economic support in using that freedom, the mass media must, on the whole, merchandise such gratifications as can be profitably cultivated under the circumstances.

The agenda of life's business as seen by the mass media is not easy for educators to perceive. Only systematic study can lift the blinders each of us wears as a matter of choice, temperament, or habit. So let me mention a few areas of concern to education in which we have scratched the surface. These have a direct relation to preparing young people for the life they will live tomorrow, as well as to the educational enterprise and profession.

THE "IDOLS OF CONSUMPTION" TRIUMPHANT

My first case study might be called a story of success. A by now classic study of biographies in popular magazines [1] traced the growth of attention devoted to personal success stories in the first half of the twentieth century. But more remarkable than the growth itself were the changes in the kinds of personalities which symbolized success. Before World War I three-fourths of these models of achievement came from political life, industry, and the professions. Forty years later the "idols of production" gave way to the "idols of consumption." Aside from political figures, in nine out of ten cases the latter-day celebrity's chief claim to fame was stardom in the world of the mass media and the markets they served.

The celebrity cult is not a simple affair. Matt Dillon outdraws the election returns. Several TV stars are more familiar to a test panel of two thousand viewers than an ex-President of the United States who does not or cannot become a TV star himself. About one out of four magazines on the newsstands can be classified in the fan-romance category. Half of the lives immortalized on *This Is Your Life*, 76 per cent of those personalized on *Person to Person*, and 69 per cent of those interviewed on *The Mike Wallace Interview* came from entertainment and the mass media. Such celebrities account for over 40 per cent of all paperback biographies in print. The size of the Hollywood press corps just about equals the combined memberships of the education writers' and science writers' associations. The much-publicized *McCall's* list of "The Most Exciting Reading of Our Time" features such famous nonauthors of nonbooks as Zsa Zsa Gabor, Fred Astaire, Arlene Francis, Keenan Wynn, Jack Paar, Art Linkletter, and poor Marilyn Monroe.[2]

So much for the story of success. My second case study might be entitled "A Study in Failure."

We know that by and large the teacher in literature is too often an inhibited, sexless prune (with apologies to Dr. Ernest Dichter, the motivation expert hired some time ago to improve the image of the prune industry). As an early student of the subject put it, the teacher in literature is likely to be "stooped, gaunt, and grey with weariness. His suit has the shine of shabby gentility and hangs loose from his undernourished frame." [3] That is, unless class is out and memory rings the school bell, when we say a tearful "Good Morning, Miss Dove," or bid a nostalgic "Goodby, Mr. Chips."

But perhaps memory fails us. So we did a study of eighty-one American movies produced since 1950 which portray teachers in leading or supporting parts.[4] Since most movies involve love, and since love has a peculiar affinity to humanity, a look at love and the teacher will give us a good measure of his human stature on the screen.

His opportunities for love were virtually unlimited. Although male teachers outnumbered female teachers (this is not surprising in the predominantly male world of the mass media), six out of ten male and nine out of ten female teachers were unmarried at the beginning of the picture. Alas, most of them were unmarried also at the end of the picture.

Not that they didn't try. They just didn't try hard enough. With so many unmarried teachers running around, it was inevitable that some of them would run into their colleagues of the opposite sex. But the encounter fizzled five times out of six.

With all the happy endings in the movies, a teacher's chances of success in love with *anybody* were fifty-fifty. The most common condition of success in love was (1) that the teacher find a partner without college education, or (2) that the teacher leave the teaching profession. The typical pattern has her quitting a New England high school and a biology teacher fiancé to "find herself" and a *man* in New York. Or it has him leaving a dull musical chair at a western college, along with a strait-laced professor girl friend, to be taught something about music and love in Tin Pan Alley.

With such a pattern of romantic success among screen pedagogues, what need be said about failure? Well, failure in love permitted teachers to remain fully dedicated to the profession. And it permitted their would-be partners to escape into the stronger, warmer arms of less educated but apparently more human creatures.

The film teacher leaving the profession usually goes to greener pastures. The still youthful screen teacher shown as going into some other specific occupation appeared to know what it takes to be really successful. Five times out of six, he became an entertainer.

In a related study we analyzed fifty-six fiction stories dealing with teachers or schools in *The Saturday Evening Post*.[5] The general character profile of teachers turned out to be quite different from the profile of other characters in the stories. Most teachers were represented as coming from the outside, as aliens to the community, often in conflict with the community. They were out of tune not only with most community activities but also with the goals and aspirations of most other story characters.

Although teachers were shown as not striving for material success nearly as much as other adult characters, teachers (as well as students and schools) were more frequently portrayed in material and financial difficulty. As one solution to this problem, about one-third of all teacher-characters quit the profession. In no story was a teacher ever given a salary raise. No student was supported on a public scholarship. No community took the initiative to build or improve schools. There was no normal way in which the financial difficulties of teachers, students, and schools could be resolved. Some degree of poverty was presented as the usual state of affairs. If any solutions were given, they were likely to be fantasy solutions such as hitting the jackpot, finding a rich donor, or holding a fantastically successful sports event. The usual practice of civilized society, financing public schools through normal tax support, was absent from the world of commercial fiction.

ARE THERE SIGNS OF IMPROVEMENT?

Maybe all this is changing. There are certainly encouraging signs in a great deal of serious and responsible news coverage and publicity given to problems of schools and education. But perhaps the cultural attitudes reflected in the selection of fiction and drama are more deep-seated than is the official thinking reflected in overtly informational materials.

Although our press devotes an increasing amount of space to news of education, and the number of education reporters is increasing, there are still only about a hundred full-time reporters specializing in education in 1,700 daily newspapers in the United States, most of which employ a sports editor. A recent study of education news in daily newspapers, by Gloria Dapper and Barbara Carter in the March 17, 1962, *Saturday Review*, found that, "in general—but with notable exceptions—the local newspaper, in the name of education reporting, concentrates on student extracurricular activities, teacher appointments and activities, school finance and buildings, scholarships, honors and awards, the school bus, PTA notes, and a variety of news about colleges." The study came to the conclusion that, "if you depend on your local newspaper for information on education, chances are you have virtually no information or perspective on the major national issues in education and only the most fragmentary view of even the local school picture."

There *are* notable exceptions. But the picture is far from reassuring.

BIASED COVERAGE OF NEA CONVENTION

In 1961 the Institute of Communications Research at the University of Illinois collected every newspaper story and editorial published about the National Education Association convention. We wanted to know how the convention's message about the state and problems of American education was transmitted by the daily press to the American people.

The federal aid bill was pending in Congress when the NEA met in Atlantic City. President Kennedy termed the bill the most important piece of domestic legislation. The extension of federal aid to schools was favored by 65 per cent of Americans in a Gallup poll. Demonstration of the need for federal aid to schools was a major effort of the convention.

But the general pattern of press coverage precluded—to put it mildly—the clear development of any such message. Senator Wayne Morse gave the major address on the bill before the convention's Legislative Commission. Most newspaper stories reported his off-the-cuff criticism of the Catholic church on the parochial issue, elicited by a question after the speech, but made little or no reference to the content of the speech.

The convention passed a resolution supporting the bill. Seven delegates spoke for the resolution, two in opposition. All news stories cited the opposition. The major wire story even cited a floor motion made which died for lack of a second. (The motion was against aid to parochial schools, thus again injecting the conflict element which had done so much to prevent passage of similar bills in the past.) But there was no place in the stories for the voices or arguments on behalf of the almost unanimously approved resolution.

Editorial comment was almost 100 per cent opposed. Editorials warned about "a well-organized pressure group" using children as "pawns" in a cynical game of power politics. They made repeated references to "the blandishments of federal munificence" spurring efforts of "teacher lobbyists" to "fast-talk" Congress "out of some cash." The school-aid bill was editorially characterized as "only the beginning" of a "sensational fiscal binge." Editorials warned of "appeals to emotion of the parents and taxpayers" as tactics characteristic of a "big teachers' lobby." One editorial drew the "logical inference" that teachers would not hesitate to blackmail parents and taxpayers; another even cautioned that they would "sell our free system into the bondage of a federal bureaucracy" to grab "another pot of gold at the end of a rainbow."

A few weeks later the bill was defeated in the House. Defeat of the bill prevented the allocation of resources for narrowing the gap of educational inequality. It perpetuated a situation in which 98 per cent of Wisconsin students complete high school while less than half of those in Virginia, South Carolina, Georgia, Kentucky, and Mississippi do so. Who will measure the cost of this "saving" of money in terms of wasted lives and squandered human resources in the richest country on earth? This is the real "gap" which—if anything—will count in the world in the long run. The gap is in ways of meeting the imperatives of the world industrial, scientific, and cultural revolutions.

This, then, is what concerns me. It is that, on the whole, the mass media play an uneasy double role as the cultural arms of a private corporate system and, at the same time, as the informal public educational agencies of modern society. The pattern of industrial behavior recorded in the media's portrayal of teachers and education exhibits earmarks of tension and of a crossing of institutional purposes.

What can we do about it?

THREE COURSES OF PLANNING AND ACTION

I would like to suggest three courses of planning and of action. One has to do with professional relations with the mass media. The second involves relations with the community. And the third is a suggestion for making our curriculum more adequate to the demands of twentieth-century popular culture and its institutions.

A recent study conducted by researchers at Stanford University about voters and their schools is instructive in these respects. The researchers wanted to know how a favorable vote on school bond issues related to the various sources of information about schools. It was found that the voters who were involved in some personal participation or contact with school representatives and who had direct access to information about schools were *twice* as *likely* to vote favorably as those who relied for information on the mass media.

These results, as well as the case studies I have reported, are food for thought. They mean that schools have much to do in the way of improving their relations with mass media and of improving the quality of educational reporting, even if there is little they can do about vested editorial interests.

Second, they mean that under our present institutional circumstances there is no substitute for the hard way of encouraging more participation, bringing more adults into the schools, trying to educate parents directly. The mass media must still be helped and encouraged to function as representatives of the public interest. But other bridges must also be built and other channels must be found to achieve a realistic public understanding of the needs and problems of modern education.

Third, our curriculum must reflect the needs and demands of twentieth-century popular culture more than it does now.

The mass media have immeasurably enriched our lives. They also have the capacity for narcotizing us, for insulating us from some realities of life, and for stacking the cards against broad public access to all information and even entertainment necessary for self-government.

These potentialities of mass-produced popular culture pose a challenge and a responsibility for curriculum builders. It makes little difference whether we think about this under the heading of social studies or citizenship or English or literature or special studies on the mass media. We must help our students to use the riches offered by the mass media. We must also help them to make some order out of the distortions, confusions, and chaos created by the mass media. Most important of all, we must help them make intelligent decisions not only as consumers but also as citizens, building and molding social institutions—including the media—for democratic human purposes.

NOTES

[1] Leo Lowenthal, "Biographies in Popular Magazines," in *Reader in Public Opinion and Communication*, edited by Bernard Berelson and Morris Janowitz (Glencoe, Ill.: The Free Press, 1950).

[2] Patrick Hazard, "The Entertainer as Hero." Paper read at the Association for Education in Journalism convention at Pennsylvania State University, August, 1960.

[3] Arthur Foff, "The Teacher as Hero," in *Readings in Education*, edited by Arthur Foff and Jean Grambs (New York: Harper, 1956), p. 21.

[4] Jack Schwartz, "The Portrayal of Educators in Motion Pictures, 1950–58," *The Journal of Educational Sociology* (October, 1960).

[5] This and the convention press study (discussed below) are as yet unpublished. I would like to acknowledge the assistance of Roger Brown in summarizing some preliminary findings of the magazine analysis cited above.

[6] See, e.g., Wilbur Schramm, "Mass Media and Educational Policy," in *Social Forces Influencing American Education*, the Sixtieth Yearbook of the NSSE, edited by Nelson B. Henry (Chicago: The University of Chicago Press, 1961).

V. Criteria for Evaluation

WHAT ARE FAIR STANDARDS to expect of broadcasting in the public interest? What are some of the problems unique to the electronic media? Where is broadcasting to get its models, its traditions, its critics?

How adequate today is the emphasis on literary appreciation, music appreciation, and other such courses in an age of television? Where and how are adequate courses, training, and standards to be found which are fair both to the public and the broadcaster?

Such are some of the issues raised here by a quartet of men whose influence and writings are well known in this country.

24. Is Literacy Passé?

by Walter J. Ong, S.J.

Father Ong, professor of English at Saint Louis University, is known as a scholar in both the Renaissance field and the area of contemporary literature, and as a prolific writer on problems of contemporary civilization. Several of his books and essays, i.e., The Barbarian Within; Ramus, Method, and the Decay of Dialogue; *and* Problems of Communication in a Pluralistic Society *represent critical explorations of literature, contemporary culture, and religion. He is also a well-known lecturer. This article is reprinted from* The Saturday Review *for November 28, 1959, by permission of the author and the magazine.*

It may be that we have come to the end of the Gutenberg era. The electronic age may not yet have made printing obsolete, but it certainly has ended the monopoly rule of published matter over our habits of thinking.

Our "typographical culture" began about five hundred years ago with the invention of the printing press, which enabled handwritten documents to be duplicated rapidly, thus facilitating the swift diffusion of ideas; but more importantly, the press changed our sense of what thinking itself is. To gauge the extent of that change we need only reflect on the communications systems used by preliterate and pre-typographical man.

Preliterate man knew no history in our sense of the term. His knowledge of the past was limited pretty much to what his parents and grandparents or great-grandparents could tell him. He lived in a voice-and-ear culture. His knowledge was stored in the mind, and when verbalized was communicated primarily by the voice and by other sounds. But sound, the ground of all verbal communication, is time-bound and evanescent. It exists only while it is passing out of existence. I cannot pronounce the last part of a word until the first part no longer exists. The alphabet—which appeared quite late in history, around 2000 B.C.—reduces the evanescence of sound in time to relative permanence in space. Pictures also do this, in a way. They enable us to recall an image or a concept, and thereby the word. But the alphabet turns the picture process inside out and upside down. It breaks the sound itself up into little spatial parts, which it reassembles on a surface in countless configurations. The alphabet thus had a revolutionary effect on our thinking processes.

However, the tug of habit is a strong one: the oral-and-aural tradition persisted long after the adoption and eventual spread of the alphabet. When, a few decades before Christ, Marcus Tullius Cicero wanted to bring Greek knowledge to the somewhat backward Romans of his day, he did not read scrolls in order to master that knowledge, but rather went to Greece to listen to the lecturers and orators of that country. Significantly, Cicero used to speak his orations aloud, then write them down as a kind of afterthought.

With the dawn of the European Middle Ages, Western man developed a "manuscript culture." Medieval scholars could not listen to Rome's learned men as Cicero had "audited" those of Greece. So they pored through handwritten manuscripts: medieval culture produced the bookworm. However, it also retained massive oral-aural commitments. We know from Saint Augustine that men of his time continued to read aloud even when they were reading to themselves. Words were worth little unless they could be vocalized. In the medieval universities written exercises beyond the level of elementary instruction were absolutely unknown. Written texts abounded, but they were used as take-off points for speeches, for lectures, or for "disputations," highly organized verbal tilts in which the students proved their mastery of logic, physics, and other subjects. Yet despite the persistence of this oral-aural cast of mind, medieval culture was pre-eminently a manuscript culture, a fact which prepared the way for printing and which perhaps made printing inevitable. For printing was and is a cheap way of producing what a manuscript culture wanted—something to read.

After the invention of the printing press, man's whole way of thinking about his own intellectual processes changes subtly. The "spatial parts" embodied in the alphabet are made even more maneuverable by the press's movable type, and are given greater permanence by its fonts. Thinking processes are now taken to be concerned with getting things into an order comparable with that observed in a printed book. Thought begins to have "content" just as books have tables of contents—a concept quite foreign to medieval and ancient man, for whom truth was not commonly associated with some kind of containment or boxing process, but rather with communications or teaching ("doctrine").

In our time a new drift, away from the "typographical" and toward the "oral-and-aural" culture, toward the world of sound, has definitely set in. As human society, despite its swelling size, becomes more closely knit around the globe, sound asserts itself more and more, supplemented and augmented by such marvels as the telephone, TV, rapid transportation systems, and earth satellites which speak to us in tiny, beep-beeping voices.

As corollary of this deep-rooted transformation, language itself is undergoing a profound overhauling. Grammar, which is based upon the written word, is giving way to linguistics, a discipline rooted in the spoken word, i.e., the word at first hand instead of at one remove.

In philosophical circles interest is veering away from logic to dialogue, from thinking conceived of as a private, silent affair, to thinking seen in its full social and public setting. There is a growing awareness that science itself at any moment is only arrested dialogue, and that the difference between what we know and what men five hundred years ago knew, or between what we know and what men a hundred thousand years from now will know, can be charted not in terms of men's private thoughts but in terms of what they have said to one another.

The new world will not forget the old. It never does. Printing is here to stay. But now, more than ever, it is only one part of a constellation of activities. Its monopoly is broken.

25. Television

by Jack Gould

Mr. Gould is the well-known television critic of the New York Times. *He has commented frequently and articulately on American television in the pages of the* Times *and elsewhere. This interview, one of a series on the American character, was conducted under the auspices of the Center for the Study of Democratic Institutions in 1961. It is reprinted by permission of the author.*

This is one of a series of interviews being conducted by Donald McDonald in connection with a study of the American Character by the Center for the Study of Democratic Institutions. The Center is the principal activity of the Fund for the Republic. Its work is directed at clarifying basic questions of freedom and justice, especially those constitutional questions raised by the emergence of twentieth-century institutions.

Jack Gould was born February 4, 1914. He worked at the *New York*

Herald-Tribune as a general reporter from 1932 to 1937. He began at the *New York Times* as a member of the drama department and is now radio-television critic. He is considered the most influential newspaper critic of television in the United States.

Harry S. Ashmore, now editor-in-chief of *Encyclopaedia Britannica,* was Pulitzer-Prize-winning editor of the *Arkansas Gazette.* A member of the board of directors of the Fund for the Republic, he has been in residence at the Center for the Study of Democratic Institutions since its establishment in Santa Barbara in 1959 and has spent much of his time in studying the role of the mass media, and particularly television, in a free society.

The Center for the Study of Democratic Institutions is a nonprofit educational enterprise established by the Fund for the Republic to promote the principles of individual liberty expressed in the Declaration of Independence and the Constitution. Contributors to publications issued under the auspices of the Center are responsible for their statements, of fact and opinions. The Center is responsible only for determining that the material should be presented to the public as a contribution to the discussion of the Free Society.

Q: Mr. Gould, from what I can understand of the television industry, it seems to me that there are half-a-dozen or so areas that are basic to any over-all grasp of the nature and function of the industry. These would include (1) the relationship of individual stations to the networks, that is, the mutual relationship between them; (2) the relationship of networks, stations, advertising agencies, and sponsors—in other words, the commercial structure of the industry; (3) the regulatory agency, the Federal Communications Commission, and its relationship with the industry; (4) the relationship of the writers to the television medium; and (5) the problem of censorship, both external censorship and the internal, anticipatory censorship by the stations and networks themselves. Would you say that any one of these was a key area, or a central problem in the sense that it explains or is responsible for the existence of the other problems?

GOULD: Well, the commercial structure of the industry is what determines the nature of the medium, and in some respects it is almost a mirror of the larger problems of the country itself.

Q: When television was established, did it merely continue the relationships and structuring that had been established by the radio industry? Did television depart in any significant way from the manner in which radio broadcasting had been conducted?

GOULD: There was practically no change from radio to television. Television inherited radio's pattern of scheduling, the breaking up of the time into hour, half-hour, and fifteen-minute segments. And, of course, the whole relationship of sponsors, advertising agencies, and networks was inherited from radio.

Q: When television was in its infancy, do you know whether any thought was given to the desirability of television's breaking away, at least in part, from the way things had been done in radio broadcasting?

GOULD: After the war, when television got the green light, the industry just adopted the existing pattern of radio, the one with which everybody was familiar. Television hasn't really changed very substantially to this day. There are so many people who have a bit of authority to say what the shape of TV should be, I don't know where you would really begin. In theory you begin with an individual television station. But the individual stations are associated with networks. And the individual station, by and large, has no more idea of what is going to be on the air than does a viewer, until it arrives from the network. So you have a kind of ridiculous situation: The government holds that the individual station must be responsible for programming, but the individual stations, in turn, assign programming over to the networks.

Q: Has the government no control over the networks?

GOULD: Only indirectly through the stations that the networks own. The FCC has exercised some jurisdiction over some of the business arrangements, the contracts between networks and individual stations, but in terms of minimum programming standards or anything like that the networks aren't required, as individual stations are required, to tell the government what general type of programming they plan to do.

Q: Who makes the determination of the type of programming that will be broadcast, I mean the real determination?

GOULD: The heads of the networks as such bear final responsibility—Paley, Stanton, Sarnoff, Kintner, Goldenson, and Treyz. They are the ones who really run American television, commercial television. But, of course, whoever has control of the dollar, the advertising program dollar, is going to make the final determination of what goes on the air in entertainment programming. It has always seemed to me very academic, this perennial controversy over whether the advertising agencies or the networks should control the programming. As a practical matter and as things stand, it doesn't make any difference. So long as the sponsor can buy an individual program, so long as he can pick and choose his programs—whether the agencies conceive the programs or whether the networks conceive them—the result, in terms of program content, will be the same because both the agencies and the networks will devise the type of shows the sponsors want. The networks like to think they have control over their programming, and they like to tell us they have control. But the real control rests with the sponsor. By the act of *not* purchasing certain kinds of programs the sponsor exercises a tremendous influence over television programming. The sponsor's negative power is enormous. The sponsor refuses, say, to buy a "controversial" drama and he thus has a power of veto over what the public sees; he simply does not buy that program.

In news and public affairs programming, the networks do exercise control. I think the fact that news and public affairs are by far the most substantial area of television programming indicates what can be done when the networks themselves really control the content of their programs. Sponsors, to their credit, have kept their hands off the news.

Q: From what you've said, it seems that the whole enormous apparatus of television broadcasting has become, aside from the news and public affairs programming, simply an instrumentality for moving the goods and selling the services of the sponsors.

GOULD: The sponsor's decision to purchase or not purchase a program is based on how many potential customers he will reach with that program. It stands to reason that he's not going to reach as many customers with Shakespeare as he is with *The Untouchables.* This is the tragedy of a mass medium because so long as it has to appeal to the largest number and its whole economic strength is based on making that kind of an appeal, then automatically there is a great limitation on what you can do in television programming.

Q: Would you say, then, that the other problems in television are more or less subordinate to the commercial orientation of the medium?

GOULD: The matter of program control involves everything else in television—content, censorship, the writing problem. They flow one from the other. Sponsors naturally want to make the best possible impression on potential customers. That's basic. No businessman is going to go out and offend the guy he wants to sell his product to. You don't slap a customer in the face; you don't offend. All the major advertising agencies have testified at FCC hearings that this is basic policy in television: *you do not offend.* The result is that you're bound to have blandness. The oil company that sponsors Chet Huntley and Dave Brinkley has already felt some boycott repercussions in the South because of their excellent pieces on integration. Other sponsors see this. They say, "The news you have to put on, but why put on a play that's going to get the Southerners sore at you?" And since sponsors still want to sell their products in the South, they just avoid the great contemporary issue of civil rights and integration so far as drama is concerned.

Q: Wasn't Rod Serling's original play about the lynching of Emmett Till a victim of this blandness? By the time the agency and network people got done with it, it wasn't recognizable.

GOULD: Yes. There's an old gag in television that the only character who is safe to portray today is a Swiss Protestant. So there is this limitation. Du Pont, for example, said very candidly that on their *Show of the Month* they wanted to project from coast to coast the image of a happy, contented company. They didn't want to offend anybody. I don't say there is anything to quarrel about in this, from a business sense. But you can see what that does. . . .

Q: What it does to art.

GOULD: To art and culture. It's disastrous. You wind up with a bland diet of mish-mash. You get Westerns, which are safe. The gangsters are always safe, you can do those. Situation comedies you do as innocuously as you can. And you can do the classics. You can have a very controversial classic and nobody screams. But get an original play and everybody looks for something sinister and evil in it. In this kind of environment a playwright finds nothing stimulating. He may have a good, original idea but if it has anything to do with a contemporary controversy of any kind he can't get it out on television.

Q: What about the financial rewards for the television writer?

GOULD: For the writer, the economics of TV are pretty terrible. A writer who has a good idea, if he makes a Broadway play of it and hits the jackpot and goes on to a movie sale, would be dealing in sums of six figures. The most that a playwright on television can hope to realize from the same idea and play would be $10,000 or $15,000 and he would have all the censorship and other frustrations of television to contend with.

Q: What kind of writers are left in television then?

GOULD: Those who are still around are trained in the taboos of the business; they anticipate them and keep them out of their scripts right from the beginning. The agency people testified to this at the FCC hearings. They said, "We don't have any trouble with writers any more. They know what we don't want and they leave it out." So you have this terrible sterility.

Q: Again, this stems from the commercial exigencies of television, doesn't it?

GOULD: It comes from the commercial nature. I don't want to be unfair to the businessman, but . . .

Q: But all these purposes—artistic, commercial, etc.—are at war with each other.

GOULD: Continually at war. The thing I object to is that the world of commerce is using the resources of the theatre, of all our culture, for sales purposes.

Q: On a medium that belongs to the public.

GOULD: Yes. I think that is short-sighted and selfish. I think they have an obligation for the good of the country and for the good of our culture to put something back into that culture rather than just bleed it white in order to sell their products. I think they have a responsibility here to reinvigorate the resources they're using, to replenish the reservoir of capable writing talent.

Q: What can a governmental regulatory agency do in this regard? The Federal Communications Commission has been rather a toothless organization in recent years, hasn't it?

GOULD: Until the last few months the FCC has been an utter disaster so far as making its influence felt.

Q: Some of the ex-FCC commissioners like Fly and Durr have said that one of the FCC's problems is that it is under-staffed, that it cannot make a realistic determination of the renewal qualifications of the thousands of radio and television stations whose licenses come up for renewal every three years.

GOULD: Sure, they are under-staffed, but guys like Larry Fly were exciting to have around. They were stirring up a fuss; they cared about broadcasting. They cared about it deeply and they didn't hesitate to wade in for their point of view. I remember when Larry Fly stood up to all the broadcasters in the country and told them off. And Clifford Durr was fearless. And both of these men knew a great deal about the power of persuasion and discussion. But in the last few years there has been no discussion of the big issues on any meaningful level. The Commission fell into a dead sleep.

Q: Well, the new FCC chairman, Newton Minow, has spoken very candidly to the broadcasters about the "vast wasteland" of television. How much influence will he be able to exert? I noted that within a few days after his famous talk to the broadcasters in Washington certain Congressmen served notice he would not be given the powers the administration asked for so far as his agency is concerned.

GOULD: Mr. Minow can exert a big influence and he already has. There was a lot of screaming over his speech but many stations and networks have begun to make changes in their schedules. The Kansas Association of Radio Broadcasters has told its members that they better make sure they are presenting educational and cultural programs. The broadcasting industry always has had an efficient lobby; politicians think radio and TV time is essential to election. But in their hearts many of the radio and TV boys know Mr. Minow is right. In public they will yell over the terrible man at the FCC but in private they will be apprehensive. If Mr. Minow doesn't get bored or discouraged, he could give TV a lift. The FCC needs a reorganization, but it is not powerless as it stands.

Q: You seemed to imply earlier that the problem of size of staff is not a major one on the FCC.

GOULD: Well, on the specific level of regulation and license renewal, sure, the FCC is under-staffed. But on the other hand there is some leverage the FCC can use that is not related to how much staff it has. A station operator makes voluntary promises, when he applies for a license, concerning what he intends to do with his station, what kind of programming he intends to broadcast. I don't see how anyone can claim that when the FCC asks him to live up to these voluntary promises the FCC is guilty of "socialism" or that it is "anti-free enterprise." I think making station owners live up to their promises should be done very, very toughly. I think it can be done, and now Mr. Minow says he will do it. Certainly there will be a lot more improvement in broadcasting if these station owners know they're going to have to be called in for an accounting of their performance. And I would like to see the FCC announce publicly in advance when the licenses of stations are about to be considered for renewal so that the public will have a chance to testify too about the quality of programming on these stations. Mr. Minow has said he intends to hold such hearings.

Q: Of course, if the public gets nothing but a steady diet of Westerns, crime stories, and bland comedies, how can it know what it has been missing? Wouldn't it be difficult for the public to specify its complaints about broadcasting standards if there is no basis for comparison?

GOULD: Yes, if you get fed anything long enough, you get satisfied with it, and then if something comes along that's just a bit better, you tend to over-praise it, really, because you're so numb from the mediocrity you've had.

Q: Well, what can the FCC do? You mentioned earlier that the FCC has no control over the networks and yet the individual stations are pretty well tied to the networks and must take so many hours of network programming each week.

GOULD: One of the things the FCC can do is to make stations owned by networks give an accounting of their performance. And then there are other things that can be done. In a democracy we want to raise standards but we don't want to get tangled up in governmental action and censorship in the process. We have always been able to accomplish a good deal through discussion and persuasion. I think if the real issues of television can be more widely discussed, if Mr. Minow continues to raise the important questions, constructive action by the broadcasters will follow. I remember years ago when Paul Porter and Larry Fly of the FCC issued their famous Blue Book on broadcasting. My heavens, the reform movement that got under way in one week-end in the broadcasting industry! Some of the effects of that have lasted to this day. It was fantastic. And yet nothing in that Blue Book was ever really implemented in any concrete, regulatory action by the FCC. But it was tremendously effective.

Q: The effectiveness of such actions, I suppose, is a matter of good timing and rhetorical technique.

GOULD: Yes, and, of course, this is part of the answer; it's one of the ways we work out our problems in this country. I think it needs to be pursued. I think there have been some bad boys in broadcasting and that the Commission has let them off. I don't believe a single station license has been denied renewal throughout the history of broadcasting, except for the few in the famous case of the "goat gland" advertising by stations in the Southwest years ago. But apparently no other station has ever done anything contrary to the public interest in broadcasting all these years.

Q: The law doesn't specify, does it, how many hours a station must give to religious, educational, and public affairs programming?

GOULD: No, it specifies no percentage. But when a broadcaster asks for a license, he lists what he proposes to do with his station. He makes voluntary promises about his performance. A case came to light just a few weeks ago in which a station owner had promised to devote 5 per cent of his schedule to religious programming and his actual percentage was zero. Well, I think the FCC has a right to call that man in and say: "Look, you haven't done what you said you were going to do. We aren't specifying what you must do, but you haven't lived up to even your minimum, voluntary promises. So we think you ought to be put on suspension for a day." One day! And, boy, the revolutionary effect that would have on broadcasting performance in this country!

Q: What a regulatory agency can do in any area of life in this country is, I would imagine, limited by what the public wants it to do.

GOULD: Yes, a regulatory agency can't get too far ahead of public opinion.

Q: Some of the old FCC commissioners who had had some good ideas about radio and television broadcasting have said that Congress never gave them enough appropriations to do the job that had to be done. In other words, there wasn't enough conviction or indignation in Congress or in the country at large about the broadcasting industry's performance to support a realistic regulatory program in the industry. Perhaps Congressmen would act more affirmatively if the people themselves would exert the necessary pressures of public opinion. I should think that the television critics of the newspapers and magazines would play an important part in the formation of such a public opinion.

GOULD: It's partly the television critic's responsibility, partly the responsibility of schools and of organizations such as the Fund for the Republic which has done a number of studies of the broadcasting industry. All this has helped. And yet there remains a huge indifference to the general problem.

Q: Aside from the taboos arising from commercial sources, sponsor anxieties, and the like, is there much of a censorship problem in the television industry? I am thinking principally of groups outside the industry. Do they exert much censorial influence?

GOULD: Yes, and, of course, the dark, dark period was the McCarthy era. That was an outrageous and shocking period in the history of television. That was the time of "Red Channels," and . . .

Q: And the grocer from Syracuse.

GOULD: Yes, the grocer from Syracuse who took sponsors' products off his shelves and thereby set off the most shocking panic I've ever seen in my life. And many people are still suffering the consequences of that panic by the industry. Well, the situation is a little better now. But you still have outside pressure groups. The networks, for example, just don't want to get into the Civil War centennial in any meaningful way because the affiliated stations in the South would get upset. And we've had Italian groups objecting to *The Untouchables*. But I don't think the problem is as bad as it was. The networks discovered that even in the most publicized case of all—the case of Jean Muir's being dropped, which was front-page copy day after day—Jello sales of the sponsor, General Foods, were not being substantially hurt. Even General Foods admitted this. So the threat of boycott on any significant mass scale has just never been transformed from threat to reality. But the networks, in the area of news and public affairs, have been far more outspoken. I don't know of any subject now that the networks would avoid in their news and public affairs programming. And here in New York the independents have been useful. David Susskind has helped to open things up. And while I don't like all the things that Mike Wallace did, he did broaden the horizons, he did expand the areas that could be discussed.

Q: You have mentioned several times the networks' comparatively better performance in the presentation of news and public affairs programs. Yet it was only a few years ago that Edward R. Murrow spoke to a professional group in Chicago and said that "this country is in grave and perhaps mortal danger, and . . . [yet] to a large extent during the hours between 8 and 11 in the evening the television audience is being fed a diet that tends to cause it to be indifferent, that tends to insulate it from the realities of the world in which it lives." Has the situation improved much since Murrow made that statement?

GOULD: Well, of course, the quiz scandals have intervened since that statement, and one of the beneficial effects of those scandals was that the networks made greater effort in news and information programming. This was the quickest way to tidy up television's image, to ease the embarrassment of the industry. Also I think that Robert Kintner, who is newspaper-oriented and head of NBC, had a genuine interest in this area and has tried to do something about it. I think the situation has improved. We've had many public affairs specials on NBC; and we've got *CBS Reports* now.

Q: Coming in on good times?

GOULD: Yes, at least on evening time. The picture is better. But Murrow was absolutely right. Now, however, the question arises: Will people who have been conditioned to look for only the exciting and the jazzy and the easily entertaining view the better type of program?

Q: A thing that never fails to intrigue me is that a good "public service" type program on television may draw five or ten million viewers, but, by commercial television standards, that is considered an insignificantly small audience and so it is usually broadcast at odd, inconvenient hours, say early Sunday morning or late in the evening.

GOULD: That's right. If you can get ten million people to read and absorb a serious treatment of a problem like the migratory farm workers in this country, or the Cuban situation, why, that's a fantastic accomplishment. It ought to make the TV industry jump up and down in delight. But in practice if your competitor puts on a gangster show opposite you and outdraws you, you're in trouble. But I do think the networks' performance in this area is better. They recognize that, in fact, public service programs are getting through, that they're getting better acceptance than they used to.

Q: I wonder if many people in the television industry are concerned with something that Walter Kerr wrote about in *Horizon* a year or two ago. Kerr emphasized the importance of television's discovering and developing its own unique form and he gave a few examples of things television had done which no other medium could do: Bernstein's seminars on music; Agnes de Mille's essays on choreography; the conversations with Susskind, Murrow, Wallace, Paar, which combined, as he said, public figures with private tone, the exceptional event and our matter-of-fact view of it. To what extent are the leaders of the industry striving to realize television's unique form?

GOULD: There was a time, back in the flowering period of *Omnibus*, when television was an indigenous art form. Its use of camera, its "visual" thinking, was enormously exciting. TV's truly creative use of the camera in those days gripped everybody. But certainly for the vast bulk of television today all this has gone by the boards, again under economic pressure.

Q: Is it because experimentation is too risky?

GOULD: Well, there's very little of it. CBS, to its credit, still does about as much experimenting as anybody, particularly on Sunday morning. I think *Camera Three* has done a brilliant job over a fairly long period of time.

Q: But such experimentation is invariably reserved for the odd and inconvenient Sunday-morning period.

GOULD: Yes. And while video tape could be a creative tool, I'm afraid it has become simply a speed-up mechanism to expedite the television operation. This has its economic, but not artistic, advantages, of course. The basic television you see every night is now video films shot in something of the order of two or three days. There just isn't time in such an operation for any real subtlety or polish. So what you have is an abortive sample of movie-making instead of live televised drama. I think the whole technique of television has sadly deteriorated.

Q: Are any people trying to do something unique and creative in this medium?

GOULD: Here in New York some of the technique on *Play of the Week* was exciting. There is a show called *Accent* that is good. I've mentioned *Camera Three*. There is *Lamp Unto My Feet* and a few others. But these are the last forlorn outposts of experimentation. Agnes de Mille and Walter Terry have had some exciting stuff, small programs. Few are seen nationally. In the old days of television the summer was the logical time for some experimentation. Now summer is given over almost wholly to re-runs of winter stuff; that helps amortize the cost of the film. And sponsors find that the re-runs pay off with much bigger audiences than does experimental TV, so nobody feels a responsibility to replenish the medium with new ideas, new artists, new material. I think Kerr is right; really creative TV could be terribly exciting.

Q: Do you know the mentality of the people who run the industry? I mean are you familiar enough with the thinking of the Stantons, Paleys, and Sarnoffs so that you can describe their own attitude to the quality of the material they are broadcasting? I've heard it said, for example, that the network officials and the advertising agency people frankly admit that they themselves can't stand to look at what they put on the air day after day, that they program the stuff . . .

GOULD: For business purposes, they look, yes.

Q: I suppose that's not hypocrisy as such.

GOULD: No, it's a business matter really. According to their statistics, this popular, day-by-day television material does gain a huge audience.

Q: Then they don't pretend that this is "great" television, do they?

GOULD: Oh, no. They make the point that in order to survive they have to appeal to the majority, that this is pure economics, and that it is unrealistic to argue to the contrary. When I was in Britain I was amused to see the same forces and logic at work there. The British apparently are going for the same popular stuff that the Americans go for: *Wagon Train* is a smash hit in Britain just as it is here. Let's face it—there *is* a huge appetite for diversion, pure diversion, and the basic economics of television—the way it is structured—demands it.

Q: Certainly such programs demand little from the viewer.

GOULD: Yes, and I don't think that the average fellow who's finished a day's work is always looking to be elevated by his television viewing at night. But my quarrel . . .

Q: Is that nobody takes care of the minority?

GOULD: Well, that, but also—well, everybody in the television industry argues that they have a "democratic obligation" to "give the most people what they want most of the time." I say that that is true only to a limited extent because the public, the disorganized public, really has no way of articulating its demands and its desires. All the public can do at this point is to select from what is offered to it.

Q: Are you, in effect, saying that the television industry has some responsibility for raising the general intellectual and cultural level of the country?

GOULD: Yes. My idea—and I'm sure that people in the industry would agree and want to do it if it could be worked out economically—is that, as trustees of the air waves, the broadcasters have an obligation to furnish leadership in all fields. You see, the way in which the public expresses its wishes, its likes and its dislikes, is, to a great extent, the way in which it responds to the leadership that is offered. I recall no survey, for example, that said that Shaw's *Pygmalion* should be turned into the musical *My Fair Lady*, and that this would be a great success and that it would delight millions. No, Lerner and Loewe and the rest of them had an idea and they had enough faith in their own judgment to go out and try it and it was successful. I think the broadcasters have to do this. I'm sure nobody told Mr. Ochs at the *Times* that there was a demand for a certain kind of newspaper. He didn't go out and make a survey. He put out the kind of newspaper he wanted to put out.

I think here is basically the weakness of the leadership in broadcasting: they're not really putting out the kind of network programs they would like to put out. They're just putting out what is economically feasible. It seems to me that this is the real test of their free enterprise method—does it just have to drift, does it have to keep drifting downward? Must the only goal be the getting of large audiences and making more money? Are they going to express, at some point, real leadership, as they have to a limited extent in the news programs? As of now, the leaders of the industry have virtually abandoned the field of entertainment.

Q: This reflects not so much a conscious contempt of the audience, I suppose, as a kind of indifference to responsibility.

GOULD: Kerr has hinted at the point; others have, too. An awful lot of people, after an amazingly short time in television—thirteen or fourteen years—are dead tired. I think they've lost the zest for the battle. They're uneasy. I think perhaps the time has come when the industry must have new blood.

Q: Those who did have the zest and tried to do something have fallen by the wayside, haven't they?

GOULD: They've fallen by the wayside. And practically all the writers have checked out. Few of the imaginative producers are around any more. And the actors! There's nothing more heartbreaking than to see what's happened to them. If you look carefully at the actors in some of the West Coast half-hour television film programs, you will see people who five or eight years ago were giving beautiful, creative, sensitive performances on live television—live originals from here in New York. And now, just to survive, they're doing these junky Westerns—not that we don't need Westerns, but not at that price! The thing that's most disheartening is that darned few people in the industry seem to want to come to grips with the problem. Governor LeRoy Collins of the National Association of Broadcasters might do something. From his first talk to the broadcasters it seems he hasn't yet learned the "rules" and I find this enormously encouraging.

Q: What do you see as alternative or supplemental forms for television? Can reform and improvement be achieved within the existing structure, the commercial structure, of television?

GOULD: Perhaps it can to a limited extent. There are some signs of that in the New York area where independent stations have gone to the minority audience. They have conceded the majority audience to the popular networks. But it's very questionable, for the reasons we've discussed, how much improvement can be expected from within the present structure. I think we might learn something from another mass medium. I think the most fascinating mass medium today is the long-playing phonograph record. This enables a manufacturer to put out Elvis Presley in 5,000,000 copies if he wants to and another manufacturer to put out the Budapest String Quartet in 50,000 or 100,000 copies. Both can make money proportionately. The public has a choice; they can play both records on a common device. But in television, with the limitation of the number of channels (all of which have long since been spoken for), and with these dominated by commercial interests, the public has no real choice of program. What it has is choice between programs of one kind. There is no diversity in American television as there is in a book store, record shop, or newsstand. It seems to me that the basic problem is the small number of channels available, and here is where the government has been at fault. It has not moved ahead fast enough to open up new channels, to get the industry away from this whole scarcity philosophy and to open up UHF channels throughout the country.

Q: Do you envision these UHF channels as operating in the same pattern as commercial television, not necessarily programmatically but economically and technically?

GOULD: What I am thinking about is that if you had sufficient channels, then you could have, in addition to existing commercial and popular television, a pay-as-you-see channel in terms of one or more kinds of program—sports, theatre, education—perhaps several pay-as-you-see channels for the several major categories. You could have educational stations as such. There's almost no limit to the diversity you could achieve if all seventy UHF channels were activated. The industry has never made any real effort to get that going and the FCC is just starting now to do something about it—thirteen years after the start of television.

Q: Would you compel certain channels to program certain kinds of material?

GOULD: I think that ultimately all television should be moved to UHF where everybody will be on a technical parity. In this country we have no local television to speak of. We could have hundreds of broadcasting stations in towns of under 100,000. If it were made technically feasible, we could have local television, coverage of local meetings, local sports contests—even local commercials. We have no regional television to speak of, either, in this country. I've heard it said that in New York City, if all the channels were opened up, we could have what broadcasting we now have; we could also have educational television not only as it is presently contemplated (through the purchase of an existing station) but as a collaborative effort by several colleges; and the city could have a channel. There would also be some pay-as-you-see television. There's no reason why we couldn't have at least ten or fifteen different channels in the big metropolitan centers; we have thirty-five or forty radio stations in those centers now. Those channels would all be doing different kinds of programming. And no one would expect a station to try to do what it is now doing, which is ridiculous—to program from six in the morning until two o'clock the next morning; nobody can program enough good material for that many hours. One of the channels could concentrate on pay-as-you-see Broadway theatre presentations. I'm sure there are people here, theatre people in New York, who would love to put on a play; they'd be perfectly delighted if they could get 20,000 viewers a night, at 50 cents each, to view their plays. That would bring a huge amount of money, more than they could gross on Broadway.

Q: Do you think this diversity of programming would follow naturally from the opening of UHF channels?

GOULD: If you had just one pay-as-you-see TV channel, then that station would have the same motivation as commercial broadcasters to get a maximum audience and it would come right around to where we are now. With a number of various kinds of operations—commercial, educational, pay-as-you-see, perhaps municipal—I believe you would get a natural diversity of programming.

Q: You are not in favor, then, of direct federal government intervention in this thing.

GOULD: For educational television, through local setups, I think governmental action can be good. But a central governmental network? No. The idea of any Congressional committee having authority over the network just frightens me to death. Bad as we may be we don't have to go to that extreme.

Q: Arthur Schlesinger, Jr., once said in a discussion of this problem that if the government would "equalize" the "disadvantages" inherent in quality programming, then maybe all the stations and networks would start behaving responsibly. He illustrated his point by referring to the child labor laws of a generation ago. Until the government outlawed all child labor, no individual manufacturer—much as he personally would have liked to do so—would stop hiring children in his plant or factory; he would have been economically penalized by other manufacturers who would continue to hire cheap child labor. Schlesinger said that if the government insisted that all networks and stations devote a certain amount of prime time to "public interest" programming, no one station or network could be economically punished by complying with the law. Of course, this comes around, again, to enforcing statutes and regulations that are already to a great extent on the books.

GOULD: And there are many ways in which you can circumvent such a system. First, you have to pick the hours, and if you are a smart network operator and you've got an hour you can't sell—say opposite a top commercial program, *The Untouchables*, or whatever—you put your "public interest" program there. You're not going to be able to sell that hour anyway, so you give it away to satisfy Mr. Schlesinger's group. I think we have part of the solution (and it has not been appreciated) in the nucleus of an alternative service in educational television. Right now we've got about forty-five or fifty educational stations. Many of them are not very good. The quality of educational television is nothing like what it could be. But this is partly due to budget limitations, partly due to lack of experience, lack of help, lack of public appreciation, lack of newspaper support. But the nucleus is there. If these could be put together for major purposes, then we would have a genuine alternative service. Undoubtedly government aid by the state would have to come into it, through educational sources probably. But I think such a service should be encouraged and developed. As I say, I also think pay-as-you-see TV should be developed and sufficient UHF channels opened up to make genuine diversity possible.

Q: What hopes do you have that any of these alternatives will develop?

GOULD: I think the pressures for this are building up. The improvement in the news and public affairs programming is a sign of a change in thinking. The talks of Governor Collins of the NAB and Mr. Minow of the FCC reflect another change in thinking. Now the danger is that the individuals and groups who could do so much to insure the success of efforts to make television broadcasting more responsible will become disheartened in the face of the real commercial pressures to keep things as they are. I recall in the days of radio that the intellectuals of the country—the university people and foundation people—had great hope in the educational and cultural potentialities of radio. But they lost interest and they lost heart when the commercial pressures in radio built up. The intellectual elements just dropped out and radio pretty much just went to pieces. It has only been in the last few years that radio has come back in the form of FM "good music" and "arts" broadcasting. Today many people who could be helpful to television are losing interest and are saying the same things and adopting the same attitude as the intellectuals twenty-five years ago did with regard to radio. They're saying that TV will never be anything more than a neighborhood movie house and why must they expect to get all they need from television, why can't they go and read a good book or just resume their other interests, whatever they may be. I just feel that if the leadership of the country does despair of television and refuses to stay in the arena, then it can't complain about what happens in this medium.

Q: I suspect that some of the sharper television critics on the staffs of newspapers have abandoned the arena, too. I can't say that I blame John Crosby of the *Herald Tribune* for leaving the field, but the fact is he did leave it.

GOULD: That's what's happening, you see. Johnny has left; others around the country have left. I think this is exactly what television would like—for all of us simply to drop out so they could go on their merry way.

Q: I'd never say that Crosby has a personal obligation to stay on as a television critic, but I would think that if he decides to get out, his paper should do everything it can to replace him with another critic of comparable acumen.

GOULD: Yes, and other papers around the country have an obligation, it seems to me, to take a critical, constructive, intelligent approach to television.

Q: What effect have you had through your writing on the thinking of the television industry? What effect have any of the newspaper critics had on the industry? Have you been able to discern any impact on any of these people?

GOULD: Yes, I think we've had some impact. An extreme case occurred in 1956 at the time of the Suez crisis when the whole world was blowing up and when not one of the three networks which were within a few blocks of the United Nations bothered to televise what was going on there. That was a most chilling thing. A local station, WPIX, did carry some of the UN proceedings and did quite a good job. But on that occasion I screamed my head off about the networks' lack of responsibility. NBC started answering the criticism over the air. John Daly, who was at ABC, got on the air and said that the broadcasters were all good boys even though they hadn't done a good job in this case. But the upshot was that within the next few days all the networks got busy and began covering the UN. A lot of other critics got on the networks for the same thing at that time. But certainly where a contract is signed for a popular entertainment show for, say, twenty-six weeks, criticism isn't going to take that program off the air.

Q: I was wondering whether critics can affect the attitude of those in the industry.

GOULD: Over-all, I think, yes. Most of the fellows in broadcasting are extremely perceptive; they know they've made a mistake and they can almost forecast what the critics are going to say the next day. But I think they lack confidence in their own judgment. As the television business has become more and more competitive, they have become more and more nervous. I remember I had one little line in one of my pieces about a small mishap in a commercial and the advertising agency people had an all-day conference about it. I think that's absurd.

Q: They might be nervous about the criticism, but I wonder whether they're moved to do anything about changing the major pattern of their programming. I gather from what you have said that you don't think too highly of the British approach which separates the sponsor from the content of the program.

GOULD: I would have some hopes in a situation where the sponsor's control truly stopped with the actual advertising—I question if it does on British commercial TV—but I don't see who could or would start that kind of broadcasting. Frank Stanton, in the wake of the quiz scandals, halfway toyed with the idea of devoting one night a week to a kind of magazine concept of programming on CBS. CBS would produce the whole thing and would simply drop in spot commercial announcements. The sponsor would pay for the announcements but would have nothing to say about the programming. But that idea died pretty fast, because let's say a sponsor wants Thursday night and if that's the night for CBS's "magazine of the air" and if the sponsor doesn't like that kind of a set-up he can just go over to Oliver Treyz at ABC. Or maybe Treyz has already been over to see him and tells the sponsor he doesn't have to worry about "all that jazz," because "over here" he can get the same old commercial stuff and a big audience. You see, you have no floor under such a thing. How do you raise standards if everybody else involved in the same industry doesn't have to follow suit?

Q: So far as getting things improved and corrected is concerned, I keep coming back, in my own mind, to this question of how much Congress and the government can do if the people themselves are not aroused or concerned about the matter.

GOULD: This really goes back to the fundamental concept of the American character which the Fund for the Republic is studying. Do we just drift and don't give a damn? Do we just accept what we have? Or do we care? Do we want to do something about it?

Q: I'm not trying to put the major responsibility on the critics, but I would think that people like yourself can do a great deal to help form and shape and sharpen the public taste for quality television. If more critics had more penetration in their criticism, if they were more perceptive and thoughtful, and if in the course of their review of a program they explained to their readers in some precise fashion why a particular program was good or bad, and if they occasionally wrote a longer piece, or essay, in which they examined the nature and function of television, what Kerr calls its "form," and then related the present performance of the television industry to its proper function and form, I think the public would gradually begin to develop an intelligent, insistent demand for better programming from the industry. Too many newspaper critics tend, I think, to identify themselves with the industry; when they say "we," they mean themselves and the industry, not themselves and the public. And I think too many merely talk about what was on the air the night before; they talk about it very uncritically. They're narrating, not criticizing or examining.

GOULD: I think that is true all across the country today. I think the newspapers have a deep obligation to appraise and judge all the phenomena of American life, including television. I remember the faculty at one college rising up in wrath because of a proposal that a study of communications be included in the curriculum. If such an attitude prevails, then the best minds won't contribute anything to communications. So far as the papers are concerned, most don't criticize radio, television, concerts, books, or movies. I wonder what they're in business for. They deluge you with words, but there's very little constructive critical writing going on. Now I think many of the people inside and outside the television industry today acknowledge that the industry has pretty well hit bottom. All of us should be asking one question: "Why?"

A Comment by Harry S. Ashmore

It may be that there are a few studio engineers and bedridden invalids who have seen more television than has Jack Gould. In the nature of their entrapment, however, none of these could have sampled the variety that has been Mr. Gould's daily fare ever since the radio waves first took on an added dimension. Some of Mr. Gould's stung detractors may consider him a common scold, but no one can question his credentials as TV's pre-eminent professional viewer.

Mr. Gould's peers fall roughly into two classes. The majority, as he notes in this interview, tend to identify themselves with the TV industry, finding compensation for their boredom in the freeloading and glamour by association that attends any form of show business. A small minority take a gingerly, intellectual approach to the medium, viewing the small screen as from an aisle seat in the theatre. Only a hardy few, of whom Mr. Gould is not only the most conspicuous but the most accomplished, combine the function of reporter with that of critic, displaying as much pleasure in turning up a fact as in turning a phrase.

Considerable stamina is required by Mr. Gould's pulpit at the *New York Times*, from which he must pay close attention to at least the major television offerings season after frustrating season. He could not be faulted if he followed the lead of his erstwhile colleague of the *Herald Tribune*, John Crosby, and turned away in search of fresher air. But the evident fact is that Jack Gould cares deeply about what happens to television—cares in a fashion not readily discernible in the industry itself except in the case of an occasional and unusual beleaguered producer, director, or writer.

For his limited but influential audience Mr. Gould performs the invaluable function of keeping the continuing controversy over American broadcasting in reasonable focus. It is not easy. Partly because of the temper of the times, partly through the conscious efforts of their skilled fuglemen in the advertising agencies, the proprietors and managers of American broadcasting have managed to reduce one of the most urgent contemporary issues to a polar exercise in black and white, as over-simplified as the character of a TV cowboy. We are told, and most of us seem to believe, that our choice is between what we are now seeing on television and government censorship, accompanied by subversion of the free enterprise system, and violation of our democratic principles. If it is pointed out that, even so, there must be something wrong with an industry that can't produce an honest quiz show, we are enjoined to have faith in the self-corrective instincts of men who are making large fortunes out of the status quo.

If this appraisal seems exaggerated, I offer the testimony of an interested witness, Mr. Lester S. Clarke, director of research at the American Broadcasting Company's station in Los Angeles. Taking issue with a statement by Frank K. Kelly of the Fund for the Republic urging a more active role for the FCC, Mr. Clarke wrote:

> Notwithstanding your denial, I feel that in effect you are suggesting censorship of and direct control over program content through the FCC. Certainly we have all seen enough governmental intervention into our business economy and the effects on our free enterprise system. We, as an industry, have met the challenge of our obligation and will continue to bear the responsibility of quality programming.
>
> I would like to close with a quote from Mr. Leonard H. Goldenson, president, American Broadcasting—Paramount Theatres, Inc., which I feel aptly evaluates our responsibilities in broadcasting.
>
> "My experience in show business has taught me that there is no substitute for quality. Television, of course, is show business—with an important added dimension. Though a primary interest is entertaining the American public, we broadcasters—network and stations alike—bear the responsibility of informing as well as entertaining."

A week after Mr. Clarke thus admonished Mr. Kelly, the *Los Angeles Times* reported:

> The [ABC] network planned a program called *Discovery* which was described as "a new concept in TV programming for children—a series that would plumb the world around us, and present an infinite variety of entertaining and informative subjects for young viewers." Earlier this month the network's announced plans had to be cancelled because so few affiliated stations would agree to run the program.

What then, of the dimension beyond *Wagon Train?* The industry concedes that it is there, or should be. But Mr. Gould documents, and no one except the professional apologists seriously questions, the fact that the added dimension, as well as program quality, actually has been diminishing year by year, apparently in direct proportion to TV's financial success. Here is the heart of the matter: show business of a fairly trivial order can command a massive TV audience and sure-fire advertising sponsorship; programming that rises above the norm is at best a financial risk, since the TV industry has not found—and does not appear to be actively seeking—a way to merchandise a selective audience, even one that runs as high as ten million.

To suggest that this is not a tolerable situation is not to trample on the flag, or even to condemn the profit motive out of hand. There is no suggestion that Mr. Gould thinks it is wrong for broadcasters to make money; on the contrary he quite evidently wishes for the industry a prosperous and expanding future. I gather that his concern, like mine, is not so much with what TV is doing, but with what it is not doing. I have no disposition to halt the wagon train, fan away the gun smoke, and stop Maverick from dealing off the bottom of the deck. I have as much concern as the next man with keeping the economy viable through the rising sales of razor blades and proprietary drugs; I, too, wish everyone used a good deodorant. My intelligence is not unduly insulted by TV's programming, but it is outraged by the pious argument that the public interest is coincident with the existing effective monopoly on the air waves.

Nor am I unsympathetic to the industry's concern over the possible consequences of government intervention. I am with Mr. Gould to the last goose-bump when he says of the proposal for a publicly supported, BBC-model broadcasting system: "The idea of any Congressional committee having authority over the network just frightens me to death." Finally, I am by instinct as well as occupation a First Amendment man, and I would not see the Constitution's immunities denied to any part of the communications media. But I insist that consideration of what should be done about broadcasting must not be cut off, as the industry obtusely and so far successfully insists, with recognition of the valid limits on government action. This is not the end of the argument, but the beginning.

The government has had a regulatory role in broadcasting ever since it first became practical to transmit sound by radio. If the operating agency, the FCC, has never seemed to be quite certain as to its duties beyond that of traffic control, the agency's authority and responsibility are nevertheless real. Only the FCC, for example, can provide the remedies discussed hopefully by Mr. Gould: more broadcast channels in the present spectrum and ultimately a shift of all TV to the UHF band; a legitimate accounting of how the stations and networks are employing their valuable licenses in the "public interest, convenience and necessity"; development of subscription TV as a means of serving minority viewers at their own expense; support in a variety of ways for the languishing experiment in educational television. Nothing here smacks of censorship, and in the unlikely event of transgression by the bureaucrats we, the people, can safely rely on being alerted by a mighty chorus of electronically amplified watchdogs.

In sum, Mr. Gould is pleading for greater diversity in television. Far from being an assault on the sacred precepts of free enterprise, this could be a blueprint of salvation for private broadcasting; diversity means competition, and competition is the life of trade. Neither is he seeking to shackle the industry; on the contrary he is seeking ways and means of freeing TV's discouraged creative spirits from the dead hand of the common denominator. He reminds us that censorship is not the exclusive province of government; it occurs in a particularly insidious way when broadcasters conclude that their final obligation is to make everybody happy, or at least make nobody unhappy. Mr. Gould cites the generally good record of the networks in the sensitive area of news and public affairs, yet it is a reasonable assumption that for every aired program dealing with a subject in contemporary controversy a dozen are still-born in executive conference. There is an inevitable cumulative effect here. All of the network commentators of my acquaintance at one time or another have been diverted from more useful pursuits to plead with a sponsor or his advertising agent that a dozen letters from the Citizens Council of Sunflower County, Mississippi, do not constitute a national boycott. Performers, if they will pardon the expression, are not inspired to walk very often on this tightrope when sponsorship means the difference between mere affluence and a Hollywood-style income.

I admire Mr. Gould's passion and his persistence. I do not think we can afford to be complacent about television. This technological marvel is one of the great accomplishments of our time. In a historical moment, broadcasting has hooked up virtually the whole of a sprawling nation in a system that enables all of us to see the same sights and hear the same sounds, and in so doing it has pre-empted the leading role in mass communication. The potential here is as great as that of the invention of the printing press. Surely we ought to be able to employ this miracle to produce something more than a fast buck—something, say, on the order of a Gutenberg Bible, which, after all, turned out to be quite a bestseller.

26. How Big Is One?

by Edward Weeks

Mr. Weeks is editor of the Atlantic Monthly, *where this article first appeared in 1958. Copyright is held by the* Atlantic Monthly Company, *and all rights are reserved. It is reprinted with permission of the author and the publisher.*

My late friend, the French writer Raoul de Roussy de Sales, who knew America intimately, used to tease me about our infatuation with bigness. "It's in your blood," he would say. "When I listen to Americans talking on shipboard, or in a Paris restaurant, or here in New York, it is only a question of time before someone will come out with that favorite boast of yours—'the biggest in the world!' The New York skyline, or the Washington Monument, or the Chicago Merchandise Mart—the biggest in the world. You say it without thinking what it means." How right he was, yet until he prodded me about it, I had never realized that this was indeed our national boast. We take pride in being big, and in a youthful way we used to think that bigness was our own special prerogative. But now we know better; now we find ourselves confronted with nations or with groups of nations which are quite as big as we are and which have the potential of being considerably bigger. This calls for a new orientation; indeed, I think it might be timely if we examine this concept of bigness and try to determine how it has affected our private lives and our thinking.

We have been in love with bigness ever since the adolescence of our democracy. The courtship began on the frontier: the uncut virgin forests, so dense and terrifying; the untamed flooding rivers; the limitless prairies; the almost impassable Sierras—to overcome obstacles like these, man, so puny in comparison, had to outdo himself. He had to be bigger than Hercules. The English live on a small, contained island, the English humor is naturally based on understatement; but an American when he is having fun always exaggerates.

Our first hero of the frontier was a superman, Davy Crockett, who could outshoot, outfight, and outwoo anyone. One day he sauntered into the forest for an airing but forgot to take his thunderbolt along. This made it embarrassing when he came face to face with a panther. The scene is described in the old almanac, as Howard Mumford Jones says, "in metaphoric language which has all the freshness of dawn." The panther growled and Crockett growled right back—"He grated thunder with his teeth"—and so the battle began. In the end, the panther, tamed, goes home with Davy, lights the fire on a dark night with flashes from his eyes, brushes the hearth every morning with his tail, and rakes the garden with his claws. Davy did the impossible, and listening to the legends of his prowess made it easier for the little guy on the frontier to do the possible.

Davy Crockett had a blood brother in Mike Fink, the giant of the river boatmen, and first cousins in Tony Beaver and Paul Bunyan of the North Woods and Pecos Bill of the Southwest. They were ringtailed roarers, and everything they did had an air of gigantic plausibility. Prunes are a necessary part of the lumberjack's diet, and Paul Bunyan's camp had such a zest for prunes that the prune trains which hauled the fruit came in with two engines, one before and one behind pushing. "Paul used to have twenty flunkies sweepin' the prunestones out from under the tables, but even then they'd get so thick we had to wade through 'em up over our shoes sometimes on our way in to dinner. They'd be all over the floor and in behind the stove and piled up against the windows where they'd dumped 'em outside so the cook couldn't see out at all hardly. . . . In Paul's camp back there in Wisconsin the prunestones used to get so thick they had to have twenty ox-teams haulin' 'em away, and they hauled 'em out in the woods, and the chipmunks ate 'em and grew so big the people shot 'em for tigers." Only an American could have invented that build-up, and I am grateful to Esther Shephard for having recaptured the legend so accurately in her *Paul Bunyan.*

Texas, with its fondness for bigness, preferred the living man to the legend: it provided the space for men like Richard King, the founder of the King Ranch. Richard King's story as told by Tom Lea is Horatio Alger multiplied by a thousand. The son of Irish immigrants, he ran off to sea at the age of eleven; a river boat captain in his twenties, he came ashore, married the parson's daughter, bought 15,000 acres of desert at two cents an acre, and went into the cattle business. His close friend and adviser was Lieutenant Colonel Robert E. Lee of the Second United States Cavalry, and it was Lee who gave King what has come to be the family slogan: "Buy land; and never sell." The King Ranch has grown to 700,000 acres in Texas with big offshoots in Kentucky, Pennsylvania, Australia, Cuba, and Brazil, and those of us who dwell in cities and suburbia have developed a kind of Mount Vernon reverence for this vast domain. It is just about as big, we think, as a good ranch ought to be.

I entered publishing in the summer of 1923 as a book salesman in New York. As I look back over the thirty-five years of my working life, I recognize that a significant change has taken place in our business community. The motorcars which I used to covet as a young bachelor, the Stutz Bearcat, the Mercer, the Simplex, the Locomobile, the Pierce Arrow—all these beauties and hundreds of the lesser breeds, like the Hupmobile, the Maxwell, the Franklin, the Stanley Steamer, and the Moon—are museum pieces today. The beauty and the originality which went into their design have been melted down and vulgarized in the models of the five major companies which survive.

In the days I am speaking of, Mr. Potts was our family grocer, and he knew the exact cuts of roast beef and lamb which would bring joy to my father's heart, just as he was prepared for my mother's remonstrance when there was too much gristle. There used to be a family grocer, like Mr. Potts, in every American community. Then some genius in Memphis, Tennessee, came up with the Piggly-Wiggly, the first gigantic cash and carry where the customer waited on himself, and in no time there were chains of these supermarkets stretching across the country. Such consolidation as this has been going on in every aspect of business, and at a faster and faster tempo.

When I was a book salesman, an American book publisher who sold a million dollars' worth of his books in one year was doing quite a prosperous business. Today a publisher who sells only a million dollars' worth of books a year cannot afford to remain in business; he has to join forces with another and larger publisher so that their combined production will carry them over the break-even point.

In the nineteen-twenties almost every American city had two newspapers, and the larger ones had four or five, and there is no doubt that this competition for ideas, for stories, for the truth was a healthy thing for the community. Today most American communities are being served by a single paper.

Of the daily papers that were being published in this country in 1929, 45 per cent have either perished or been consolidated. This consolidation, this process of making big ones out of little ones, is a remorseless thing, and it may be a harmful thing if it tends to regiment our thinking.

We Americans have a remarkable capacity for ambivalence. On the one hand we like to enjoy the benefits of mass production, and on the other we like to assert our individual taste. Ever since the Civil War we have been exercising our genius to build larger and larger combines. Experience has taught us that when these consolidations grow to the size of a giant octopus, we have got to find someone to regulate them. When our railroads achieved almost insufferable power, we devised the Interstate Commerce Commission, and we eventually found in Joseph Eastman a regulator of impeccable integrity who knew as much as any railroad president. We have not had such good luck with our other regulatory agencies, as the recent ignoble record of the FCC makes clear. What troubles me even more than the pliancy of FCC commissioners to political pressure is their willingness to favor the pyramiding under a single ownership of television channels, radio stations, and newspapers. Isn't this the very monopoly they were supposed to avoid?

The empire builders, who were well on their way to a plutocracy, were brought within bounds by the first Roosevelt. Then under the second

Roosevelt it was Labor's turn, and in their bid for power they have raised the challenge of what regulations can be devised which will bring them to a clearer recognition of their national responsibility. In the not far future we can see another huge decision looming up: When atomic energy is harnessed for industrial use, will it be in the hands of a few private corporations or in a consolidation which the government will control? My point is that in the daily exposure to such bigness the individual is made to feel smaller than he used to be, smaller and more helpless than his father and grandfather before him.

I realize, of course, that twice in this century our capacity to arm on an enormous scale has carried us to victory with a speed which neither the Kaiser nor Hitler believed possible. But it is my anxiety that, in a cold war which may last for decades, the maintenance of bigness, which is necessary to cope with the USSR, may regiment the American spirit.

In his book, *Reflections on America*, Jacques Maritain, the French philosopher, draws a sharp distinction "between the spirit of the American people and the logic of the superimposed structure or ritual of civilization." He speaks of "the state of tension, of hidden conflict, between this spirit of the people and this logic of the structure; the steady, latent rebellion of the spirit of the people against the logic of the structure." Maritain believes that the spirit of the American people is gradually overcoming and breaking the logic of their materialistic civilization. I should like to share his optimism, but first we have some questions to answer, questions about what the pressure of bigness is doing to American integrity and to American taste.

Henry Wallace has called this the century of the Common Man. Well, the longer I live in it the more I wonder whether we are producing the Uncommon Man in sufficient quantity. No such doubts were entertained a century ago. When Ralph Waldo Emerson delivered his famous address on "The American Scholar" to the Phi Beta Kappa Society of Harvard in 1837, he was in a mood of exhilaration, not doubt, and he heralded among other things a change which had taken place in American literature. It was a change in the choice of subject matter; it was a change in approach, and it showed that we had thrown off the leading strings of Europe. Here is how he described it:

> The elevation of what was called the lowest class in the state assumed in literature a very marked and benign an aspect. Instead of the sublime and beautiful, the near, the low, the common, was explored and poetized. . . . The literature of the poor, the feelings of the child, the philosophy of the street, the meaning of household life, are the topics of the time. It is a great stride. It is a sign—is it not?—of new vigor when the extremities are made active, when currents of warm life run into the hands and the feet. . . . This writing is blood-warm. Man is surprised to

find that things near are not less beautiful and wondrous than things remote. The near explains the far. The drop is a small ocean. A man is related to all nature. This perception of the worth of the vulgar is fruitful in discoveries.

This change from the appreciation of the elite to the appreciation of the commonplace, or as Emerson called it, the vulgar, has been increasingly magnified under the pressure of numbers. But were Emerson able to return to us for a short visit, I am not sure that he would be altogether happy about what we have done to elevate the vulgar in literature or in television.

In contemporary literature, new books—the best we can produce—are still published in hard covers and sold to a discriminating body of readers. If I had to guess, I should say that there are about one million discriminating readers in this country today and what disturbs me as an editor is that this number has not increased with the population; it has not increased appreciably since the year 1920. What has increased is the public for comic books, for murder mysteries, for sex and sadism. This debasement, especially in fiction, was most noticeable in the early stages of our paperbacks, when the racks in any drugstore were crowded with lurid, large-bosomed beauties who were being either tortured or pursued. Recently there has been an improvement, both in quantity and in seriousness, thanks to the editors of Anchor Books and the New American Library, thanks also to a feeling of outrage which was expressed in many communities. But it still seems to me regrettable that after a hundred years of public education we have produced such a demand for the lowest common denominator of emotionalism.

Am I, I sometimes wonder, a minority of one when I shudder at certain photographs in our pictorial magazines? The picture of a Negro being lynched; the picture of an airliner which has crashed and burned, with that naked body to the left identified as an opera singer whose voice we have all heard and loved; the picture of a grieving mother whose child has just been crushed in an automobile accident? Am I a minority of one in thinking that these are invasions of privacy, indecent and so shocking that we cringe from the sight?

Television, for which we once had such high hope, is constantly betrayed by the same temptation. It can rise magnificently to the occasion, as when it brought home to us the tragedy in Hungary, yet time and again its sponsored programs sink to a sodden level of brutality, shooting, and torture. And is there any other country in the world which would suffer through such incredible singing commercials as are flung at us? Does the language always have to be butchered for popular appeal, as when we are adjured to "live modern" and "smoke for real"? Am I a minority of one in thinking that the giveaway programs, by capitalizing on ignorance, poverty, and grief, are a disgrace? These are deliberate efforts to reduce a valuable medium to the level of the bobbysoxers.

There was a time when the American automobiles led the world in their beauty, diversity, and power, but the gaudy gondolas of today are an insult to the intelligence. In an era of close crowding when parking is an insoluble problem, it was sheer arrogance on the part of the Detroit designers to produce a car which was longer than the normal garage, so wasteful of gasoline, so laden with useless chromium and fantails that it costs a small fortune to have a rear fender repaired. I saw in a little Volkswagen not long ago a sign in the windshield reading, "Help Stamp out Cadillacs!" There speaks the good-natured but stubborn resistance of the American spirit against the arrogance of Detroit.

Is it inevitable in mass production that when you cater to the many, something has to give, and what gives is quality? I wonder if this has to be. I wonder if the great majority of the American people do not have more taste than they are credited with. The phenomenal increase in the sale of classical music recordings the moment they became available at mass production prices tells me that Americans will support higher standards when they are given the chance. I stress the aberration of taste in our time because I think it is something that does not have to be. The Republic deserves better standards, not only for the elect, but straight across the board.

I wish that our directors of Hollywood, the heads of our great networks, and those who, like the automobile designers in Detroit, are dependent upon American taste—I wish that such arbiters would remember what Alexis de Tocqueville wrote a hundred and twenty-five years ago in his great book, *Democracy in America.* "When the conditions of society are becoming more equal," said Tocqueville, "and each individual man becomes more like all the rest, more weak and more insignificant, a habit grows up of ceasing to notice the citizens to consider only the people, and of overlooking individuals to think only of their kind."

It seems to me that our tastemakers have been guilty of this fallacy ever since the close of World War II. They have ceased to notice the citizens and consider only the people, just as Tocqueville warned. They no longer plan for the differences in individual taste, but think only of people in the mass.

In the years that followed the crash of 1929, Americans began to transfer their trust from big business to big government; if big business and banking, so ran the reasoning, could not be trusted to keep us out of depressions, perhaps big government could. Gradually in this emergency we began to shape up our version of the welfare state, a concept which was evolving in many parts of the Western world and to which both Democrats and Republicans are now committed.

A welfare state requires a big government with many bureaus, just as big government in its turn requires big taxes. We embarked on big government with the idea of safeguarding those segments of American society which were most in jeopardy, and now after twenty-five years of experimentation we are beginning to learn that the effects of big government upon the individual are both good and bad. It is good to provide the individual with security, and to give him the chance to adjust his special claims; another and perhaps unsuspected asset has been dramatized by Edwin O'Connor in his novel, *The Last Hurrah,* in which he showed us how President Roosevelt had diminished and destroyed the sovereignty of the city boss. It is Washington, not Ward Eight, that has the big patronage to give today.

The maleffects of big government are more subtle. Consider, for instance, the debilitating effect of heavy taxation. I remember a revealing talk I had with Samuel Zemurray when he was president of the United Fruit Company. Born in Russia, Zemurray made his start here by pushing a fruit cart through the streets of Gadsden, Alabama. Then he set up his own business as a banana jobber by selling the bunches of bananas the fruit company didn't want. He sold out to United Fruit and continued to acquire shares until he controlled the majority of the stock. In the autumn of FDR's second term, when we were sitting in adjoining Pullman seats on the long run to Washington, Mr. Zemurray began talking about the President's promises to "the forgotten man." "He made three promises," Zemurray said, "and he has kept two of them: the promise to labor and that to the farmer. The promise he has not kept is to the little businessman. Under today's taxes it would be quite impossible for a young man to do as I did—he would never be able to accumulate enough capital."

Some years after this talk, in 1946 to be exact, I was on a plane flying West from Chicago. It was a Sunday morning and the man who sat beside me at the window seat had the big bulk of the Chicago *Tribune* spread open on his knees, but out of politeness' sake he gave me the proverbial greeting, "Hello, where are you from?" And when I said, "From Boston," his face lit up. "Do they still have good food at the Automat?" he asked. "Boy, that's where I got my start and it certainly seems a lifetime ago." And then in a rush out poured his life story in one of those sudden confidences with which Americans turn to one another: How he had become a salesman of bedroom crockery, and his Boston boss had refused to raise him to thirty dollars a week. In his anger he had switched to the rival company, and under their encouragement he had simply plastered Cape Cod with white washbowls, pitchers, soap dishes, and tooth mugs. "Seven carloads I sold in the first year," he told me. The company called him back to its head office in Chicago, and then came the crash. The company owned a bank and lake shore real estate, and when the smoke had cleared away and recovery was possible, he found himself running the whole shebang. His wife hadn't been able to keep up with it all, he said, shaking his head sadly. He had had his first coronary, and what kept him alive today was his hope for his two sons, who had just come out of the Navy. "But, you know," he said to me, his eyes widening, "they neither of them want to come in with me. They don't seem to want to take the chances that I took. They want to tie up with a big corporation. I just don't get it."

Security for the greatest number is a modern shibboleth, but somebody still has to set the pace and take the risk. And if we gain security, but sacrifice first venture and then initiative, we may find, as the Labor Party in England did, that we end with all too little incentive. As I travel this country since the war, I have the repeated impression that fewer and fewer young men are venturing into business on their own. More and more of them seek the safety of the big corporations. There are compelling reasons for this, the ever-shrinking margin of operating profit being the most insistent. But if we keep on trading independence and initiative for security, I wonder what kind of American enterprise will be left fifty years from now.

A subtle conditioning of the voter has been taking place during the steady build-up of big government. During the Depression and recovery we took our directives from Washington almost without question; so too during the war, when we were dedicated to a single purpose and when the leadership in Washington in every department was the best the nation could supply. And for almost twenty years local authority and the ability to test our political initiative in the home county and state have dwindled. About the only common rally which is left to us is the annual drive for the community fund. Too few of our ablest young men will stand for local office. Their jobs come first, and they console themselves with the thought that if they succeed they may be called to Washington in maturity. We used to have a spontaneous capacity for rallying; we could be inflamed, and our boiling point was low. Our present state of lethargy, our tendency to let George do it in Washington, is not only regrettable, it is bad for our system.

I remember one of the last talks I had with Wendell Willkie. He was still showing the exhaustion of defeat, and he spoke with concern as he said, "One of the weaknesses in our democracy is our tendency to delegate. During an election year we will work our hearts out, and then when the returns are in, we think we have done our part. For the next three years what happens to the party is the responsibility of the national committeemen. Have you ever looked at them?"

The decision having been made to drop the atomic bomb on Hiroshima, President Truman tells us that he retired and slept soundly. But those in authority in these days are less sure. The delegation of so much authority to those in Washington and the difficulties of dealing with an opponent so ruthless and enigmatic as Russia seem to have developed in our most responsible officials a secretiveness and an uncertainty which make it hard for the citizen to follow. This administration has practiced a policy of nondisclosure toward the press and the electorate which has left the average citizen in a state of constant uncertainty. I have nothing but admiration for the dedication and stamina of Secretary Dulles, but I wish with all my heart that he had made our purpose and our commitments clearer for our allies and for our own people to understand. When we pulled that dam out from under Nasser's feet, we projected a crisis which must have come as a great shock to France and Britain. And how can we blame the young leaders of Hungary for misunderstanding the words "dynamic liberation" when we at home had no clear notion of what they meant? It was inexcusable not to have warned the American people that the Sputniks were coming and that greater exertions must be expected of us. This is no time for remoteness or for lulling slogans or for the avoidance of hard truths. The volume of material, the thousands of articles dealing with the great issues of today which are pouring into my office from unknown, unestablished writers, testifies to the conscientiousness and the courage of American thinking. The pity of it is that such people have not been taken more fully into the confidence of their own government.

I have said that the concept of bigness has been an American ideal since our earliest times. I pointed to our propensity to build larger and larger combines ever since the Civil War, and how the process of consolidation has speeded up during the past thirty-five years. I suggested that we cannot have the fruits of mass production without suffering the effects of regimentation. And I ask that we look closely at what the pressure of bigness has done to American taste and opinion. Is the individual beginning to lose self-confidence and his independence? In short, how big is one?

Surely, in an atomic age self-reliance and self-restraint are needed as they have never been before. See with what force Van Wyck Brooks expresses this truth in his *Writer's Notebook:*

> Unless humanity is intrinsically decent, heaven help the world indeed, for more and more we are going to see man naked. There is no stopping the world's tendency to throw off imposed restraints, the *religious* authority that is based on the ignorance of the many, the *political* authority that is based on the knowledge of the few. The time is coming when there will be nothing to restrain men except what they find in their own bosoms; and what hope is there for us then unless it is true that, freed from fear, men are naturally predisposed to be upright and just?

As we look about us, what evidence can we find that in an atmosphere overshadowed by Russia and made murky by the distrust of McCarthyism there are citizens who will still stand forth, upright and ready to speak the hard truth for the public good? How big is one?

One is as big as George F. Kennan, who believes that we cannot continue to live in this state of frozen belligerency in Europe. We do not have to accept all of his proposals before applauding his thoughtful, audacious effort to break up the ice.

One is as big as Omer Carmichael, the superintendent of schools in Louisville, Kentucky, who led the movement for voluntary integration in his border state; as big as Harry Ashmore, the editor of the Little Rock, Arkansas, *Gazette*, for his fearless and reasonable coverage of the Faubus scandal.

One is as big as Frank Laubach, who believes in teaching the underdeveloped nations how to read their own languages, and then in supplying them with reading matter which will aid them to develop their farming and health.

One is as big as Linus Pauling, Harold C. Urey, Robert Oppenheimer, and the other editors and sponsors of the *Bulletin of the Atomic Scientists*, who have never underestimated Russian scientific capacities, who have always believed in the peaceful value of scientific exchange and never ceased to struggle against fanaticism in secrecy and security.

One is as big as Edith Hamilton, the classicist, the lover of Greece and of moderation; and as Alice Hamilton, her younger sister, who pioneered in the dangerous field of industrial medicine.

One is as big as Sheldon and Eleanor Glueck, who for years have been guiding lights in the resistant field of juvenile delinquency.

One is as big as Ralph Bunche and Eleanor Roosevelt.

One is as big as Louis M. Lyons, whose interpretation of the news and whose judgment of the popular press have provided, in the words of the Lauterbach Award, "A conscience for a whole profession."

One is as big as I. I. Rabi, a brilliant scientist and a passionate humanist, who, on being asked how long it would take us to catch up to Russia and to safeguard our long-range future, replied, "A generation. You know how long it takes to change a cultural pattern. The growing general awareness of this need will help us, but nevertheless we will have to work hard to succeed in a generation."

One is as big as Frederick May Eliot, president for twenty-one years of the American Unitarian Association, who worked himself to the bone for the deepening of faith and for reconciliation.

One is as big as you yourself can make it.

27. Television: The Dream and the Reality

by Robert Shayon

Mr. Shayon is professor of communications, Annenberg School of Communications, University of Pennsylvania, and a radio-television critic and writer for The Saturday Review. *This speech was delivered by Mr. Shayon on November 16, 1959, in Milwaukee, Wisconsin. Mr. Shayon's appearance was cosponsored by the Marquette University Press Club and the Milwaukee County Radio and Television Council. It is reprinted by permission of the author.*

Dream and Reality is the title of a very enlightening book on American foreign policy by Louis J. Halle, published by Harper Brothers. It is a study of how our foreign policy has blundered along through the decades since the nation was founded, because the United States has dreamed that we were isolated by two oceans from the troubles of the rest of the world. In reality, however, we have, from the beginning, been inextricably bound up with the destinies of other nations.

While I was reading this book, the quiz show scandals were approaching their climax in Washington, and William H. Stringer, the Washington Bureau Chief of *The Christian Science Monitor,* wrote a front-page column suggesting an American BBC as a competitive alternative to our commercial networks, I was impelled to write the following letter to this newspaper. I congratulated Mr. Stringer for his article and I said:

> It is heartening to find a thoughtful journalist of a responsible newspaper at last coming around to the notion that perhaps a government sponsored alternative to the commercially sponsored broadcasting pattern would widen the spectrum of television fare available to the American people.
>
> The nation's experience with broadcasting has been marked by a conflict between dream and reality. The dream has been that private enterprise, in the traditional American fashion, could somehow solve the dilemma of how to harmonize the demands of the marketplace in TV-radio with the rights of the alleged cultural minorities and with the much greater and more important requirements of the whole nation's communications vision. Twenty-five years of experience with TV-radio since the Communications Act of 1934 set the pattern, have shown us the reality.
>
> Desirable as a self-regulated industry solution would be, it has proven to be inadequate, for compelling and obvious reasons. Opinion-leaders, however, have been reluctant to let go the dream and acknowledge the reality.

The general public, as a consequence, has never been led to an honest and searching consideration of possible competitors to commercial TV-radio but has always been kept dead-center on the road marked: "This is the best of all possible worlds." Mr. Stringer's forthright suggestion for an American BBC is the first grappling with the meaningful specifics of the problem that I have seen in a leading newspaper. It may be useful to your readers to know that in 1922 Herbert Hoover, then Secretary of Commerce, speaking at the first Annual Radio Conference in Washington, expressed what was then the fashionable attitude toward sponsored broadcasting. Said Mr. Hoover: "It is inconceivable that we should allow so great a possibility for service . . . to be drowned in advertising matter."

Congress, in 1934, when it was rewriting the Communications Act into its present form, was seriously concerned about maintaining non-sponsored types of broadcasting activity. It asked the FCC to study the proposal that 25 per cent of the available radio frequencies be reserved for non-profit broadcasting. The FCC rejected the proposal.

Congress, too, may have dreamed the great dream—that the evils inherent in government control were greater than the risk in the abilities and will of the commercial broadcasters to use their franchise for private gain and at the same time to serve well the public need.

The issue today is much deeper than service to cultural minorities. It is the pervasive erosion by the possibly best-intentioned but inevitably economic-bound broadcasters of vital qualities of mind and spirit. No broadcasting system could be perfect.

A government network is but one of several proposals which have been kept in obscurity in the wings of public awareness. Others include federal support of the burgeoning educational broadcasters, listener-viewer subscription support of stations, the recently-blocked pay-TV and a public authority to make an annual critical audit of broadcasting behaviour. But all of us must begin to think about possible modifications of the dead center present; and this we will not do generally and with influence, unless courageous reporters like Mr. Stringer show us the way.

I suggest to you that you are dreaming if you think that in the wake of the quiz exposé, television is bound to improve. The reality to which you should be alert is that the prospect for any real improvement is slim indeed. The dream is that you can relax. Congress, the sponsors, the networks, the agencies, the FCC, the public—everybody is going to take it from here. If you buy this dream, you are buying disappointment. The work of the listener councils is far from done. In truth, it has become more important than ever.

Let's look at the situation. The new Congress meets in 1960. The general impression left by Representative Oren Harris and his colleagues of the House Special Subcommittee on Legislative Oversight is that the TV industry has a period of grace in which to clean its own house. If it fails to do so, it will face legislation.

Will the industry clean its own house? NBC has promised to throw out the rascals who duped them in the quiz shows, but it wants to keep the quiz shows and assure their integrity by an NBC FBI. CBS abjures a trade Gestapo for creative artists, but promises to take matters more into its own hands, not only where honesty of claims and authenticity behind appearances are concerned, but also in the more difficult area of program balance—of getting some informational and cultural shows in prime evening time.

As I wrote in a recent *Saturday Review* piece, public concern about any future rigging of quiz shows is, as the humorist S. J. Perelman might say: "flogging a dead cat." The fix is out in TV—for the foreseeable future, at any rate. Public sensitivity to corruption seems to describe parabolic arcs. It rises from exposures of rottenness to heights of morality and sinks after a time to indifference again. There *are* more responsible elements in the broadcasting industry, who are genuinely hoping that they can use the present climate as an opportunity to win some real gains for broadcasting "in the public interest, convenience and necessity." But there are also elements who are genuinely hoping to ride out the storm. They expect the American public to lose interest, to switch attention to the next big headline story that comes along, to lapse again into apathy, boredom on the subject, or even to the feeling of "Well, there's no business like show business," and "What's wrong with a little fun anyway?"

These people will hedge, drag their feet, temporize as long as they can, hoping for the storm of public indignation to blow itself out. And just as in real life, there are not only "the good guys" and "the bad guys," there are also the "gray guys," the people in between, the people who may say as Paul said: "The good that I would I do not, and the evil I would not, that I do." These are the men and women enmeshed in TV's complex dilemmas. The real problem is not how to polish up the minor surfaces of television's blandishment and deception, but how to refashion its architecture so that the image it presents to thoughtful individuals is mostly positive instead of largely negative.

Let us assume that, in its period of grace, responsible elements in TV will want to do something about improving the situation to prevent legislation. What is the heart of the dilemma which they face?

The essence of the matter is "circulation." Circulation is what networks sell. Circulation is what the sponsor wants. Circulation is what is offered by the individual station owner who holds the licence. NBC's first advertising rates on radio were $120 per hour from 6 P.M. to 11 P.M. and $60 from 8 A.M. to 6 P.M. Now they are more than $50,000 per hour in the evening on TV and more than $25,000 an hour during the day. What brought about the jump? Increased circulation. *The Saturday Evening Post* sells circulation. So does *Life Magazine*. So do the *Reader's Digest* and the *Saturday Review*. The articles, the stories are merely means to achieve that circulation for the advertisers. It is the same in TV. TV is the biggest advertising medium of all because it can supply the largest circulation at cost per thousand. In order to achieve maximum circulation, the networks and the sponsors seek programs that will attract the greatest number of people at economically significant times. Even the broadcasters will agree that as you achieve greater circulation, you flatten out the curve of sophistication. You reach the commonality of most minds most of the time.

Now the popular mind is the least informed about the realities of the external world in which it lives: it is the most given to the gratification of the moment and to the impulses of the short run in life. It tends to be the most immature. All this is familiar broadcasting jargon; not even the commercial broadcasters deny it. They argue that you have to lead the common mind to the uncommon things in the life of the mind, but that you cannot do this without at the same time holding their attention—and to hold them you must follow them.

Paradoxically, following and leading at the same time is what the commercial broadcasters claim they are successfully doing for the good of the nation and, of course, for the good of the industry. There is no better broadcasting system in all the world, they tell us. People who speak of changing the system really wish to inflict the snobbery of the intellectually elite, a minority snobbery, on the healthy, normal, mass, democratic cultural illiteracy of the majority. Some broadcasters go further. They contend that any hope we may have for uplifting audiences culturally in this country depends directly on the maintenance of the commercial system. Eliminate the high-rated Westerns and the crime shows and the situation comedies, they say, and you destroy all significant potential for getting information and enlightenment and cultural uplift to the masses. A network executive urged this point very recently at a meeting of educators.

Now TV's big circulation periods, TV's bread and butter, are in prime evening hours. Last year, in his notable address before the Chicago meeting of the Radio and Television News Directors, Edward R. Murrow urged the sponsors to pick up part of the tab for the presentation of serious informational and cultural programs in prime evening hours. He acknowledged that the networks couldn't afford to pay the whole freight themselves. Shortly thereafter, Mr. Murrow left CBS for a year's sabbatical. Perhaps there was some connection between the lack of response to his appeal and his sabbatical. At any rate, give or take, more or less, there have been no more prime-time informational specials this year than in the year before.

Competitive mass advertisers are committed to circulation. They *cannot* lead and follow at the same time. They can only follow where the circulation leads them. They have no message for their audience except their sales message. In fact, they must go looking for a message in program terms. They must shop for a program. And in buying a program message for the common mind, they follow the conventional method of telling the common mind what it wishes to hear, giving it the familiar, the pleasant, the nondisturbing, the unsophisticated. A leader appeals to the common mind's awareness of the real world. A follower appeals to its dreams.

Now what are the sponsors saying about circulation, since the quiz scandals compelled them to make some public statements about the matter—something they have been traditionally reluctant to do? In Hot Springs, Virginia, the Association of National Advertisers met on November 9. This organization has 653 members, including the nation's leading television advertisers. They did not absolve sponsors from the blame of the quiz shows. They announced, "It is our responsibility to see that every aspect of television with which we are connected meets our obligation of fair play to the public." They spoke of an immediate inventory of advertising, including factual support for accuracy and the techniques used in its preparation. Then, according to the *New York Times,* someone introduced the subject of turning complete control of program content over to the networks.

If the networks were to have control, the sponsors would merely purchase ads in the same manner they do in newspapers and magazines, without having any voice in the shaping of the medium's content. Henry Schachte, executive vice-president of Lever Bros., was quoted as commenting: "You could get a heck of a debate going among people right here on this subject." Edwin W. Able, vice-president for advertising for the General Foods Corporation, asserted that certain conditions would have to be met by the networks before he would agree to such an arrangement. "Under the present method of television programming, the advertiser takes the financial risk," he added. "If the networks are to take over show selection, then we would want certain guarantees we do not now have." He indicated that such revisions might include some type of guarantee as to the size of the audience watching the sponsored shows. This would be comparable to the circulation guarantees offered to advertisers by newspapers and magazines.

What these sponsors were saying, in other words, was that they are buying circulation. They pay the going rates and take the risk of achieving circulation at present because they have a voice in the shows. But if they were to be denied this voice, if they were allowed merely to buy space, they would have to insist on circulation guarantees—the rates according to the circulation delivered, rebates and all. This means that if they sponsored a show and its rating fell below the network's guarantee, they would get part of their money back. Are networks in the business of returning money to advertisers any more than newspapers and magazines? Rates are based on guaranteed circulation. Will the networks willingly agree to cut back their profit potential in order to put on informational and cultural broadcasts in prime times? Networks are not only concerned with single time periods, say the half hour between 8:00 and 8:30, but also with the periods immediately before and after, with cycles of time. Networks have been known to reject a sponsor's chosen program because it lost part of the audience which it inherited from the preceding program.

Networks sell circulation—not fragmented, but "back to back" and as continuous as possible and as high as possible. On November 11, in the *New York Times*, there was this story, also from Hot Springs, where the National Advertisers were meeting:

> Advertisers are not interested in relinquishing their present role in television network programming. This was the consensus among leading company advertising executives interviewed here. . . . Although some major advertisers acknowledge that they are giving this proposed revised operation "some thought," virtually all said they wanted to stick with the present system.
>
> A major advertiser, who asked that his name not be used, said: "Actually, if you are running your company in an ethical manner, you should take more of a role in programming, rather than less."
>
> John Barlow, who is in charge of corporate advertising at the Chrysler Corporation, was emphatic in his insistence that there be no change in the present manner in which companies buy TV programming.
>
> "Surely there are a lot of things wrong," he said, "but getting the advertiser out of the picture is not going to help.
>
> "Our top executives are vitally interested in seeing that nothing in bad taste gets onto a program we sponsor and I am on hand at the network to see that necessary changes are made before the show goes on the air."
>
> Al Hollender, a vice-president of the Grey Advertising Agency, who was a speaker, denied that "advertisers control shows." "There is a big difference between involvement and control of a show," he asserted.
>
> "We think it is important for an agency and advertiser to be involved to make certain that they are getting a return in terms of the corporate image they want to project, in relation to the money they are spending."
>
> Advertisers generally contended that since they were bearing the financial burden for the shows, they should have a voice in what was put on the air.

The advertisers, then, apparently are willing to tidy up the house, but they are not willing to give up their circulation voluntarily. Thus far, we have been mentioning the networks and the sponsors, neither of whom have any legal responsibility under the law for what is broadcast, except for obscenity, libel, and lottery. What of the man who has that responsibility—the licensee, the individual station owner? In all the reams of copy written during and following the recent hearings, the local station owner was Mr. X, the missing man. No one called him as a witness. He made no statements to the general press. He was monitored, however, by the alert weekly tradepaper of show business, *Variety*. On November 3, while the Washington quiz show hearings were warming up to Charles Van Doren, the Broadcasters' Promotion Association convention met in Philadelphia. These are the gentlemen who publicize the programs which are broadcast locally. They develop the audiences in their local communities for what the networks originate. They are vital control-points of information. Newspaper readers in all the cities learn what's on the air mainly from their local listings and ads. The convention was addressed by Louis Hausman, director of the newly formed TIO, Television Information Office. This organization, in *Variety's* phrase, is the industry's "new propaganda wing." It's out to shape TV's corporate image more affirmatively in the public mind.

Hausman praised TV's over-all performance, labeled the quiz shows "a single, narrow area of programming," and spoke of television's good intentions. I quote now from *Variety:*

> At the same time, he took the station men to task for taking public affairs shows for granted. "Sometimes I think many of us are too prone to consider our public affairs and cultural programs as laudable in their way, but not very important. . . . There is no question that as far as the immediate profit and loss statement is concerned, these programs may not be as impressive a factor in your financial statement as the staples of entertainment."
>
> He asked for the re-examination of the amount of publicity being given to public affairs and other prestige shows.
>
> "Obviously, as far as network-originated programs are concerned, you can't promote these programs unless your station is carrying them. It is of little value for a network to create and originate programs of this kind if only a relatively few stations clear time for them."

The president of a regional network, at the Philadelphia convention, also said "the local stations must face up to the clearance problem on quality programs from the webs."

What was the reaction from the local station representatives to this inter-family conflict among the broadcasters? A *Variety* reporter quizzed the promotion men and he reported as follows:

> The talks and proposals centered around TV's dark hour of the quiz scandals seemed to stir wide disinterest among members of the Broadcasters' Promotion Association.
>
> The general feeling seemed to be, "I'm just out there at the local station helping to pump it through. Let the boys in New York worry about it" (which is exactly the way someone put it).
>
> The relative calm of the promotion men, compared to the sensitive concern of networks and Madison Ave. could stem from public reaction.

> Stations are getting mail on the quiz mess. Most of it is for forwarding to the webs. A good deal of it is asking for the return of axed shows. This, of course, may only be a reflection on the quality of the folks who write in.
>
> There's not much chance that promotion managers will be diverting funds from tight budgets in competitive situations to plug the small-audience public service and quality web shows, as was suggested. Few if any of the BPA members believed that better clearances for culture would in any way clear the muddy impression left by quiz fixing—prime time or any other time.
>
> The confession of Van Doren was a good conversation piece in the hospitality suites. But most of the promotion men were hoping to get away tomorrow with a couple of ideas that would raise their station's audience, not its image.

The circle of circulation is complete. The sponsors *must* have circulation. The networks have to plead with their affiliates to risk circulation in presenting informational and cultural programs, and the gentlemen at the local switches are worrying today about the same things they've always worried about—circulation. Before we point the finger at the circulation-minded broadcasters, let us ask ourselves what we would do in their place.

The situation was mournfully put in philosophical perspective in the Sunday *New York Times* of November 8, by that wise Washington observer, James Reston, who wrote on the editorial page:

> Charles Van Doren, brooding on the mysteries of life at his Connecticut farm this week-end, can scarcely be more puzzled or gloomy than the capital of the United States.
>
> There is an overwhelming feeling here that somehow we have lost our way. Nobody seems to know just how or why, but everybody feels something's wrong.
>
> It is not only the TV quiz scandal, but the steel strike that has given an impression of haphazard greed, and a system debased and out of balance.
>
> The problem is not primarily the weakness of a Charles Van Doren. It is that the struggle for power and money has become so savage that even the leaders of the institutions concerned are trapped in the system.
>
> Frank Stanton of CBS and Robert Kintner of NBC are high-minded men, but they lost control of the boys who will do anything for a fast buck. Dave McDonald of the steelworkers' union has to get more money for his men every time a contract ends or lose control of his union; or so he thinks. The leaders of big steel are fighting for bigger profits and control of their mills, and in the whole process the public is manipulated like a bunch of boobs.

How ironic! The nation looks to the TV industry for discipline, for self-restraint, when every other special interest group in the nation, caught in the competitive web, seems incapable of exercising self-restraint. In all this, television says: "We are young and immature. Give us time. We will grow up. Why should we be expected to behave differently from all the other media which worship circulation?" Why, indeed? Except perhaps because the air is federal and the spectrum is limited. The broadcasters constitute a quasi-public utility, using what belongs to all of us for their corporate gain. The press, to which they compare themselves in social responsibility, is not spectrum bound and does not use federal property. And the one thing the broadcasters are reluctant to recognize is the fact that as you feed the popular taste, you perpetuate it at the level you find it.

A revelatory key to the whole affair may be found in The Television Code of the National Association of Radio and Television Broadcasters. Under the heading of "Advancement of Education and Culture," paragraph 3 states: "Education via television may be taken to mean that process by which the individual is brought toward informed adjustment to his society." Now I ask: What is the precise meaning of "informed adjustment to society"? The word "adjust" suggests to me "being in harmony with," and the word "informed" means "knowing the facts, being educated and intelligent." Educated, intelligent, and in harmony with the prevailing order of things in television—"informed adjustment." Is that not an accurate description of the mental and moral experience on TV of an American citizen known as Charles Van Doren?

"Informed adjustment" means to me the commonality of ideas, the commonality of ideas means circulation—big circulation—and big circulation means commercial broadcasting. You must understand I speak more in sadness over the state of things than in condemnation. My entire adult life has been given to working in radio and television—as a writer, director, producer, and critic. I began before World War II in a strictly commercial phase of the industry. I produced wild comedy shows, participation shows, and quiz shows. As the war advanced upon us, I became involved in information and propaganda documentaries and bond drive shows. I experienced intensely and creatively the power of broadcasting for a nation gripped in the consensus of a global struggle for survival. Those were the days when Ed Murrow thrilled the United States with his nightly greetings from flaming England: "This is London." When the war ended, many hoped that broadcasting would maintain its war-time public affairs concern and go on to mature consideration of the realities of the post-war world. But the return was to circulation and the commonalities of broadcasting. TV did not change radio's patterns, it merely intensified them under the pressure of the most compelling circulation medium of all, sight and sound and color in the home.

As a parent of two little girls, I watched with growing concern the impact of TV's parade of violence, frivolity, and informed adjustment on my own daughters. I could merely wonder what its impact was on other children in homes where parents made no effort and had no capacity to counteract its influence. As a citizen, I observed the realities of the external world in Europe, Asia, and Africa, and contrasted them with the general vision of life in television's dream. A courageous program was. broadcast last week by CBS, called, *The Population Explosion.* It dealt with this grim problem of overpopulation chiefly in India. It was neither pleasant, popular, nor frivolous. It was an attempt to deal maturely with the spectre that stretches over our propensities for entertainment and luxury. It was presented in the East at the margin of the prime-time period. I wonder how many local stations across the nation carried it, even though it was sponsored courageously by B. F. Goodrich and General Electric.

The networks tell us that audiences who are concerned about such programs should be willing to watch them in the times they can be scheduled. This is the Carnegie Hall theory: that you can't expect to go into the public market of entertainment at your time and find good music. You must go to Carnegie Hall. Of course the obvious fallacy is that at the time someone else is going to the Roxy theatre, I can go to Carnegie Hall. The two forms are competitive. On television, they are not competitive for time. But this is more than a matter of serving the tastes of the alleged cultural minority. Does not the popular, dreaming mind need, for reality's sake, to know the potential violence residing all over the world in masses which are hungry? Eighty-five per cent of India's children (I believe the figure was) were going to sleep hungry every night—while an advertising agency man in New York was telling the producer of *The $64,000 Question* that the blonde who opened the doors of the isolation booths for contestants, "should look svelte and be dressed in white from head to foot, a look that should be long, thin and sleek and create talk." The best intentions of the broadcasters under the commercial system may be dreams—and the implacable reality may continue to be circulation.

So much for the possibilities of reform within the TV industry's own house. What are the realities underlying the talk of federal legislation to compel changes? The record of the FCC is plain, and so are its character and sympathies. At this moment, it is still wrestling with the question of whether or not it has the power to rule over matters of program content. According to a newspaper report on November 11, "the FCC plans to make a survey of its own powers . . . a broad, new inquiry into its authority to control programming." Let us assume it discovered, what many people think it should have found out a long time ago—that it does have powers in this respect. The rule-making procedure of the FCC is well known. It announces it is going to rule in a matter and invites testimony from all interested parties. This takes a few years. Then the FCC goes into conference and considers what to do. This takes more time. Then it announces its proposed rules and gives the industry an opportunity to comment on them. This takes still more time. Then it rules, and the possibilities of court action appealing its rule are wide open, as they surely would be if the commission ever ruled on drastic reform. How long would it take to get any change? And would it all be obscured by the march of bigger headlines anyway? The FCC files are filled with self-study material, including the celebrated Barrow report proposing licensing of the networks, which was never acted upon.

As for Congress, bills *may* be offered in the coming session. But when moral posturing gives way to sober considerations of introducing sanctions into the realm of traditional free enterprise, especially where free speech is allegedly concerned, there are likely to be second thoughts. And many Congressmen are themselves either owners of radio and television stations or connected with newspapers which in turn own stations. It may be noted that Representative Oren Harris, chairman of the House Committee on Legislative Oversight, which is running the present investigations, was the Congressman who introduced a resolution opposing a trial run for pay television. One newspaper stated recently, "There is evidence that the House Committee under Representative Harris is almost embarrassed by its success. . . ." It fired its energetic first counsel, Dr. Bernard Schwartz (who laid the basis for both the Adams and the FCC inquiries). Since then it has tended to concentrate on the immediately sensational rather than on the long-range inquiry into the regulatory agencies.

Nor is it only the traditional bias of Congress for keeping government out of the realm of ideas which is likely to slow down legislation of meaningful character. There is also the very real problem of writing good legislation in this highly charged business of the communication of ideas. Much of the early New Deal legislation ran into trouble in the Supreme Court not only because it represented a major shift to the left in American politics and economics, but also because, under the stress of the emergency situation which President Roosevelt and the Congress inherited—the need for getting things done quickly—the early big laws were poorly drawn with respect to the precision of their language. Can Congress decide what proper program balance should be in sponsored television? Does it have the necessary wisdom and experience in such difficult matters? Legislation reform will take time. To leave matters ambiguously worded in public utility operation is a tradition. The phrase "in the public interest, convenience and necessity" is an old legislative chestnut, not invented for the Communications Act but inherited from time-honored and multiple state laws. It is deliberately worded that way because Congress had no experience on which to base anything more specific. And the intensity of Congressional efforts is at all times related to public interest. Politicians tend to swim with the flood tides, not in the backwashes of public attention.

This is the reality. There will be resistance to change—strong and natural. Resistance of the powers of reason, vested interest, and emotion. And the general public, without the information or the compulsion to act, may continue in apathy or even in satisfaction with the present situation. TV ratings are just as big as ever. There is no sign that because of the quiz exposé, the public has cut down its regular viewing habits. But organizations like yours have accepted the job of leadership. You keep yourselves informed and you inform others. In spite of all the negatives I have deliberately sketched out, the public climate *has* changed with respect to TV. It can never be what it was before the storm broke.

Against this changed climate, your work will be inevitably more significant. And the public will be more receptive. But the times present a challenge to you, too. It is no longer enough to criticize the broadcasters in general terms. History, unexpectedly, has made your point for you, made it more spectacularly than you have ever dreamed it could be made. It is time you attempted, along with the rest of us, to decide on some answers. This is why your forthcoming summit meeting in spring can be a great contribution. From Santa Barbara, California, comes a report that the Fund for the Republic, in its Center for the Study of Democratic Institutions, has set up a project to study and appraise the mass media. There are other such projects going up. Ferment is great. But you people have been studying for a generation. You ought to be ahead of the parade. Precisely what do you want? You can help by making up your mind. And there are a few basic questions to which you must address yourself.

1. Are you willing to have the federal government enter the broadcasting picture—and on what terms? Are you prepared to accept the proposition that government sanctions will mean placing limits on the profits of the broadcasters?

2. Are you prepared to answer the charge that any cutback in national advertising over television seriously endangers our consumer-oriented economy? That more than ideas and entertainment is at stake, but also jobs, prosperity, taxes, and even national security?

3. Do you want a BBC rival to the commercial networks? Should the Communications Act be rewritten? Or do you wish merely to patch up the commercial networks?

4. How shall we go about upgrading the character of the FCC commissioners? By getting not only lawyers, but men and women of cultural distinction into the picture? Should this body be changed by law to separate its quasi-judicial authority in determining channel awards from its rule-making authority?

5. Do you wish Congress to give the FCC authority over the most important road block of all—namely, the use of the VHF and the UHF channels? We talk about limitations of channels, and yet the truth is that only about one-fourth of the more than two thousand channels available in

the United States are being used today. We do not have a true, nation-wide television service.

You must give leadership in deciding what to do about this. Hindsight has shown that the FCC made a mess of the original channel allocations. Millions are invested in present VHF licenses. Should these be taken away—and under what conditions—from present owners? The big obstacle is that receivers are not being manufactured in enough numbers to receive UHF as well as VHF stations. The FCC has authority to regulate the standards of transmitters, but not of receivers. Shall the FCC be given this authority? Shall manufacturers be required to build all sets with UHF as well as VHF? Can we take channels away from the military, who grabbed up the cream in the early days?

These are very real, unpublicized problems, but they stand in the way of getting the kind of television to which you are committed.

The country needs a master plan of its broadcasting service, in radio as well as TV. You must take a hand at leadership. The fault of the broadcasters is that they allow themselves to be directed by an opinion which they do not attempt seriously to direct. The job of the leader is not to harmonize his followers, to obtain a consensus among them, and then give expression in action. The job of the leader is to see that the *right* consensus is reached. What is the right consensus? I don't know. But I have some philosophical guideposts. When you sum it all up, I think that what I don't like about commercial broadcasting is that it addresses itself to what I am, instead of to what I could be. It assumes that I am smaller than my possibilities. It manipulates me instead of giving me the warm, human respect of addressing me from its complete moral and intellectual manhood. It lessens itself when it talks to me, and it lessens me generally.

In 1946, when American TV was still an infant gleam in General Sarnoff's eye, a textbook was published called *Here is Television—Your Window to the World*. The introduction was written by an educator. He agreed that when you watch television, "*You are there*, to all intents and purposes." But to what intent, to what purpose, he wondered. And he proposed as follows:

> Let it be our intent, in the words of Joseph Conrad, that ". . . one may perchance attain to such clearness of sincerity that at last the presented vision of regret or pity, or terror or mirth, shall awaken in the hearts of the beholders that feeling of unavoidable solidarity; of the solidarity in mysterious origin, in toil, in joy, in hope, in uncertain fate, which binds men to each other, and all mankind to the visible world."

This is a good dream. I thank you for working to make it a reality.

VI. Effects

Of all the contradictory claims regarding effects which are to be believed, how is the evidence of educational broadcasters (i.e., that television is a great teaching instrument) to be reconciled with the statements of many industry spokesmen that television does not teach crime techniques, anti-social behavior, etc., even while showing them?

Under all the claims and counterclaims there must be some common ground for discussion. For the basis of analysis, several thought-provoking statements are available. Those presented here appear to raise some of the most serious questions. For contrary evidence readers are referred to the readily available books and studies of Joseph Klapper, Wilbur Schramm, and other research scholars. Particular emphasis is drawn here to the *Summary Report on Findings and Recommendations of the Conference on Impact of Motion Pictures and Television on Youth*, originally presented in New York City in 1964, and "Human Dignity and Television," one of several position papers prepared by Dr. Arthur J. Brodbeck.

28. Puritanism Revisited: An Analysis of the Contemporary Screen-Image Western

by Peter Homans

Mr. Homans is Assistant Professor of Religion and Personality at the University of Chicago Divinity School. This article appeared in Studies in Public Communication *in 1961, published by the University of Chicago Press, and was popularized by* Look *magazine, where it appeared on March 13, 1962. It is reprinted with permission of the author.*

One of the most noticeable characteristics of popular culture is the rapidity with which new forms are initiated and older, more familiar ones revitalized. While narrative forms of popular culture, such as the detective story, the romance, and the soap opera, have generally been less subject to sudden losses or gains in popularity, the Western has within the last few years undergone a very abrupt change in this respect. Formerly associated with a dwindling audience of adolescents, who were trading in their hats and six-guns for space helmets and disintegrators, the Western has quite suddenly engaged an enormous number of people, very few of whom could be called adolescent.

This new and far-reaching popularity is easily established. Whereas before, the Western story was told from four to six in the afternoon, on Saturday mornings, in comic books, and in some pulp fiction, now it is to be seen during the choicest television viewing hours, in a steady stream of motion pictures, and in every drugstore pulp rack. At present, on television alone, more than thirty Western stories are told weekly, with an estimated budget of sixty million dollars. Four of the five top nighttime shows are Westerns, and of the top twenty shows, eleven are Westerns. In addition to this, it is estimated that women now compose one-third of the Western's heretofore male audience.

Such evidence invariably leads to attempts to explain the phenomenon. Here there has been little restraint in trying to analyse the unique status which the Western has gained. Some have suggested that it is the modern story version of the Oedipal classic; others find it a parallel of the medieval legends of courtly love and adventure; while those enamoured of psychiatric theory see it as a form of wish-fulfillment, and "escape" from the realities of life into an over-simplified world of good and evil.

Such theories, I suppose, could be described at greater length—but not much. They not only betray a mindless, off-the-top-of-the-head superficiality; they also suffer from a deeper fault characteristic of so many of the opinions handed down today about popular culture—a twofold reductionism which tends to rob the story of its concrete uniqueness.

This twofold reductionism first appears as the failure to attend fully and with care the historical roots of any form. For example, to say that the Western is a re-telling of chivalric tales is partly true. There is some similarity between the quest of the knight and the quest of the Western hero—they both seek to destroy an evil being by force. However, the tales of chivalry grew out of medieval culture, and any effort to account for them must consider their relationship to their culture. Similarly, the Western must be seen in relation to its culture—eastern American life at the turn of the century. To relate the two forms without first considering their historical contexts is what may be called historical reductionism.

The second form of reductionism is the failure of most theories to attend the unique details of the story which set it apart from prior forms. This can also be seen in the idea of chivalric tales retold. Holders of this theory notice that both heroes are engaged in a quest, the destruction of evil, and that they both earn some kind of special status in the eyes of the communities they have served. But what is not noticed is that the modern tale betrays an intense preoccupation with asceticism and colorlessness, while the medieval one dwells upon color, sensuousness, and luxury; or, that the medieval hero exemplifies tact, manners, elaborate ceremony, and custom, while his modern counterpart seeks to avoid these. Again, the Western rules out women; the older story would not be a story of chivalry did not women play an important part. The refusal to attend with care specific and possibly inconsequential details is a form of reductionism which may be called textual reductionism.

Both types of reductionism rob a particular form of possible uniqueness and independence. They force it to be merely a dependent function of some prior form, whatever that form may be. Together, they have become the two main errors which have obscured analysis of many present-day forms of popular culture.

However, these two foci are more than pitfalls to be avoided. The textual and historical aspects of any popular art form are the very points which should be scrutinized most carefully and elaborately. If these points are properly attended, they will yield the greatest insight into the meaning and significance of the story.

TEXTUAL ANALYSIS

Any effort to analyse a particular form of popular culture must begin with the problem of text. Each of us, in thinking and talking about the Western, has in mind an over-all understanding of it—an ordered vision of character, event, and detail shaped by all the hundreds of different versions which he has seen. Therefore, one must first set forth and defend precisely what it is he thinks the Western is, before indicating what it means. Indeed, disagreements as to meaning can often be traced to disagreements as to text.

But we cannot simply lump together everything that has ever happened in every Western, fearful of omitting something important. Nor can we refuse to include anything which does not appear in each and every version. For there are Westerns which omit details which all critics would agree are characteristic of the story, just as there are others which include details which all would agree are of no consequence. The task consists in selecting, from the endless number of Westerns we have all seen, a basic construct of narrative, character, and detail which will set forth clearly the datum for subsequent analysis. This critic's basic construct can be set forth as follows:

BACKGROUND

The Western takes place in a stark, desolate, abandoned land. The desert, as a place deprived of vitality and life as we know it, is indispensable. The story would not be credible were it set in an equatorial jungle, a fertile lowland, or an arctic tundra. As the classical versions have told us again and again, the hero emerges from the desert, bearing its marks, and returns to it. Already we are instructed that our story deals with a form of existence deprived of color and vitality.

This desert effect is contradicted by the presence of a town. Jerry-built, slapped-together buildings, with falsefronts lined awkwardly along a road which is forever thick with dust or mud, tell us that the builders themselves did not expect them to endure. And of these few buildings, only three stand out as recognizable and important—the saloon, the bank, and the marshal's office (hero's dwelling). Recent Westerns have added stores, courthouses, homes, and even churches. But for the classical versions such contrived togetherness has never really been necessary.

The saloon is by far the most important building in the Western. First of all, it is the only place in the entire story where people can be seen together time after time. It thereby performs the function of a meeting-house, social center, church, etc. More important, however, is its function as locus for the climax of the story, the gunfight. Even in today's more fashionable Westerns, which prefer main street at high noon, the gunfight often begins in the saloon, and takes place just outside it.

The bank, we note, is a hastily constructed, fragile affair. Poorly guarded (if at all), it is an easy mark, there for the taking. Its only protection consists of a snivelling, timid clerk, with a mustache and a green eyeshade, who is only too glad to hand over the loot. Has there ever been a Western in which a robber wondered whether he could pull off his robbery? There is a great deal of apprehension as to whether he will elude the inevitable posse, but never as to the simple act of robbery. The bank is surprisingly unprotected.

The marshal's office appears less regularly. Most noticeable here is the absence of any evidence of domesticity. We rarely see a bed, a place for clothes, or any indication that a person actually makes his home here. There is no mirror, an omission which has always intrigued me. The over-all atmosphere is that of austerity, to be contrasted sharply with the rich carpeting, impressive desk, curtains, pictures, and liquor supply of the saloon owner or evil gambler. Such asceticism is not due to the hero's lack of funds or low salary; rather, because of his living habits, there is no need of anything else. Indeed, we are led to suspect that such austerity is in some way related to our hero's virtue.

The town as a whole has no business or industry. People have money, but we rarely see them make it. And we are not concerned as to how they got their money—unless they stole it. This town and its citizens lead a derivative, dependent existence, serving activities which originate and will continue outside the town. It is expendable, and will disappear as soon as the activities it serves no longer exist.

Home life, like economic life, is conspicuous by its absence. There simply are no homes, families, domestic animals, or children. The closest thing to a home is a hotel, and this is rarely separated from the saloon. Recent Westerns have included homes, along with cozy vignettes of hearth, wife, kitchen, etc. Such innovations do little more than indicate how harassed script writers have become, for these scenes do not contribute to the basic action and imagery of the story. Classically, home life in the Western simply isn't.

SUPPORTING PEOPLE

As in any good form of popular culture, the number of important people is small. Such people I prefer to call "types." A type is an important figure recurring again and again, whose basic actions and patterns of relationship are relatively enduring from one version of the story to another. The particular vocation, clothing, mannerisms, personal plans, names, are all conventions—concessions to plausibility—which seemingly identify as new someone we know we've seen before. Such conventions I would like to call "role." When we refer to a particular person in a story with the preface "the"—e.g., "the" hero, or "the" good girl—we have penetrated beyond the role and identified a type.

One of the most interesting types is the "derelict-professional." He is one who was originally trained in one of the traditional eastern professions (law, medicine, letters, ministry), but who has, since his arrival in the west, become corrupted by such activities as drink, gambling, sex, or violence. Most celebrated is Doc Holliday, who trained in the east as a dentist, then came west to practice medicine whenever he was sober enough to do so. The derelict-professional sometimes appears as a judge or lawyer; sometimes as an ex-writer; in other instances he is a gun-toting preacher. The point is the same: the traditional resources of society (healer, teacher, shepherd, counselor) cannot exist in an uncorrupted state under the pressures of western life.

Somewhat similar is the "nonviolent easterner." He often appears as a well-dressed business man, or as a very recent graduate of Harvard, although the roles, as always, vary. Constantly forced to defend himself, he is simply not up to it. Indeed, he is usually thrashed shortly upon his arrival in town. Sometimes this is so humiliating that he tries to become a westerner. It never works. He is either humiliated even more, or killed. Another role for this type is the pastor (a recent addition) who, when the chips are down has only a prayer to offer. The east, we soon note, is incapable of action when action is most needed.

The "good girl" is another supportive type. Pale and without appetites, she too is from the east. Classically represented as the new schoolmarm, she also appears as the daughter of a local rancher, someone en route to a more distant point, or the wife of a cattleman. She has her eye on the hero. While any dealings between them come about as the result of her initiative, she is rarely flirtatious or coy. She does not allow any feminine allure to speak for itself—surely one reason why she ends up doing most of the talking. The good girl fails to understand why men have to drink, gamble, punch and shoot each other, and she spends a good deal of time making this point to the hero. Usually she has some kind of protection—brother, father, fiancé, or relative—which makes it possible for her not to work. She is never independent, out in the world, with no attachments.

The "bad girl" is alone in the world, unattached, and works for her living, usually in the saloon as a waitress or dancer. She too has her eye on the hero, attracting him in a way her counterpart does not. She is often flirtatious and coy, but rarely takes the initiative in their meetings. She doesn't try to make him put away his guns and settle down. She is friendly with other men, and, like her counterpart, is unhappily stalemated in her relation to the hero.

The "attendant" is another type. The most enduring and easily recognizable role for this type is the bartender, although the snivelling bank clerk is a close second. The attendant observes the action, provides the instruments of it, but never becomes centrally involved with it. Like a child following adults from room to room, he remains passive, deferring again and again to the principals, performing the important function of appearing unimportant.

One final type, of which there are many—"the boys," those bearded, grimy people who are always "just there," drinking and gambling in the saloon, without any apparent interest in anyone or anything, except their cards, whiskey, and the occasional songstress. Their function is that of an audience. No hero ever shot it out with his adversary without these people watching. Isolated conflicts between hero and adversary are always postponed—sometimes at considerable inconvenience to both—until the "boys" have had a chance to gather. The "boys" are passive functions of the action, important primarily for their presence.

PRINCIPALS AND ACTION

The action of the screen-image Western takes place in three phases: the opening, the action, and closing phases; or, everything before the fight, the fight, and everything after the fight.

The opening phase first of all introduces us to the story's setting, to the supporting types (through their roles), and principals. In doing so, however, it not only supplies us with information, but also provides the very important illusion that we are to see for the first time something which we know, in the back of our heads, we have seen many times before. It is important to believe that we are not idiots, watching the same story night after night.

Secondly, the opening phase prepares us for the action by delineating the hero. He is, first of all, a transcendent figure, originating beyond the town. Classically, he rides into town from nowhere; even if he is the marshal, his identity is in some way dissociated from the people he must save. We know nothing of any past activities, relationships, future plans, or ambitions. Indeed, the hero is himself often quite ambiguous about these. There are no friends, relatives, family, mistresses—not even a dog or cat—with the exception of the horse, and this too is a strangely formal relationship.

His appearance further supports this image. In the pre-action phase the hero sets forth a contrived indolence, barely distinguishable from sloth. Lax to the point of laziness, there appears to be nothing directional or purposeful about him. Take that hat, for instance: it sits exactly where it was placed—no effort has been made to align it. His horse is tied to whatever happens to protrude from the ground—and remains tied, although little more than a lazy nod would free it. Clothes and gunbelt also betray the absence of any effort towards arrangement and order. With feet propped up on the hitching rail, frame balanced on a chair or stool tilted back on its two rear legs, hat pushed slightly over the eyes, hands clasped over the buckle of his gunbelt, the hero is a study in contrived indolence.

I have used the word "contrived" to indicate another quality—that of discipline and control—which remains latent, being obscured by apparent laxity. His indolence is merely superficial, and serves to protect and undergird the deeper elements of control which will appear in the action phase. Now he has time on his hands; but he knows his time is coming, and so do we.

The hero's coupling of laxity and control is seen in those recurrent primary images which are ordinarily referred to simply as "typical scenes." With women there is no desire or attraction. He appears somewhat bored with the whole business, as if it were in the line of duty. He never blushes, or betrays any enthusiasm; he never rages or raves over a woman. His monosyllabic stammer and brevity of speech clearly indicate an intended indifference. In the drinking scenes we are likely to see him equipped with the traditional shot-glass and bottle. The latter becomes his personal property, and therefore he is never questioned as to how many drinks he has taken. We rarely see him pay for more than one. While drinking he usually stares gloomily at the floor, or at all the other gloomy people who are staring gloomily at each other. He gulps his drink, rarely enjoys it, and is impatient to be off, on his way, hurrying to a place we are never told about. In the gambling scenes his poker face is to cards what his gloomy stare was to drink—a mask serving to veil any inner feelings of greed, enthusiasm, fear, or apprehension. We note, however, that he always wins, or else refuses to play. Similarly, he is utterly unimpressed and indifferent to money, regardless of its quantity or source, although the unguarded bank is always just around the corner.

The action phase opens with the threat of evil, and extends up to its destruction at the hands of the hero. Although evil is most often referred to as the "villain" or "bad guy" or "heavy," I prefer the terms "evil one" or "adversary."

Of the many hundreds of seemingly different versions, each is unshaven, darkly clothed, and from the west. Little is known about him. We are not told of his origins, his relationships, habits, or customs. Like the hero, he is from beyond the town, rather than identified with the interests, problems, and resources which characterize it. All details of his personal life are withheld. We can only be sure that the evil one unhesitatingly involves himself in the following activities: gambling, drink, the accumulation of money, lust, and violence. They are his vocation; with respect to these, he is a professional man. It should be noted, however, that he is inclined to cheat at cards, get drunk, lust after women who do not return the compliment, rob banks, and finally, to shooting people he does not care for, especially heroes.

The impact of this evil one on the town is electric, as though a switch had been thrown, suddenly animating it with vitality, purpose, and direction. Indeed, it is evil, rather than good, which actually gives meaning to the lives of these people—his presence elicits commitment to a cause. The townsfolk now share a new identity: they are "those who are threatened by the evil one." Unified by a common threat, the town loses its desolate, aimless quality. It becomes busy. Some hasten to protect others; some to protect themselves; some run for help; some comment fearfully. Nevertheless, they all know (as do we) that they are of themselves ultimately powerless to meet this evil. What is required is the hero—a transcendent power originating from beyond the town.

Notice what has happened to this power. Gone are the indolence, laxity, and lack of intention. Now he is infused with vitality, direction, and seriousness. Before, the most trivial item might have caught his attention; now, every prior loyalty and concern are thoroughly excluded—he drops everything—in order that he may confront with passion and single-mindedness this ultimate threat. Once this radical shift has been accomplished, the hero (and audience) are ready for the final conflict—the central part of the action phase, the climax of the story.

While the fight can take many forms (fist-fight, fight with knives, whips, etc.—even a scowling match in which the hero successfully glares down the evil one), the classical and most popular form is the encounter with six-guns. It is a built-up and drawn-out affair, always allowing enough time for an audience to gather. The two men must adhere to an elaborate and well-defined casuistry as to who draws first, when it is proper to draw, when it is not, etc. The climax also reflects much of the craft of gunplay, of which both hero and evil one are the skilled artisans (cross-draw versus side-draw, fanning versus thumbing, whether two guns are really better than one, etc.) While these issues are certainly not the main concern of the action, the prominence given them by the story as a whole tends to prolong the climax.

Although the hero's presence usually makes the fight possible—i.e., he insists on obstructing the evil one in some way—it is the latter who invariably attacks first. Were the hero even to draw first, the story would no longer be a Western. Regardless of the issues involved, or of the moral responsibility for what is to follow, the hero's final, victorious shot is always provoked by the evil one. With the destruction of the evil one, the action phase is completed.

In the closing phase the town and its hero return to their pre-action ways. The electric quality of alarm and the sense of purpose and direction recede. People come out of hiding to acclaim their hero and enjoy his victory. He too returns to his pre-action mode of indolence and laxity. At such a moment he is likely to become immediately absorbed in some unimportant detail (like blowing the smoke from his gun), indicating for all to see that he has survived the crisis and is once again his old self.

One more event must take place, however, before the story can conclude. The hero must renounce any further involvement with the town which his victory may have suggested. In some way the town offers him the opportunity to identify with it, to settle down. Traditionally, this means marrying the schoolmarm and settling down. The hero always refuses. He cannot identify himself with the situation he has saved. He forfeits any opportunity to renounce his "beyond the town" origin and destiny. When this forfeiture has been made clear, when both savior and saved realize that it cannot be abrogated, then the story is over.

ANALYSIS

The Western is, as most people by this time are willing to acknowledge, a popular myth. And by myth I mean three things. First of all, it is a story whose basic patterns of character, plot, and detail are repeated again and again, and can be so recognized. Secondly, the story embodies and sets forth certain meanings about what is good and bad, right and wrong—meanings regarded as important by those who view and participate in the myth. And thirdly, some of these meanings are veiled by the story, so that one can affirm them without overtly acknowledging them. Some part of the story (or all of it, perhaps) serves to conceal something from the participant—i.e., there is an unacknowledged aspect to the story. There is, therefore, an embarrassing question which never occurs to those in the sway of the myth—the posing of which is precisely the critic's most important task.

The meanings which the Western sets forth center upon the problem of good and evil. Evil, according to the myth, is the failure to resist temptation. It is loss of control. Goodness lies in the power and willingness to resist temptation. It is the ability to remain in the presence of temptation and yet remain in control of one's desire. Five activities make up the well-known content of temptation: drinking, gambling, money, sex, and violence.

Whenever any one of these activities appears it should be seen as a self-contained temptation episode. Such an episode first of all presents an object of temptation which can be indulged, should the hero so choose; and secondly, it sets forth the hero in such a way that he can indulge the temptation in a preliminary way without becoming absorbed in it—i.e., without losing control. And, of course, it sets forth the evil one in precisely the opposite way.

In the drinking scenes the hero possesses not one drink, but a whole bottle—i.e., he has at his disposal the opportunity for unlimited indulgence and its consequent loss of self-control. Gambling is a situation over which one has rather limited control—you can lose; but the hero does not lose. He wins, thereby remaining in control (cheating simply signifies the failure to acknowledge loss of control). Wealth is not seized although it is available to him through the unguarded bank; and both good and bad girl seek out the hero in their various ways, but to no avail—he remains a hero. However, each temptation is presented in its peculiar way in order to set forth hero and evil one in their respective functions.

The temptation to do violence is more problematic, so much more so that the climax is given over to its solution. Furthermore, in the climax we find the key to the meaning of the myth as a whole—i.e., it can tell us why each type appears as he does, why the temptation episodes have their unique shape, and why certain fundamental images recur as they do.

We perceive in the evil one a terrible power, one which cannot be overcome by the ordinary resources of the town. However, he has acquired this power at great price: he has forfeited that very control and resistance which sustains and makes the hero what he is. The evil one represents, therefore, not temptation, so much as "temptation-unhesitatingly-given-into." He is the embodiment of the failure to resist temptation; he is the failure of denial. This is the real meaning of evil in the myth of the Western, and it is this which makes the evil one truly evil. Because of this he threatens the hero's resistance (and that of the townsfolk, as well, although indirectly): each taunt and baiting gesture is a lure to the forfeiture of control. This temptation the hero cannot handle with the usual methods of restraint, control, and the refusal to become absorbed; and it leads to a temptation which the hero cannot afford to resist: the temptation to destroy temptation.

The evil one's dark appearance is related to this threat. It tells us two things. First, that to lose control and forfeit resistance is (according to the story) a kind of living death, for black signifies death. In terms of the moral instruction of the story, and speaking metaphorically, we know that the evil one has "lost his life." But his black appearance also tells us that, speaking quite literally, this man will die—because of what he is, he must and will be executed. We are therefore both instructed and reassured.

The embarrassing question can now be posed: why must the hero wait to be attacked, why must he refrain from drawing first? Why does he not take his opponent from behind, while he is carousing, or while he is asleep? Anyone in the power of the myth would reply that the gunfight takes place the way it does because this is the way Westerns are; it's natural; this is the way it's always done—or, in the language of the myth itself, it was self-defense. But if one moves beyond the grasp of the myth, if one is no longer loyal to its rules and values, the gunfight is never inevitable. The circumstances which force the hero into this situation are contrived in order to make the violent destruction of the evil one appear just and virtuous. These circumstances have their origin in the inner, veiled need to which the story is addressed. This process, whereby desire is at once indulged and veiled I call the "inner dynamic." It is the key to the Western, explaining not only the climax of the story, but everything else uniquely characteristic of it. What is required is that temptation be indulged while providing the appearance of having been resisted.

Each of the minor temptation episodes—the typical scenes setting forth hero and evil one as each encounters drink, cards, money, and sex—takes its unique shape from this need. Each is a climax-less Western in itself, a play within a play in which temptation is faced and defeated, not by violent destruction, as in the climax, but by inner, willed control. Or, reversing the relationship, we may say that in the gunfight we have writ largely something which takes place again and again throughout the story. It is precisely for this reason that no Western has or needs to have all these episodes. Therefore Westerns can and do depart radically from the composite picture described earlier. We are so familiar with each kind of temptation, and each so reenforces the others that extraordinary deletions and variations can occur without our losing touch with the central meanings.

The inner dynamic affects the supporting types as well. The derelict-professional is derelict, and the nonviolent easterner is weak, precisely because they have failed to resist temptation in the manner characteristic of the hero. Their moderate, controlled indulgence of the various temptations does not conform to the total resistance of the hero. Consequently they must be portrayed as derelict, weak, and deficient men, contrasting unfavorably with the hero's virtue. In this sense they have more in common with the evil one.

Because these two types both originate in the east, they have something in common with the good girl. We note that everything eastern in the Western is considered weak, emotional, and feminine (family life, intellectual life, domestic life, professional life). Only by becoming western-ized can the east be redeemed. The Western, therefore, is more a myth about the east than it is about the west: it is a secret and bitter parody of eastern ways. This is all the more interesting, since it was originally written in the east, by easterners, for eastern reading. It really has very little to do with the west.

Woman is split in the Western to correspond to the splitting of man into hero and evil one. Primarily, however, the double feminine image permits the hero some gratification of desire while making a stalemate ultimately necessary. To get the good girl, the story instructs us, our hero would have to become like those despicable easterners; to get the bad girl, he would have to emulate the evil one. In such a dilemma a ride into the sunset is not such a bad solution after all.

The attendant sets forth the inner dynamic by being infinitely close to the action (temptations) while never becoming at all involved in it. It is his task to provide the instruments of temptation (drink, money, cards, guns) while never indulging them himself. He is at once closer to temptation than any other type, and yet more removed than any other type.

The boys function to facilitate the action without becoming involved in it. Without them hero and adversary might find other ways to settle their differences. The boys serve to remind them of their obligations to each other and the story as a whole, thereby structuring the myth more firmly. While they are around nothing less than the traditional gunfight will do. On the other hand, because they never participate in the action, but only coerce and re-enforce it, they are thoroughly resistant to this temptation as well.

In summary, then: the Western is a myth in which evil appears as a series of temptations to be resisted by the hero—most of which he succeeds in avoiding through inner control. When faced with the embodiment of these temptations, his mode of control changes, and he destroys the threat. But the story is so structured that the responsibility for this act falls upon the adversary, permitting the hero to destroy while appearing to save. Types and details, as well as narrative, take their shape from this inner dynamic, which must therefore be understood as the basic organizing and interpretive principle for the myth as a whole.

CULTURAL IMPLICATIONS

The Western, I believe, bears a significant relationship—both dynamic and historical—to a cultural force which, for lack of a better word, I would call "puritanism." Here I simply refer to a particular normative image of man's inner life in which it is the proper task of the will to rule, control, and contain the spontaneous, vital aspects of life. For the puritan there is little interpenetration between will and feeling, will and imagination. The will dominates rather than participates in the feelings and imagination.

Whenever vitality becomes too pressing, and the dominion of the will becomes threatened, the self must find some other mode of control. In such a situation the puritan will seek, usually unknowingly, any situation which will permit him to express vitality while at the same time appearing to control and resist it. The Western provides just this opportunity, for, as we have seen, the entire myth is shaped by the inner dynamic of apparent control and veiled expression. Indeed, in the gunfight (and to a lesser extent in the minor temptation episodes) the hero's heightened gravity and dedicated exclusion of all other loyalties presents a study in puritan virtue, while the evil one presents nothing more nor less than the old New England Protestant devil—strangely costumed, to be sure—the traditional tempter whose horrid lures never allow the good puritan a moment's peace. In the gunfight there is deliverance and redemption. Here is the real meaning of the Western: a puritan morality tale in which the savior-hero redeems the community from the temptations of the devil.

The Western is also related to puritanism through its strong self-critical element—i.e., it attacks, usually through parody, many aspects of traditional civilized life. Self-criticism, however, does not come easily to the puritan. Like vitality, it functions through imagination; and it too is in the service of the will. Therefore, if such criticism is to appear at all, it too must be veiled. The Western assists in this difficult problem, for the story is well-removed from his own locale, both geographically and psychically. Because it is always a story taking place "out there," and "a long time ago," self-criticism can appear without being directly recognized as such.

It is tempting to inquire how far certain historical forms of puritanism, such as mass religious revivals, may have actually produced the Western. Was it only a coincidence that the same period of 1905–1920, which saw the early emergence of the Western myth, also witnessed the nationwide popularity of a Billy Sunday and an Aimee Semple McPherson? Their gospel was a radical triumph of will over feeling and vitality, through which the believer could rely wholly upon his increasingly omnipotent will for the requisite controls. And here too was the familiar inventory of vices, with its characteristic emphasis upon gambling and drinking.

Recently there has been an even more remarkable religious revival. Beginning in the early 1950's, it reached its point of greatest intensity in 1955. Here the gentle willfulness of the Graham gospel, and the more subtle (but equally hortatory) "save-yourself" of the Peale contingent permitted many respectable people to go to church and become interested in religion, without actually knowing why. However, like its earlier counterpart, this was not so much a religious movement as it was a renewed attack of the will upon the life of feeling and vitality.

That a re-appearance of the Western should take place precisely at this point is certainly suggestive. For the upsurge in its popularity did occur just five years ago, beginning in the same year that the religious revival reached its height. Perhaps the present Western revival has been more extensive and pervasive because the recent religious revival was equally so.

Presently, however, the religious revival has subsided, but the Western remains almost as popular as ever. This could mean one of two things. On the one hand, the many changes which the Western is presently undergoing—in its narrative, its types, and in its recurrent, primary images—could indicate that the religious recession has permitted the myth to be altered radically, such that it is on the way to becoming something entirely different. On the other hand, should such changes remain responsible to and be contained by the classical version, it could be that our puritanism is simply being expressed through nonreligious sources: most notably through the social sciences (indeed, in the sociologist's and psychologist's denunciation of the violence, historical inaccuracies, etc. in the Western, do we not hear echoes of the puritan hero himself?).

29. Television for Children, the FCC, the Public, and Broadcasters

by Ashbrook P. Bryant

Mr. Bryant is Chief of the Network Office for the Federal Communications Commission in Washington, D.C. His remarks here were delivered at the Children's Film Festival of the Wisconsin Association of the American Council for Better Broadcasts on April 27, 1962. They are reprinted here with permission of Mr. Bryant.

I am sure I don't need to emphasize to a gathering such as this, that in our country television ranks with the home, the church, and the school as a most potent influence on children. It has recently been estimated that 70 million hours per day are spent at the television set by 27,000,000 children under 12. I'll leave it to you to compute that on a yearly basis per child hour. So I think we all must recognize that television inevitably has great effects on the young. None of us can be entirely sure what those effects are.

There are, of course, a variety of so-called "expert" opinions as to how and why television affects the young. . . . Certainly the so-called experts are not in agreement. But I do think that the considered opinions of informed parents, and others who are concerned with the welfare of our children, are certainly entitled to great weight in the decisional process which results in the selection of our television fare.

Also, I think parents have a responsibility to place some reasonable limitation on children's viewing. TV cannot be entirely gauged to children's needs. For instance, I would think that broadcasters are entitled to assume that the old American institution of bedtime, while unpopular with the young, is still a reality in most homes. Believe it or not, a survey has shown that an average audience of 50 million, from 10:20 to 11:00 P.M., included just five million children under 10. If true, that is certainly an interesting commentary.

Your distinguished chairman has suggested that you "would like to hear the FCC's views on children's programming, with primary emphasis on what can be done and how it can be accomplished."

Before I attempt to answer those questions, however, I should remind you that there are rather precise limits to the authority of the government over broadcast subject matter. The Commission may not "censor" programs, nor may it dictate program schedules. Indeed, the Commission plays no direct part in the selection of programs—be they children's or otherwise. However, through its licensing process, it does have a duty to review over-all community program service, including, of course, that part primarily intended for children.

Now, as that may seem somewhat contradictory, perhaps I had better review for you briefly the basic concepts of the American system of broadcasting and point out where the Commission fits into the program picture.

1. Our system of broadcasting is based on private enterprise at the community level and is developed and operated by individual initiative.

2. Unlike other governments of major nations, your government does not directly participate in the selection of broadcast fare, nor may it prescribe what the detail of that fare shall be.

3. This system was adopted, not because Congress did not appreciate the importance of broadcasting as a social function. Rather, the reverse was true.

4. Those who devised the original pattern of government regulation for broadcasting back in 1927, and which is essentially unaltered today, were very conscious that radio, with its ability to penetrate the walls of our homes, involved a unique opportunity for those who operated it, to serve our needs and influence our thoughts and actions. As it was then expressed by Herbert Hoover, radio was recognized as a business "affected with the public interest" and that the people of the nation "had a direct and justifiable interest in the manner in which it is conducted."

5. To avoid centralized control of content of broadcast service, either by the government or by large private interests, the authors of the original act sought to diversify selection and censorship of broadcast service in many hands at the community level, but, at the same time, to encourage private enterprise—rather than the government—to develop and operate our national broadcast structure.

However, Congress did not leave broadcasters entirely free of social responsibility. It created a regulatory pattern designed to assure the public that their interests would be considered and served by those citizens who were chosen to operate commercial broadcasting.

1. In the first place, broadcast channels were declared to be public property. Neither prior appropriation, private investment in broadcast facilities, nor continued use creates individual property rights in the spectrum. This was a most important step. Through it broadcast channels became a great national resource and the government retains ultimate control of their use.

2. These channels are licensed by public authority for limited terms to persons who qualify under prescribed rules and who agree, as part of their permission to use public property for private profit, to take the interests of all the public into account in operating their businesses. Periodically, they are called to account for their stewardship.

3. Within these limits, full responsibility to choose and supervise the programs and other matter broadcast to the public rests on local broadcasters. They may not avoid it by redelegation to a network, to program producers, advertisers, or other persons.

4. On the other hand, the Commission may not designate either the source or content of station programming. Its function, in this area, is first to select qualified licensees and then to attempt to see to it that they serve the public interest.

Thus, it is a fundamental tenet of our system of broadcasting that the government does not design or determine, except in a very general sense, the social characteristics of broadcast service.

The essence of the obligation of your broadcaster is to act as a "trustee" in the operation of your property and to make a "diligent, positive and continuing effort to discover" and reasonably to fulfill your "needs and desires" for broadcast service. In a country of the size and variety of ours, it is assumed that the needs and desires of all communities are not the same in detail, so that the inquiry of the broadcaster must be something more than *pro forma*. If he wishes to continue to represent you as a broadcaster, he must consult with significant groups in your community to find out what they require of him and then reasonably meet those requirements through his programming.

These are not pious words. They are the rules of the game. Parenthetically, I might call your attention to the recent action in the United States Court of Appeals for the District of Columbia in which the Commission's right to require licensees to make the kind of audience inquiry I have outlined, was directly challenged by a broadcaster on the grounds that it constituted "censorship," and violated the guarantee of free speech in the First Amendment of the United States Constitution. The Court broadly sustained the Commission's position and specifically upheld its right to: ". . . . impose reasonable restrictions upon the grant of licenses to assure programming designed to meet the needs of the local community."

The broadcaster, as you see, is not entirely free to program as he will. The interposition of the public interest as the standard for broadcast service places an over-riding responsibility on licensees to temper their commercial interests with social consciousness.

But this is not a simple task for the broadcaster—and here is where the public comes in.

Our system of broadcast regulation assumes a constant interplay between licensed broadcasters and the members of his audience, with the Commission acting as sort of a referee, or perhaps catalyst. The licensee's duty can be made much easier and more realistic if the public assumes the obligation to make their wants and needs for broadcast service known. It is the job of the broadcaster to make informed judgments as to the relative importance in his community of the various kinds of programming and to serve them.

Early in our program inquiry a number of earnest persons, including some broadcasters, urged the Commission to set up more precise standards to determine the proportion of various kinds of programs necessary to an adequate community service—so much public affairs, so much entertainment, so much children's programming, etc. To have done so would have tended further to standardize broadcast service and to substitute the Commission's judgment for that of its licensees. However, we carefully reviewed our policies and the Commission issued a statement in which it reaffirmed the licensee's duties, and called their attention to fourteen elements of community service, generally considered by the industry and the Commission as items to be considered in designing a community program schedule. Programs for children are near the top of that list.

Now, quite probably, not all of these elements are needed in every community—certainly not in the same proportion. But I strongly suspect that a fair inquiry in any community would disclose a significant need for adequate service for children. It is the job of the broadcaster to determine and evaluate the importance to his community or service area of these various elements of service.

The Commission has recently taken a number of other actions designed to bring home to its licensees an awareness of their responsibilities as trustees in the operation of their stations. For instance, we have completed our long inquiry into the whole network television process. We are reworking our broadcast application forms to require much more information as to exactly what a broadcaster does to discover and serve the needs of his community. More renewal hearings are being held in local communities. Recently, investigative hearings have been held in Chicago to give local interests the opportunity publicly to state wherein they think their needs were not being adequately served and an equal opportunity to the stations to respond. These hearings were set after the Commission had received a great many complaints and protests from citizens and groups in Chicago. They were a very considerable undertaking and in the nature of an experiment. Whether there will be others has not as yet been decided.

Doubtless you are familiar with the work that Senator Dodd's committee has been doing with regard to crime and violence in television and its effects on children. My office has been cooperating with his staff in this endeavor. Also, as a result of matter which came to light during our program inquiry, Senator Pastore, Chairman of the Communications Subcommittee of the Senate Committee on Interstate and Foreign Commerce, is presently engaged in a series of discussions with the networks, the National Association of Broadcasters, and the Commission, seeking means through which television licensees may be effectively represented in the review of network programs.

Now I would like to return briefly to the specific area of your interest here this week. I cannot be of any help in evaluating children's programs. But I can tell you that there is almost uniform agreement—both within the industry and among public groups—that much needs to be done in the kind and quality of children's programs. During the sessions of our program inquiry in New York last year, many top creative professionals in television programming, as well as the representatives of some of the largest television advertisers, repeatedly identified children's programming as one of the principal deficiencies in network schedules.

The networks agreed that there is much to be done in this area, but pleaded that programming to meet the needs of children of varied age groups presents problems of some magnitude. However, they assured us that they are actively trying to solve these problems and improve their offerings for children. Perhaps efforts such as yours can point the way.

Also in our program inquiry a large number of persons—many of them prominent in television programming, expressed concern at the effect on children of excessive crime and violence on television. Perhaps you may wonder why this aspect of broadcasting has not been the subject of formal action by the Commission directed to broadcasters. The reason is this: to make the nice literary distinctions necessary to regulate, at the source, the amount and type of crime and violence in television programming would require the Commission to examine the content of each program and to make appropriate judgments. In the first place, we are not equipped to do it. But more importantly, to do so would involve grave questions of censorship in the form of direct prior restraint, which is specifically prohibited by the Communications Act. However, the fact remains that an increasing number of responsible people feel that constant depiction of crime and violence in television programming does produce damaging social effects. As we pointed out in our Interim Report to the Commission, which was made public in June, 1960, licensees should take careful account of community opinion in this area in designing program schedules. Since then, many television broadcasters, including all three networks, have established or expanded regular procedures to pre-screen programs for this purpose.

Perhaps you know that very recently, at Senator Dodd's suggestion, Secretary Ribicoff has called a conference which will soon be held in Washington to consider "the impact on children of the amount of crime and violence they view on television." We will of course follow this with great interest.

In the brief sketch I have given you, I think you will realize that the subject of broadcast programming is under serious consideration by the government. There are limits to what we can do or what you would want us to do. However, the fact remains that there is much that you and your broadcasters, acting together, can accomplish. As I have emphasized, television broadcasters are licensed to serve you—and I might add that the opportunity to do so is highly profitable. Your broadcaster, who is permitted to use your channels for private profit, must make it his business positively, diligently, and continuously in good faith to find out significant preferences for broadcast services in your community and serve them.

Now, the Commission sits in Washington. Most of the information we have regarding programming in particular communities comes from the broadcasters themselves. . . . If you feel that there are significant areas of community need which your broadcaster is not making an effort to provide, or that he is shirking his responsibility, it seems to me that you have a duty to tell him and to tell us. The Commission will not substitute its judgment for that of an informed broadcaster who is acting diligently and in good faith; but must, however, appraise his performance against his promises and judge him accordingly. Information as to how a broadcaster has performed—especially from those whom he is bound to serve—is certainly helpful in this process.

There are a number of other ways by which you may make your views count. The first thing, perhaps, is to let your broadcaster and his advertisers know what you think (this includes the network and its advertisers) not just once in a while, but frequently. And I assume you would want to tell him when you think he has done a good job as well as when you are displeased with his efforts. For instance, there are an increasing number of organizations such as yours throughout the country who regularly monitor and evaluate programs. If this is to be done, you must do it. Under the law we cannot—and I doubt that you would want us to.

When you tell your broadcasters of your pleasure or displeasure, you should also tell us. We forward your communications to the licensee and ask for an explanation. It would surprise you how effective this can be. Basically, broadcasters are trying to please you. They live by your good will.

Now, if simple protest doesn't work, there are other avenues open to you. Broadcast licenses are renewable at least every three years. Your broadcaster has no prescriptive rights to represent you. To continue, he has the burden of showing us that he has made, and will continue to make, diligent efforts to ascertain your needs and desires for programming and reasonably to fulfill them.

Applicants are *now* required by law and the Commission's rules to inform you, through the press and their own stations, when they have filed application with the Commission for renewal of their licenses. Normally, there is a sixty-day waiting period. The applications are public, and copies of them may be obtained at a nominal charge by writing to the Commission in Washington. One purpose of this recent amendment to the Act and the Rules was to afford local communities the opportunity to review a broadcaster's proposals and to express their views to us as to their adequacy.

If, after reviewing the application, or on the basis of your knowledge of a broadcaster's performance, you feel that he does not measure up, you may ask to be heard, either in writing or in person. If the Commission feels that your complaints are of sufficient weight and specificity to raise serious questions as to the public interest, it may set the renewal for hearing and place the burden of proof on the applicant to answer in detail. The Commission, under the law, may not grant or renew a broadcast license unless it finds on the basis of an adequate record, that the public interest will be served. The station's service area composes his public and it is their interest which is to be served.

And finally, it seems to me that the entire broadcast industry will be greatly benefited by the kind of activity you have been engaged in here this week. Broadcasting is a community function. Ultimately, its form and function must be determined by interaction between enlightened broadcasters and interested members of the public.

30. Summary Report on Findings and Recommendations of the Conference on Impact of Motion Pictures and Television on Youth

by the National Council on Crime and Delinquency

The following report is an unpublished draft. The Conference was held June 15–18, 1960, at the Barbizon-Plaza Hotel in New York City under the auspices of the National Council on Crime and Delinquency. It is published with permission of that organization.

THE PROBLEM

1. *Motion pictures and television are a new and powerful educative force affecting the young.* They are more powerful than just hearing and reading. There is a television set in almost every home. Some studies show today's children spend an average of 20 to 24 hours per week at TV screens and that the time spent in viewing is on the increase. The movie houses attract large numbers of teenagers who often return several times to see the same film. Many children learn the language of pictures before they learn to speak. The new media elicit a high degree of emotional involvement on the part of the viewer during their most vulnerable years.

The force of the new media is acknowledged by the television industry in advertising, in instruction, and in propaganda. This same force is also powerful in affecting youth both for good and for harm. These modern media represent a significant contribution to culture and entertainment. New windows to the world have opened to children and adults. Knowledge generally has been advanced for millions.

The conference believes this educative factor intensifies the responsibility of parents, schools, the TV and motion picture industries (including the sponsors and their agents), the FCC and all citizens who are concerned with our nation's youthful human resources, to see the maximum of good and the minimum of harm result from their exposure. The entrance of television into homes places a special responsibility on the television industry to insure reasonable safeguards against harm to children. Virtually all products which come into the home for consumption by children carry the warranty of the maker as to their harmlessness. Therefore, it would appear that the burden of proof regarding harmlessness of the entertainment product should be upon the producing industry rather than primarily upon the parents, educators, or others concerned about the welfare of children.

2. *The amount of time and number of television programs involving crime, horror, violence is conspicuous.* Rather than diversification and balance of "menu" or "diet" there appears a preponderance of such programs. This is considered by the conference as part of the problem.

3. *Public concern has been expressed about the bad effects of the horror, crime, violence, and sex movies and TV shows on the youth of the nation and about its relationship to juvenile delinquency.* This concern is growing. It is reflected in public opinion polls, opinions expressed in "letters to the editor," in editorials, and by columnists, by statements and actions undertaken by various citizen organizations, and by testimony before the United States Senate Subcommittee to Investigate Juvenile Delinquency.

4. *This conference believes that there are bases for concern about what children are learning through movies and TV.*

(a) The Effect of Creating and Reinforcing Certain Stereotypes, Antisocial Attitudes, and Behavior Patterns.

These are intensified by dramatization and glamorization. For example:

1. The weak, comic, or brutal father and his representatives, the stupid, ineffectual, callous police as well as other authority figures such as judges, law enforcement personnel, and institutions in our society.

2. The well-meaning but ineffectual or naive mother and her representatives, the teacher or social worker.

3. The teenager devoted only to pleasure-seeking and depicted largely as a noncontributing, if not actually dangerous, member of society.

4. The criminal who has prospered for many years before his downfall in the last three minutes of the program.

It was recognized that such models for identification, delinquent sociocultural standards, values, and mores are often representations of the kind of reality with which the child is only too familiar. Nevertheless, for the child, the dramatization of his reality in these programs implies acceptance and even tacit approval by the adult world which creates and brings them into his home. It is the accumulated effects of such programs over long periods of time and during the most influenceable years of the child's life which is of most concern.

(b) Effects of Reinforcement of Reality-Fantasy Confusions.

Whether because of a normally immature capacity to separate reality and fantasy or because this function of their mental development is poorly developed, the constant viewing of programs which present these children with too large quantities of material which stimulates aggressive and sexual fantasies by means of either symbolic or real visual representation, may result in:

1. Overwhelming amounts of anxiety.

2. Imitative or reactive behavior based upon a confused concept of the social-reality consequences of such behavior.

3. Further confusion of a child's developing perceptions about reality. These programs frequently present blurred notions of what really constitutes strength or weakness, goodness or badness, intelligence or stupidity, masculinity or femininity, love or eroticism.

4. Lowered sensitivity to the value of human life and dignity and reinforcement of attitudes of self-centered indifference.

(c) Effects of Reinforcing the Already Existing Observable Tendencies of Younger Children to Adopt the Behavior and Mores of the Next Older Age Group.

The general tendency of young children to imitate or take over the attitudes and behavior of the next older age group has long been observed in the group activity of children. This tendency appears to be accelerated and reinforced by the content of many films and TV programs. Imitative identification with delinquent attitudes and behavior, or with those levels of adolescent patterns of behavior generally associated with delinquency, is of special concern. It was felt that there may well be a connection between this tendency and the increasing incidence of the more serious types of delinquencies now being committed by younger-age children and by girls.

(d) Children Have a Need for Passive or Less Active Kinds of Entertainment.

This often provides both relief from outer and inner pressures as well as release of tensions in vicarious experience. It is very important that this need not be simply exploited for commercial interests, or for relief and release purposes alone. The child and his parents should be provided with the opportunity to select programs appropriate to the age of the child, his real life situation, his intellectual and social growth, as well as pleasure needs. This means that it is imperative that a diversification of programs be available so that self- and parent-directed selection is at least possible.

It was felt to be generally true that mass media can reinforce positive social values, patterns of behavior, actions leading to problem solving, deepening of understanding and empathy in children as well as adults. Furthermore, because of the powerful emotional appeal and capacity to elicit emotional involvement in the content of the shows provided by the media itself, such reinforcements and reaffirmations of positive and constructive aims in child-rearing and education could be, and sometimes are, well served by this media.

However, at the present time, such aims and intentions are either misdirected or entirely neglected because of an over-emphasis upon the kinds of program content which have a high degree of emotional involvement but which too infrequently utilize this involvement for intellectual and cultural growth.

5. *The conference observes that the National Association of Radio and Television Broadcasters in their "Television Code"* and the "Standards of Practice for Radio Broadcasters," and the Motion Picture Association of America in their "Code to Govern the Making of Motion and Talking Pictures" have set *many highly* desirable standards for their product, such as "Responsibility toward children . . . in avoiding material which is excessively violent or would create morbid suspense, or other undesirable reactions in children, . . . in exercising particular restraint and care in crime or mystery episodes involving children or minors."

The reality however is different: many movies and TV shows clearly violate the standards set by the codes.

6. *The conference observed that excessive crime and violence depictions in American movies and TV shows are well known abroad and represent a serious threat to the image of America.* As one article has stated this: ". . . the true image of America is being dirtied by shoddy, garish exaggeration of its worst aspects. . . ."

FOR THESE REASONS THE CONFERENCE SUGGESTS THE FOLLOWING PROGRAM OF ACTION:

1. The establishment of a National Commission on Television and Motion Pictures (sponsored either by the federal government or independently by some foundation) to inquire into the role, function, and responsibility of the mass media. Such a commission should be composed of leaders of recognized stature and include representation from the motion picture and television industries. Such a commission was also recommended by the U.S. Senate Subcommittee to Investigate Juvenile Delinquency and by the 1960 White House Conference on Children and Youth.

2. Support the regulatory agencies (FCC) in the performance of their functions which the federal government must assume in the public interest; specifically—monitoring of the TV programs; and more diversified and flexible sanctions should be emphasized in connection with the licensing of the TV channels.

3. Call to the attention of television stations the commitments about programming to which they agreed when their license was granted. Not all stations have maintained the balance of programs specified in their licensing application.

4. The attention of the television and motion picture industry should be urged to correct the considerable gap between their established codes regarding children and the extent of horror, crime, violence, and sex programs.

5. Education of the public through professional, civic, and service organizations regarding (a) the effects of audio-visual programs on children, (b)

the need for improvement, and (c) the means by which the public can register protests and commendation.

6. Encourage professional educational organizations (such as the National Association of Education, the National Council of Teachers of English, etc.) in their efforts to integrate into the primary and secondary school curriculum, materials and units of study which will increase understanding of the new media and improve standards of taste among children. Also encourage more creative educational use of motion pictures and television by those working with children and for improving communication between parents and children.

7. Encourage universities, foundations, and professional associations to offer annual awards for outstanding motion pictures and television scripts for children's programs.

8. Enlist the interest of national leaders of industry who use television in the nature of the impact of television shows with which their products are becoming associated.

RESEARCH

In addition the conference strongly recommends further research regarding the presumable delinquency-inducing effects of crime, violence, sex, and horror shows:

1. The conference recognizes the difficulty in establishing a causal nexus between delinquent acts of a particular child with some specific audiovisual experience, except in *modus operandi* and the psychological "trigger effect." This, however, does not preclude the study of present program content as a component of our culture and recognizing that individual and group behavior is influenced by the climate created by a heavy diet of violent and criminal exploits of juveniles.

2. The conference feels that an integrated program of research, including basic, applied, and demonstration research of interdisciplinary nature should be developed. For the purpose of working out such a comprehensive program a conference of leading research experts in this field is indicated.

3. Because of the public responsibility of the television and motion picture industries, they should be urged to provide funds for objective and independent research. In addition, all other available research resources should be utilized.

31. Human Dignity and Television

by Arthur J. Brodbeck

Dr. Brodbeck is Senior Research Psychologist with the Center for Urban Education in New York City. This article appeared in the National Association for Better Radio and Television Quarterly *for Fall, 1962. The editors refer the reader interested in a more technical approach to "Television Viewing Patterns and the Norm-Violating Perspectives and Practices of Adolescents: A Synchronized Program of Scope and Depth Policy Research," by Arthur Brodbeck and D. B. Jones in* Television and Human Behavior *(Appleton-Century-Crofts, New York, 1963). The article is reprinted by permission of the author.*

We have all been repeatedly told of the advantages and power of positive thinking. From many different quarters of current social science research, it does appear that accentuating the positive produces a better and more stable result than concentrating on the negative. Censure and blame, so the evidence begins to show, can unpredictably produce almost anything, and nothing very stable by way of result, whereas encouragement and the clear depiction of what is wanted has the effect of lifting people up toward stable and positive forms of reaction to us, when positive goals are provided.

How can we think and act positively about TV? There is a concept which man has been continually clarifying over his entire historical span of life. It is the concept of "human dignity." Each age and each generation has added something more to it; and every period has to approach it anew. As it has been handed down to us, it seems to contain one fundamental idea. It is this: *No man or group should pursue his values so as to overdeprive or overindulge any other man or group.* Hence, the concept has always been a part of democracy, starting with the half-slave democracy of the Greek city-states. Today, our world is bipolarized and world tension is so high that the prospects of violence are real and unremittingly with us night and day. Violence is a last resort for those who believe in human dignity. Yet, estimation of murders shown on TV each year exceeds the number of murders committed in America per annum. More intelligent strategies are required. Each age has sought to discover them, and we must do so in ours.

What is television doing by way of substituting intelligence for sheer violence? Is it helping people at all levels and ages to think more creatively about the world they live in? Is it raising the level of consciousness about that world, exploring new ways to look at it so that it becomes more possible to see the value of every human being in it? Mr. Minow has called it "a vast wasteland" and one suspects he has done so because it contributes so little to furthering and sharpening the concept of human dignity needed for our time.

If we think positively, what recommendations can we give to those who produce, write, direct for (and act in) the television dramas that daily come into the American living room? Here are a few to which many more could be added:

(1) *More Science Fiction:* The scientific outlook has grown up slowly in the world and is still rising to a coherent world view. Science stands for the development of effective intelligence and creativity and innovation. It is true it has produced instruments of violence that threaten to make violence itself obsolete, since the use of it may end human life altogether. But science also has capacity for increasing human dignity. Our youngsters, and we adults in authority over our youngsters, need to explore all the positive and negative ways in which the development of science can in the present and future affect human dignity. Science fiction offers a way of doing this. By imagination, we are able to plan for an ideal future, inroads into which science can begin to make now. We are also, by anticipating the future developments of current scientific discovery, able to see some consequences that we wish to avoid. There is much evidence that our brightest youngsters "tire" of television claptrap as they reach puberty, but they do remain glued to the set to continue to watch the science fiction programs, even such diluted and sometimes inferior programs as *Way Out* or *Twilight Zone*. Why aren't our television sets giving us more of them, of better and better quality? Everyone is interested in science now—and concerned about the human consequences of it. Why not use dramatic programs about science to raise the levels of public consciousness about the present and future human consequences of scientific developments so as to preserve or challenge human dignity?

(2) *More Programs that Promote Understanding of Violent Personalities.* Expectations of violence have been growing in the world ever since Hiroshima, fanned by the world tension that has resulted from the division of the globe into two great powers. There is no way of denying that we live with this tension and with such expectations. What do our television programs do to help us, and primarily our children, face this contest in the world today in which human dignity is severely threatened?

News analysis is one thing. It is usually quite impersonal, removed from everyday concerns. But the drama the world is witnessing today between the free nations and the totalitarian ones is not something outside of our daily lives. It affects our personalities. And since it does, it acts upon all of our day-to-day interpersonal relations. It is fine to have news programs, especially those which make us think, but it is also required that the thoughts move our hearts, relate to ourselves. Much of the violence in television today teaches us very little about what makes for personalities addicted to violence. It, therefore, teaches us almost nothing about the world situation which affects us, and our children, and in which we are all caught. Violence is more dangerous today than it has ever been before; the need, therefore, is to exercise every potential we have in a democratic community to learn how to deal with violent personalities without ourselves succumbing to the disease. The need, in short, is to exercise all our intelligence, our inventiveness, our creativity, our ingenuity to convert violent personalities to a way of life more congenial to human dignity.

The violent personality is characterized, above all, by extreme egocentricity. An egocentric individual has difficulty in being able to extend the self to include the perspective of others. He has, through his training or culture, come to believe there is more to be gained by shutting himself off from communication and collaboration with others than by opening himself up to it. Along with egocentricity go a host of related traits. Some of these are: paranoia, an unconscious low estimate of the self, a marked rigidity of thought and deed, and among the innumerable others, a deep-lying (and often covered-up) cynicism and pessimism, rather than sanguinity about the future. There is a great deal known about such traits—and psychotherapists every day have had to deal with them and attempt to modify them in patients who suffer from them.

Much of the fare we have on television which deals with violent personalities seems to be written by authors who have no contact with such scientific knowledge. Instead of increasing our understanding of such violent personalities, they contribute to and satisfy our own world-created violent impulses by depicting violence itself as the major weapon against the violent personality. It is almost as if the majority of TV writers were so impressed with violence that they could not creatively conceive of how violence arises from weakness rather than strength. Violence is sickness—both in those who use it and those who allow it to be used. A democratic personality succumbs to it only as a last resort, and it indicates a failure of the brain potentiality of the democratic personality to persuade and successfully communicate with the violent personality. Why doesn't some of our current knowledge of the sources of violence begin to appear in the dramas that currently flood the TV screen? Have TV writers cut themselves off from the very science which is our one hope as a democratic nation against the totalitarians?

(3) *More Dramas about Creative Personalities.* In the last ten years, an enormous amount of research has been done about what has made people creative and what a creative act consists of and feels like to experience. There is as much—indeed, *more*—of "thrill" in such creative experiences than anything now depicted by way of "thrills" in sadism or lusty sensuality. To be suddenly illuminated after facing a problem in the dark, as it

were, is one of the most exhilarating human experiences that one can dramatize. Isn't it about time that we had them conveyed to our children and to those in our communities who have not yet felt the rewards of being creative? Creativity is not something that exists among a tiny segment of the population. Each child in growing up experiences it in some form by becoming a part of his culture. No matter how humble the station in life—i.e., shoemaker, housewife, what have you—there is still room for further innovation. There are creative shoemakers and housewives and there are thoughtless ones.

When I have suggested the need of depicting the creative act on the stage to a famous playwright, he replied: "But, surely, you know how difficult it is to dramatize creativity?" It is also difficult—and in principle no more difficult—to suggest, by a host of special film devices, how a sexual act has been consummated between lovers without actually showing it. Yet, the films put enormous ingenuity of effort into getting around censorship so as to suggest a sexual act by indirect symbolic devices. One suspects the writers knew something about the feelings and experiences associated with sexuality; hence, they were able to "symbolize" these in all sorts of remarkable ways that left nothing to the imagination even though the imagination was fully exploited. Do such writers know nothing about a creative experience? If they do not, it is time they studied some of the recent investigations into the creative mind, familiarize themselves with the subjective lives of those who manifested great creative innovations. Of course, there is always a problem in depicting what few have yet "standardized' in the language of the film. If one looks at films made in the thirties, he sees how current film language is ever so much more sophisticated because it was built upon the "experiments" with communicating through film made then. What has not yet been tried remains a problem. To repeat a formula involves us in a repetition compulsion that is more like the outlook of rigid, egocentric personalities than those who enjoy the freedom of a democracy to make continual new advances with so rich a resource as television.

There is much truth to the idea that we become like that which we study or write about. To dramatize creativity might have the liberating effect of making the television industry itself more creative. The need to be commercially successful is only a further challenge—not an excuse to avoid facing and mastering a problem.

These are positive suggestions made to the television industry. Here are three simple kinds of things they can do. Indeed, it is puzzling why they have not done them. Part of the reason may be that the public media have cut themselves off from the communications they most need to be in touch with—i.e., namely, the current developments in natural and social science. Science is leaping ahead and is now producing a rich variety of concepts and ideas and concrete findings about nature, personality, and community. The public remains out of touch with these when television lags behind these developments breaking out in our democratic culture. An unevenness in society develops, always dangerous when it goes to extremes. It is an unevenness regarding man's enlightenment of himself. Social and natural science today has so much to give the public by way of encouragement and stimulation about human dignity that it is a wretched condition when the public's "window out into the world" contains almost none of it, especially in the dramatized form in which it can be best understood by the average person. And scientific ideas are dramatic. Worst of all, our children are deprived of exposure to the best we are producing, and it is they who most need to be in constant and intimate touch with the trends of ideas in the scientific world which is now expanding to touch all of life. They are the hope of the free world. Surely, the power of positive thinking is a power that the television industry needs to share with the community in which it functions and which supports it. That community is increasingly being forced to move toward scientific outlooks. Does TV want to lag rather than lead in the movement? If it does not lead, how can it maintain any sense of self-respect, a fundamental feature of human dignity?

VII. News and Public Affairs

PERHAPS the most agreed-upon statement of critics and industry spokesmen alike is that radio and television's finest accomplishments have been in the area of news and public affairs.

Yet there are nagging questions that have been raised by many of the most distinguished practitioners themselves. In conventions of newsmen, discussions of problems are more common than self-congratulation.

Some of the problems of television, particularly, as an instrument of news, reporting, and political orientation are so serious in their implications that they require more time and space for discussion. Several of the most basic and thought-provoking of problems are raised in the selections included here—from the Mike Wallace interview with Sylvester Weaver to John Day's provocative analysis of television news as a reportorial or entertainment function. Some highly articulate comments are also included here on the role of the news commentator in television by Eric Sevareid, Martin Agronsky, and Louis Lyons.

32. Radio: Dollars and Nonsense

by Hans V. Kaltenborn

Without question, the name of Hans V. Kaltenborn belongs in the literature of broadcasting. When he made his first radio broadcast in 1921 he brought to it an experience of more than a quarter century in newspaper reporting. The rise of the United States to dominance at the end of World War I found Kaltenborn well-equipped to interpret that role at home and abroad. He became a world traveler and an accomplished linguist, interviewing in the next decades scores of leaders including such notables as Adolph Hitler, Benito Mussolini, Chiang Kai-Shek, the Emperor of Japan, Mahatma Gandhi, and scores of others. The vast amount of information he gathered furnished bases for newspaper reports, numerous magazine articles and books, and radio editorials. Eventually, broadcasting became for him a full-time occupation and won him the title, dean of American radio commentators. This article appeared in Scribner's Magazine *for May, 1931, and is reprinted with permission of Mr. Kaltenborn's estate.*

Millions of average Americans still regard radio as a toy, and a few superior Americans continue to scorn it altogether. They will admit that it offers "some good things," but will not bother to look them up or tune them in. Many who read this magazine will tell you: "I would not have one in the house." Asked why, they answer: "I hate canned music," "We have too many distractions now," "The phonograph is bad enough," "My friends bore me with it."

Business-minded Americans think of radio as a leading industry and a new medium of advertising. They know that 13,000,000 families own radios, that Americans spend $8,000,000 a year on radio sets, and that big advertisers spend millions on radio entertainment. But they give little thought to radio as an instrument of government, of education, of open or disguised propaganda. Few people realize that radio has already become a subtle and powerful factor in shaping our public opinion, that it rivals the press in directing public thought, because the average citizen is much more responsive to what he hears than to what he reads.

As a newspaperman writing daily signed and unsigned editorials during the nine years in which I have also broadcast from one to four times a week on news topics, I have had unusual opportunity to compare the effect of the spoken and of the written word. There can be no question about the superior persuasive power of speech.

Radio listeners are in a more pliable and responsive mood than newspaper readers. Listening becomes a social function shared by several members of the family. Listeners chat about the things said with neighbors and business associates, who have heard the same speech. Thousands of radio friends have written me that they always ask people in to hear my talk and to discuss it afterward. Few newspaper editorials are read aloud, but speeches that come in over the radio stir debate at home and abroad. Controversial radio material receives an emphasis which deepens its effect.

The radio audience loves vigorous expressions of opinion on public questions. My own broadcasts on current events have gained steadily in popularity largely because they present a personal point of view on such live issues as prohibition and foreign relations. No day passes in which I do not receive letters of vigorous condemnation or hearty approval. Up to a few years ago many of the condemnatory letters were addressed to the broadcasting station with the suggestion that a speaker who voiced such views be left off the air. The writers expressed personal resentment against a voice which penetrated the privacy of their homes to contradict their beliefs. Since the 1928 campaign, in which radio was widely used for political controversy, there has been steady growth in the tolerance of the radio audience and in the courage of the broadcasters. Opinion and propaganda find their way over the air more easily with every passing year.

As a factor in public education and in moulding public opinion, radio has now assumed large importance. Routine musical programmes are giving way to spoken features. Radio dramas, interviews, news summaries occupy an increasing number of hours. In the New York City area, radio now provides some forty distinct daily "information" features. Many of the new programmes sponsored by advertisers feature talks instead of music.

The distribution of spot news via radio has become so universal that some newspapers have abandoned election and sporting extras. Street crowds get results through the loud speakers, and the press is no longer "first with the news." We hear the latest long before we read it. Editorial writers in many states have learned to depend on radio to give them first information on important news developments.

The London Naval Conference of 1930 was effectively "covered" by American radio reporters. Both the National Broadcasting Company and the Columbia Broadcasting System, the two corporations which monopolize coast-to-coast hookups, sent special representatives to the London meeting to talk across the Atlantic to American listeners about what was going on. These two American radio chains were in keen competition in London to secure radio addresses from leading delegates. Secretary of State Stimson gave no interviews to the press during the negotiations, but delivered several frank, explanatory talks over the air, which led the gentlemen of the press to file vigorous protests against this new way of appealing to America. But our delegates rejoiced. They still maintain that radio defeated press pessimism.

At future international meetings radio will rival the press even more effectively. This year important addresses at the League of Nations meetings in Geneva and big events in the European world will be routine features of American radio programmes. Already we receive regularly several weekly international programmes from Europe.

The competition for political material between our leading radio chains, which was evident in London, is always apparent in Washington. Both the National Broadcasting Company and the Columbia Broadcasting System have established elaborate studios in the capital to which they lure men and women in the public eye. By getting the right people before the microphone they impress radio listeners, radio advertisers, and the Radio Commission. Radio forums, international hours, farm hours, public holidays, and all sorts of special occasions are utilized to obtain speakers whose names will add prestige to radio programmes. All politicians, from the President down, welcome the opportunity to "sell themselves" to the radio public. Some even refuse to speak at banquets unless assured that the occasion will be broadcast. After you have talked to millions "on the air," a roomful of food-stuffed celebrants means nothing.

From the public point of view this competition between leading radio systems is most important in preserving "freedom of the air." Monopoly would be disastrous. Yet the danger of monopoly is ever present. It could come through a secret understanding or open alliance between the two dominant radio chains. It already exists in patent control, as anyone who enters the broadcasting field will soon learn.

The Radio Corporation of America, which already controls 3,800 separate patents, is now engaged in the following activities, directly or through related corporations:

Broadcasting.
Ship-to-shore communications.
Motion-picture and talking-picture production, distribution, and exhibition.
Manufacture and marketing of phonographs and phonograph supplies.
Vaudeville entertainment.
Manufacture of vacuum tubes (for which 173 separate uses have already been developed).
Manufacture of broadcasting equipment for sending and receiving.
Telegraph, cable, and telephone services.

Such varied activities, many of them involving an almost monopolistic control, make possible a highly centralized and powerful pressure on public opinion. It does not mean that such pressure can be exercised too directly or without discrimination. It does mean that the policies and purposes of those who control this corporation are reflected in many radio programmes.

A group of important newspapers recently entered the field of news transmission via radio under the name of Press Wireless, Inc. The purpose of this new public-service corporation, which is backed by the *Chicago Daily News, Los Angeles Times, Christian Science Monitor,* and the *San Francisco Chronicle,* is to "accept for reception or transmission press despatches filed by any correspondent or newspaper, provided the subject matter of the despatch is intended for publication in the press." It seeks to eliminate the wire costs, which are such a large factor in editorial expense.

After a vigorous fight before the Radio Commission, this new corporation was awarded twenty short-wave bands for its special purposes; but when the company sought to purchase transmitting equipment from the Radio Corporation it found itself in conflict with the powerful "radio trust." According to the president of the Press Wireless, the license agreement, which the Radio Corporation sought to exact, involved severe restrictions. Press Wireless was not to be permitted to operate as a nonprofit-making enterprise. Its communications were to be restricted to points within the United States. It was asked to agree to route all traffic to and from foreign countries over the Radio Corporation of America's circuits, at regular rates. The Radio Corporation also insisted on the right to examine books and accounts and required the surrender of all patents owned or claimed by Press Wireless.

The attempt of the Radio Corporation to impose such conditions has led to the charge that it is operating as a monopoly, in defiance of the laws of the United States. Various radio manufacturers have brought suit on this ground. The federal government has also proceeded against the corporation and haled it into court charging that it is "an unlawful combination and conspiracy in restraint of trade and commerce among the several States and with foreign nations in radio communications and apparatus." The outcome of this suit will be of great importance to the future of radio.

In foreign countries the danger of monopoly assumes a different form. Almost everywhere outside of the United States governmental supervision of broadcasting is rigid and complete. Advertising over the air is barred or much restricted. Controversial material is avoided, and until recently political broadcasts were rare. Because there is no need to guarantee returns to advertisers by appealing to the lowest common denominator of public taste, foreign programmes have a higher average of cultural appeal.

The chief reproach against American broadcasting as against the American press is that its dominant purpose is commercial. Just as most newspapers are published to make money for those who buy and sell advertising, most radio stations are operated to bring financial returns to those who buy and sell time. Radio stations do those things which help them to make money and leave undone whatever interferes with immediate business success. Practically all the more important stations are trying to find out what the public wants and to satisfy that want whenever this can be done with profit. Raising the standard of public taste or catering to more discriminating listeners is no part of a broadcaster's function.

Favorable public reaction to an advertiser's programme is his test of merit. If a speaker brings in shoals of mail he is a good programme feature. He can talk about his preference in neckties, the love life of the Eskimo, the way he licked the Germans, or the home life of prizefighters—it is all the same to the station and to the advertising sponsor. What they demand is that the sales talk which follows bring in enough leads from listeners to produce profits. It is sad but true that the very radio features which are sneered at by most of those who read these lines bring in the best results for those advertisers who use them. Radio response proves once again that in the United States money and good taste are associated in only the best—and the fewest—families. I have been called a missionary to the radio morons because I broadcast regularly on important problems of current history. But it is disillusioning to inquire too closely why people tune me in. Often the answer is: "I love the way you talk," "Your voice sounds so nice," "I adore your accent; you are English, aren't you?" "I am a stenographer and need practice," "You save me the trouble of reading the newspapers."

Directors of radio stations have learned their lesson from the tabloid press. Educational features are necessary for publicity purposes, to impress members of the Radio Commission, or as a talking point with so-called advisory committees. But they are rarely heard on the air during the popular evening hours when they might conflict with the sock seller, who sponsors *Dick & Mick*, the Wear Forever Duo.

President Merlin H. Aylesworth, of the Radio Corporation, is a veritable Merlin in this publicity game. He is a past master in the subtle art of public relations. Before assuming his present position he helped persuade the American public that government ownership and control of public utilities was of the devil, and that all virtue and advantage lay with unchecked private exploitation. The Senate Investigating Committee has exposed the lengths to which this propaganda was carried. The work he did for these privately owned public-service corporations was direct and valuable preparation for his present job.

The publicity connected with the creation of the so-called Radio City in the heart of New York City is an excellent illustration of Aylesworth skill. This is a huge construction enterprise, involving a total expenditure of

$250,000,000, undertaken for a commercial purpose. The idea of including an opera house was a later development. The plan contemplates a conglomerate of buildings occupying three long blocks—from 48th to 51st Streets, between Fifth and Sixth Avenues—suited to the diverse functions of the various entertainment and communications enterprises related to the Radio Corporation.

These plots had been assembled for John D. Rockefeller, Jr., as a possible location for the new opera house which directors of New York's Metropolitan have long contemplated. When that project fell through, Mr. Rockefeller found himself carrying a lot of valuable real estate for which he had no particular use. It was then that the persuasive Mr. Aylesworth and his able associates stepped into the situation. They pictured the Radio City project to the public-spirited Mr. Rockefeller in the same public-service light in which it was later announced through the press. Here is the eight-column streamer headline over the full-page story carried by the *New York Times* on June 22, 1930:

> Science brings to us a unique Radio City. . . . Great broadcasting centre to be built in New York by John D. Rockefeller, Jr., and associates promises to open a new era in the fields of amusement and education, and by television to carry throughout the country glimpses of the stirring events of the day.

To anyone who read the story that followed with some knowledge of the facts, it seemed that there was a shrewd use of the Rockefeller name to give public-service glamour to the establishment of a high-grade amusement centre which would house and unify the varied commercial undertakings of the Radio Corporation. I wrote to Mr. Rockefeller and voiced my disappointment that his first important contribution to radio should be made through such a purely commercial undertaking. I expressed the hope that the Aylesworth publicity had not given a true or complete picture of his contribution to radio.

Mr. Rockefeller was about to leave for the West and asked his associate, Mr. Thomas M. Debevoise, to reply to my letter. Here is what he wrote me under date of June 24:

> The publicity was unfortunate. Mr. Rockefeller is not making any contribution to Radio. The company which he caused to be incorporated for the purpose of taking over the lease of the Columbia tract (Columbia University owns most of the land) is only subletting portions of the tract to the Radio Corporation of America and affiliated companies. The transaction is purely a commercial one, made necessary because of the failure of the Opera House plan, on the basis of which Mr. Rockefeller was drawn into the lease. Something must be done with the property

and it was fortunate to find a group of such desirable tenants willing to take over so large and important a part of it.

Mr. Rockefeller's denial will never catch up with the original story. As late as October 16 the *New York Times* began an article on the project as follows:

> The $250,000,000 radio and amusement centre to be built by John D. Rockefeller, Jr., in the three blocks bounded by Fifth and Sixth Avenues, Forty-eighth and Fifty-first Streets will start about January 1, it was learned yesterday at the offices of Todd, Robertson and Todd, general contractors for the Rockefeller development.

This story also indicated that part of the site might after all be reserved for a new opera house. In any case, Mr. Rockefeller will not be permitted to escape from the project. Calling one of the auditoriums Rockefeller Hall and other means of flattery may serve their special purpose, even if "the publicity was unfortunate."

Mr. Aylesworth is similarly skillful in making it appear that radical speakers are not barred by the National Broadcasting Company. His organization is extremely conservative. Yet every now and then he takes particular care to allot radio time to some well-behaved liberal or radical speaker like Norman Thomas, and then advertises this concession widely and vigorously. The National Broadcasting Company has practically no more time for sale during the valuable evening hours, which run from eight to eleven. This provides a convenient excuse for excluding undesired or undesirable features.

As a business radio has become such a huge success that it can now afford to show some independence. Broadcasting is becoming more and more profitable as the number of listeners increases and each existing station can claim a larger circulation. The Columbia System in 1930 showed 58 per cent increase over 1929 in gross sales. For three years the number of stations in this country has not increased. The Federal Radio Commission feels that in the present state of technical development not more than 650 broadcasting stations can serve the "public interest, necessity and convenience." Only 89 wavelengths or channels are available. Of these, 40 are cleared channels used by only one station, while the other 49 are shared by several stations operating at different hours or on different days. Technical limitations make it impossible to permit any one to establish a broadcasting station anywhere. Control of a good wavelength in an important population centre has a high money value. One minor New York City station recently on the market valued its government-granted license at more than $1,000,000.

Until recently the Federal Radio Commission favored commercial broadcasting at the expense of all others. Almost every decision went against special groups seeking to serve some part of the listening public. "There is no place for a station catering to any group," the Commission declared in denying a license to the Chicago Federation of Labor. "All stations should cater to the general public."

Applied to the newspaper field such a decision would rule out everything except the tabloids, and this was the practical effect of the Radio Commission's attitude. But since certain groups were already in the field and had developed a public following, certain additional special licenses to particular groups had to be granted to avoid the charge of unfairness. Organized minorities in a position to exert political pressure in Washington occasionally break through the Commission's general rule. Important newspapers which can stir up the political animals also win an occasional favorable decision. In Buffalo, N.Y., the Buffalo Broadcasting Corporation secured a monopoly by the simple expedient of buying up all the local broadcasting stations. The *Buffalo Evening News*, which had long vainly sought to obtain a broadcasting license, successfully exploited the monopoly charge and thus won the right to establish its own station.

Newspapers are supposed to be very much alive to the significance of new developments, but many of them have been lamentably slow in realizing the close relation between the informing, entertainment, and commercial aspects of radio and the press. Half the newspapers that began operating stations when radio was still a toy closed them up or sold them because of the expense. Today several newspapers are making more money out of broadcasting than out of publishing, and scores of publishers rue the day when they missed their radio opportunity. The conviction is growing that a newspaper and a broadcasting station are both more profitable and more effective when they unite their efforts under the same management.

Educational institutions do not have the power of the press when they appeal to the Radio Commission for a license, for high power, or for a more favorable position on the wave band. Believing that "the entire listening public within the service area of a station is entitled to service from that station," the Commission does not look with a kindly or tolerant eye upon stations that eschew entertainment for a more serious purpose. The Payne Fund, after a thorough investigation of the use of radio by educational institutions, reported as follows:

> Colleges and universities that own and operate broadcasting stations devoted primarily to education find it difficult and sometimes impossible to secure from the Federal Radio Commission licenses permitting them to operate during the evening hours, or with power enough to cover their territory or on desirable wave lengths. These restrictions limit their possible audience to a minimum.

The Advisory Committee on Education by Radio, appointed by Secretary of the Interior Wilbur, presents in its 1930 report detailed information about the unhappy experiences of educational institutions without commercial broadcasters. The authorities of Columbia University reported to the Advisory Committee that their early experience

> began to be more and more unsatisfactory chiefly because the items which were wanted for broadcasting were more sensational in character and had little regard for sound educational presentation. It was a great disappointment to those at Columbia who were interested in developing the possibilities of radio to find out that they could not carry out what they regarded as a significant development in the general field of education.

Columbia University's experiment was made over Station WEAF, one of the two New York City key stations of the National Broadcasting Company.

The Massachusetts Department of Education had a similar experience. James J. Moyer, Director of the Division of University Extension, reports as follows:

> The broadcasting stations on the national chains have always agreed to co-operate, but in many cases when time on the air for educational broadcasting was requested in consecutive weeks for educators of national reputation, having subject matter of unusual interest, the reply has been that the national broadcasting companies have contracted for all the available time and that, therefore, the educational broadcasting must be set for some time in the day, which is not commercially useful.

Dean R. B. Smith, of the Extension Department of New York University, which started its experiment with high hopes, reported in June, 1929, that the period given the university for broadcasting had been so reduced in duration and value (finally to fifteen minutes during the Thursday lunch hour, and fifteen minutes of the Friday dinner hour) that the department dismissed the radio director and expected to abandon broadcasting. "Commercialism is pushing educational broadcasting to the wall," was his significant comment.

Apart from a government-controlled system of license fees imposed on set owners, private endowment of radio education seems to be the only remedy for this difficulty. Radio stations have become more liberal in accepting certain kinds of information material. But their first response will always be to the paid-programme sponsor and to the taste of the average listener. Only an endowed station can or will meet the high standards of the trained educator.

Occasionally a first-class educational feature like the Damrosch concerts has an opportunity to win a popular following, which can then be sold to some national advertiser. Or a well-known figure like Heywood Broun is permitted to express radical or unconventional opinions. An individual who has a national reputation and who is thoroughly established with the radio audience is allowed to go pretty far in the expression of personal views. Will Rogers has said things over the air that would not have been tolerated coming from almost any one else.

Restriction went much further in the early days of radio. In 1924 Station WEAF, in New York City, ruled me off the air because of my expression of liberal opinions. This station was then owned and operated by the American Telephone and Telegraph Company. Each time I criticised a federal judge (who might have to pass on telephone rates), a labor leader (who supervised the company's labor contracts), or a Washington official (whose influence counted in the issue of a broadcasting license), one of the vice-presidents became frightened and protested. Finally the much-harassed vice-president in charge of broadcasting decided that he would be happier without my spoken editorials, even though the radio audience continued to enjoy them. The policy adopted at that time was to bar all controversial material. This is still the rule at many minor stations. Station WHEC, at Rochester, N.Y., recently barred a "wet speech" by ex-Senator James W. Wadsworth on this ground.

Since 1924 there has been steady development in "freedom of the air." New York City officials once brought indirect pressure to bear on Station WOR, over which I was speaking, in an effort to modify my adverse comments on Mayor Walker's frequent vacations. The method used was to suggest that the municipality might be willing to co-operate more freely in granting Station WOR broadcast facilities on important public occasions if I were more charitable toward the Mayor. The station took no steps beyond transmitting the suggestion and I paid no attention to it.

During the Wall Street panic of 1929, when I described conditions in the market without mincing words, there was a good deal of protest by brokers who felt that in a time of crisis such comments should not be allowed. The officers of the Columbia Broadcasting System were appealed to and asked me to submit the text of my speech, to see just what I had said. This was impossible, since all my news talks are extemporaneous, and the matter was dropped without further comment. A few months ago Claudius Huston, ex-chairman of the Republican National Committee, got very angry when I cited some of his activities as revealed before the Senate Investigating Committee. He instructed his lawyer to take the matter up, and it looked like a libel suit. Good judgement or sound legal advice seems to have intervened, since no further steps have been taken. Perhaps the threat of legal intervention was intended to forestall further comment. If so, it failed in its purpose.

Except for speakers with known radical opinions or for discussion of a highly controversial topic, the old rule about submitting copy in advance has been abandoned by some stations. The well-known astrologer, whose nightly reading of the stars stimulates the sale of a certain toothpaste, must submit her copy to three separate censorships. But since the stars are more reckless in predicting unhappy events than advertisers or radio stations, this precaution is only natural. One astrologer told me that she was not permitted to broadcast President Hoover's horoscope because of the dire events which it foretold.

Various religious sects whose doctrines are offensive to some people have obtained control of radio stations or have managed to purchase radio time to disseminate their doctrines. Medical quacks are using the radio to such an extent that Doctor Shirley Wynne, Health Commissioner of New York City, made a vigorous effort, only partly successful, to eliminate that type of advertising from the air. Many small stations persistently violate the rule against direct advertising and constantly offend good taste in their programme material. Even when there is loud protest from the listening public the Radio Commission assumes a tolerant attitude in all these cases. So many troubles are deposited upon its doorstep that its representatives take good care not to raise difficulties on their own account. "We have no authority to exercise any censorship" is the Commission's established reply to all protestants.

If political quackery produced cash with the same speed and ease as the sale of cure-alls, radio would be more generally used to reach gullible voters. Entertainment and propaganda could be mixed so judiciously that the interest of the listener would not flag. Thus far no religious or political group has spent enough money on mechanical equipment or programmes, or has displayed sufficient showmanship to develop a really popular station. Yet the recent gubernatorial campaign in Kansas demonstrated what a first-class ballyhoo artist can accomplish in the way of vote-getting with radio's aid.

When the Socialist Party station, WEVD, was established in New York City some years ago as a memorial to Eugene V. Debs, great things were expected from it. Yet the average New York listener does not even know that such a station exists. Probably not one set-owner in a thousand ever tuned it in. It missed its opportunity so completely that the liberal press remained almost indifferent when the Radio Commission revoked its license. It resembled New York's municipally owned station, WNYC, in filling the bulk of its time with third-rate material.

The one political group that has taken full advantage of radio for propaganda purposes is the Communist Party in Russia. Millions of Russian peasants, unable to read or write, listen to the radio newspaper which is published on the air between six and seven o'clock each evening. Each time a Red leader delivers an important address the loudspeakers in peasant reading-rooms and village squares all over the land blare forth his words. Each one of the many Communist organizations has its hour on the day's programme. Every musical or dramatic feature, every bulletin, every educational talk, carries its Communist message.

The powerful Comintern (Communist International) station in Moscow sends out these programmes so that all the world may hear. In order that the world may also understand, many of them are repeated in French, German, Swedish, Polish, Rumanian, and Esperanto. The Comintern station has a range of 1,250 miles and is frequently heard all over Europe. Rumania bars this Red propaganda by actuating a disturbance transmitter whenever Moscow puts a Rumanian programme on the air. Germany filed a formal protest with the Moscow government against the German programme. There are 4,560,000 Communist voters in Germany, many of them eager to keep in touch with the Moscow headquarters of the Third International, to which they also belong. The German government charged that the Moscow speeches incited German soldiers and police to revolution, and that they concluded with the words: "Long live the German Soviet Republic." The Russian reply ignored the specific German charges and declared that the speeches were only intended to reach German minorities in Russia, such as those in the German Republic on the Volga.

Obviously the time is coming when international radio conventions must deal with this issue. Propaganda carried across frontiers can be just as powerful a weapon in time of peace as it was in time of war. Sentinel lines, barbed wire, and anti-aircraft guns provide no defense against vocal persuasion which wings its way over oceans and continents on invisible ether waves.

Before long our Senate chamber and House of Representatives, our state legislative halls, and even our aldermanic chambers will be wired so that selected proceedings can be carried to the radio audience. Thanks to the microphone and loudspeaker, interest in government is growing among millions of citizens. Radio has done more than any other agency to make women realize what it means to have the vote. Intelligent women voters are often shocked to hear the uncultivated voices, the bad grammar, and the poor logic of the men they have chosen to represent them in public office. Many thousands of women voted against Al Smith in 1928 because they "did not like the way he sounded over the radio."

The prospect of broadcasting legislative debates would make the competition for speaking time much keener. Speeches would be shorter and more meaty. Legislators who do not now take the trouble to participate in debates would be more inclined to do so. One can well imagine hearing from the Senate chamber in Washington a series of compact, well-delivered fifteen-minute addresses on both sides of some important public question. The daily radio hour might easily become the most stimulating and valuable feature of Senate proceedings.

It is evident from all this that radio, which Edmund Burke might call the Fifth Estate, is just beginning to exercise its all-persuasive influence. Already broadcasts from mountain tops, from ocean depths, from polar regions, from planes and dirigibles spanning oceans and continents, have become a matter of daily routine. Through radio we participate personally and instantly in man's great adventures. Space has disappeared and all mankind speaks and listens in unison.

Radio is a magic instrument of unity and power destined to link nations, to enlarge knowledge, to remove misunderstanding, and to promote truth. But it will not achieve these things unless we keep a more watchful eye on those who use it and those who control it. We must stir in them a greater sense of their responsibility for the proper employment of this modern miracle.

Today radio's chief purpose is to make money for those who control and use its mechanical devices. It threatens to prove as great a disappointment as the moving-picture for those who sense radio's undeveloped power as an agency of education, culture, and international good-will. There is a great opportunity through endowment to divorce a few first-class stations from commercial control. Federal supervision must also receive a different emphasis. The public is entitled to a more ideal interpretation of that "public interest, necessity and convenience" which broadcasting is supposed to serve under the Radio Law. We can and should avoid the crippling restrictions of complete government control and the unhappy alternative of abject subservience to the profit motive.

33. Interview with Sylvester L. Weaver, Jr.

by Mike Wallace

This is one of a series of thirteen Mike Wallace interviews, produced by the American Broadcasting Company in association with the Fund for the Republic for the purpose of stimulating public discussion of the basic issues of survival and freedom in America today. Mr. Weaver was connected with the National Broadcasting Company from 1949 to 1956—as vice-president in charge of television from 1949 to 1953, as president from 1953 to 1955, and as chairman of the board 1955–1956. He has operated his own firm since that time. His professional career has been entirely in the fields of advertising, radio, and television. The interview is reprinted here with Mr. Wallace's permission.

WALLACE: This is Sylvester "Pat" Weaver, one of the great creative forces in the brief history of television. Formerly president of the National Broadcasting Company, Mr. Weaver created *Wide Wide World, Today,* and *Tonight* and originated the television "spectacular." He now questions whether television is failing to fulfill its role in a democratic society. He says: "The television set will become like a juke box in the corner of the room, to keep the kids quiet." We'll find out why in a moment.

ANNOUNCER: *The Mike Wallace Interview,* presented by the American Broadcasting Company in association with the Fund for the Republic, brings you a special television series discussing the problems of Survival and Freedom in America.

WALLACE: Next to working and sleeping, we Americans spend more time watching television than doing anything else. What we see on television—the drama, the comedy, the newscasts, and the documentaries—all help to shape the way we think and feel and act. Tonight we'll try to find out how we're being shaped and by what. Our guest, Sylvester "Pat" Weaver, former president of the National Broadcasting Company, now a television consultant, recognized as one of the most creative minds in the industry.

Mr. Weaver, first of all let me ask you this. Several years ago when you were head of NBC you wrote a memo which said in part that television must be the instrument which prepares us for progress into tomorrow's good society or steels us to fight for our democratic way of life. How well do you think that television is living up to that idea?

WEAVER: I'm disappointed in what's been happening in the last couple of years.

WALLACE: How, specifically, is it failing?

WEAVER: I think it is going from open forums to closed forums. It is lacking in balance. It is really reducing its over-all mission to doing nothing but being largely a story-telling medium. That is, all the shows are really either game shows or story-telling shows.

WALLACE: And its function should be what?

WEAVER: It should reflect as a communication medium the whole richness and pluralism of our society. In other words, we should have all the magic of live performances in the New York theatre. . . . the great issues in documentaries and telementaries presented . . . we should have all of the people passing across our sets. It is a port, you know, through which you can look out on the entire world, but if you aim it only at a film projector and show the cans out of Hollywood, together with some game shows that can be presented cheaply and get pretty good audiences on a commercial value, you are degrading a service, and I am afraid that's what's happening.

WALLACE: Whose fault is it?

WEAVER: Well, of course, no matter which group I discuss I'll be speaking of my closest personal friends, so I guess I can just start making enemies from the top. But as a former advertiser, a former agency head of the radio and television department twice, and a network man twice, both in early radio and television, I know the needs of different units and I am afraid that I must point the finger at the managements of the television networks.

WALLACE: Why the managements?

WEAVER: An advertiser spends his money to sell goods. An agency is his agent; that is, they are not free to play any very important role although they have a very positive and affirmative influence, I think. But basically the needs of the advertiser do not involve any ability to balance programming. They have specific needs. The agency cannot actually run the schedule, although I must say in radio we took the ball away from the network and did a fair job but not anything like the potential radio offered. In television I am afraid the only force that can balance programming and give us what we should have in the home and be responsible for the influence the set has is the management.

WALLACE: In a recent issue of *TV Guide,* Ed Murrow places the blame elsewhere. He says: "Look at our recent 'See It Now' show on radioactive fall-out . . . because it was not sponsored, we had only a limited Sunday afternoon network" instead of having a large audience at a better time. And Mr. Murrow blames the advertising agencies which are reluctant to buy shows that don't "pull a mass audience." What about that?

WEAVER: It is certainly true that you can take the position that if the management of television will only program those things that get big audiences and that can be sold commercially, the advertiser can share part of the blame. But we have already had on television a whole range of experiments with commercial sponsorship. They proved you can get advertising support for fine things and for information programs and for controversial things. They involve certain formulas, problem-solving formulas, and you start with the real needs of the advertiser and try to solve the problem of protecting his interest and at the same time give the public a balanced diet of programming.

WALLACE: Of course this is the nub of the issue. Is it possible to give the public a balanced diet which will include public affairs shows, news shows, documentary programs, controversial shows, and at the same time serve the needs of the advertiser who doesn't want to offend anybody—who searches, we are told, for bland things that will make people happy and make friends for their products?

WEAVER: Not all advertisers search for bland things. We have had live dramas in television—all of which will be off the air except two on CBS; when I left NBC in the fall of 1956 we had eleven live hour dramas—all gone now or will have gone by this summer. We have a change going on in program schedules that reflects a retreat before what we knew when we started in 1949 . . . that the advertiser wanted game shows and cheap shows. They didn't want to spend any real money. But management is supposed to overcome those things and get what they think is the right thing as far as television is concerned. Now in terms of public affairs programming, we built the telementary series which was less successful here, but we were able to find clients who would present telementaries that did have a certain element of controversy in them and present them at night in pre-empted time on NBC, and we got big audiences and very favorable, overwhelmingly favorable results. In other words, it can be done. Ed Murrow was on at 10:30 Tuesday nights for a couple of years, and sponsored.

WALLACE: And is no longer there.

WEAVER: But why?

WALLACE: That's what I would like to know.

WEAVER: I just can't believe that there wasn't a solution, a more intelligent solution than putting him on Sunday afternoon once a month.

WALLACE: Perhaps the issue boils down to this: You talk about Ed Murrow. CBS Vice President Richard Salant, in a recent speech, says: "We've let ourselves get pushed into agreeing that 'public interest' really means that kind of program in which not much of the public is really interested . . . let's admit we're in business to entertain. If we do that, perhaps that will end this attempt to make us over in the image of the British Broadcasting Company, but with advertisers." What about that?

WEAVER: I am sure that he would like to withdraw it. At least I think he would. You can't really have in your hands the power that television has in this country in this time of crisis and be agreeable to solving the problems by letting it become the jukebox in the corner of the room to keep the kids quiet and just pile on one crime or Western show or game show after another—moving news out of network time as has now been done by all three networks, and gradually abdicating any responsible role on a whole series of assumptions about the public. The American public is 170,000,000 people, different people with different backgrounds, but you can gather most of them for great attractions of many kinds. You can get hits that have very high quality in them as hits in all show business have had high quality, not low quality. You can do special things with which most Americans have had no experience at all, like ballet. At NBC we got 30,000,000 viewers watching the Sleeping Beauty ballet. You couldn't have done that if you couldn't put on a ballet. I doubt if there were three million people who had ever seen a ballet.

WALLACE: You say that television has an important social function that it's failing to perform properly, and since our most important job right now probably is to survive and to remain free, let's try to focus if we can for a while on that. What do you believe television should be doing in the field of news and current affairs that it is not doing now to keep us as a people informed and properly concerned?

WEAVER: We should have a great important report to the nation at least once a month by each of the networks, at night, in premium time, that is, network time. I think we should have a news service that really spends a lot of money in developing a coverage of this country and everything that happens in it, live and with tape. That is far beyond what we are presently doing. I think that beyond the information programs, and there should be all sorts of informational telementaries, we should be going into the cultural field and showing all the good things that we know people, when they have a chance to learn about them, will become interested in . . . their tastes upgraded and their standards elevated.

WALLACE: What about the reluctance of networks to permit news commentators to editorialize? Do you feel there is sufficient editorial comment on television currently?

WEAVER: No, but I have always felt again that television is a communications instrument. Frankly, I would rather hear what Walter Lippmann has to say than most television commentators, if they'll pardon me. I think that we should use television to bring in the opinions of men who have done enough and said enough to have stature and the respect of most thinking people regardless of the shift of their opinions from left to right. In other words, just to use the right to editorialize, to let somebody write an editorial who may not have the stature of most of the major reporters or commentators, is less important, it seems to me, than to use the mechanism of television to bring into every home in the country the product of the best minds on the situation as they see it.

WALLACE: A Harvard professor, W. Y. Eliot, has written that "there might be real merit" in Congress passing a law requiring television stations to devote at least one half-hour in the evening—that is, at popular times—"for political discussion . . . preferably by debate." What would you think of that?

WEAVER: I am usually against the law stepping in to try to solve all problems. I don't think that really is the way they get solved. Of course in this case a half-hour political show undoubtedly would get no rating at all on whichever station had it, whereas under the impact of a dedicated management solving problems you will come up with information shows that do get ratings. I'll point again to ones we had on NBC that reached thirty or forty million people. You won't do that by putting on a show every week at the same time that everybody finds out is pretty dull. There are other ways to do it.

WALLACE: According to a *New Yorker* magazine article about you back in 1954, you once wrote a memo in which you said that for the good of everybody television must tackle what you called "great issues"—"great themes," but you didn't say what they were. What do you think they are? . . . Perhaps we have been talking about some of that although in that particular piece you did not say what they were. What do you think they are?

WEAVER: There are many, but there are several vital ones right now, in which television under dedicated management can be the catalyst, the driving force, and the most important element. First of all, most people I don't think fully appreciate what's going on, that is not only that we are in a revolution but that it is a good revolution, that all the trouble we are having comes from the fact that we are not any longer blinding ourselves to unnecessary human suffering and misery on which most societies have been based and are still based in most countries today. The American society, having its political equality and the American dream which started back when the country started and electrified the whole world, is still going on and getting stronger all the time, if we don't fail it. We have economic abundance now in this country, we have strength of political equality and we have the beginning of a social equality; that is to say, fewer and fewer second-class citizens, so that you can have a family and have any goals at all for the child. In other words, there is almost nothing that children have to feel is forbidden to them. Now this is not true in any other societies, with few exceptions like Canada, in the world or in history. Television, an adult medium that gives a balanced diet, that has a charge to make people learn about the history of mankind and the arts and culture, couldn't help but accelerate this march into the future.

WALLACE: What you say makes sense. The thing that comes to mind is this. Television is a business that needs a profit to continue to operate. Are you perhaps not saddling it with too much social responsibility?

WEAVER: Of course the record is very clear as far as I am concerned. We made an awful lot of money at NBC television starting back in 1951, and I started running it in 1949. I am sure they would love to have the profits that I made, today. I think that a business must be an upward-thinking, going business and when you try not only to program down but to solve everything on a business basis you are liable to get into some rather difficult things. Let me explain what I mean. You build a show for heavy viewer interest to get just a rating, as they have been doing. Now here you'll get a pretty good rating and make pretty good sales and then all of a sudden advertisers find out the light viewers aren't watching and then that the slightly light viewers aren't watching and while the rating is there the people at the sets are drifting away. Parents won't fight kids any more at 7:30 or 8 o'clock to make them watch a show acceptable to the whole family but rather will drift off and just let the kid and maybe one person reading stay in the room. All these things affect commercial vitality. They affect circulation, and circulation is what you sell, what you buy. As this comes about and the service is degraded, its actual appeal becomes less because the cumulative audiences are low. Result, the ratings break. They have broken already. That's pretty well understood.

WALLACE: What you are suggesting is that not only must there be more statesmanship, if you will, more social responsibility in television, but you're also suggesting that the businessmen in television don't really understand their medium as well as they might and the future of their medium?

WEAVER: I certainly am saying exactly that and have been saying it right along. I have been saying right along that unless you understand advertising and its real uses, unless you keep the uses for all advertisers, you have national problems solved by innnovations and patterns, not by the radio formula everybody's drifting back to—the things they started at NBC, the special audience usage, the prestige usage, the in-and-out, the magazine concept, all this stuff. Instead of having a broad base of support you limit it, you narrow it. That's true of circulation. If you don't appeal to light viewers, you lose them.

WALLACE: Until now we have been talking about what television can do to strengthen our society and preserve freedom, but to do this, of course, television must have certain freedom itself. Freedom from censorship, pressure groups, from fear of serious controversial issues perhaps offending some people. If your career as a television executive what restrictions on freedom have you seen?

WEAVER: I haven't seen very many, although you mentioned one to me before we went on the air, so I guess we do have some sort of controls. I am aware of the pressure as an advertiser and agency man, as I said earlier, who is in business to sell goods and who is not going to go out and offend people. But the groups of people who try to have their way with management, the special organized groups, are just groups that must in large part be shown, because most of them have a very limited viewpoint that's so overconcerned about a stereotype of some kind that I think they're in error. I think the networks are and have and will continue to resist those pressures.

WALLACE: Let's be a little specific. The South, for instance, is very sensitive about the race issue. Many religions will oppose portrayal or discussion of what they consider touchy subjects. Various groups bring pressures to bear, and perhaps justifiably, on the networks. The networks want to protect their sponsors, and certain sponsors simply don't want to offend various groups.

WEAVER: There are two parts. The sponsor part I have explained. The other part—the pressure group part—I am aware of some of it. But as head of NBC from 1949 to 1956 I have very few if any positive recollections of any attempt made by the pressure groups to influence us where it was sufficiently important to pay attention to it. I mean I know that all kinds of kicks are made. You'll hear from everybody all the time about something. You know, lawyers don't want lawyers put in the villain's role and national groups don't like this or that.

WALLACE: Mr. Weaver, the *New Yorker*, October, 1954, talks of your job at Young and Rubicam producing the Fred Allen shows and said: "What particularly endeared Weaver to Allen was his agility at fending off or sidestepping the crippling three-way censorship that the sponsor, the advertising agency, and the network were forever trying to impose. 'It was fearful,' said Allen. 'But Pat fought the censors like a pioneer fighting Indians.' " So you must be aware. . . .

WEAVER: This is the third kind of censorship, really. We were talking about advertisers who are afraid of offending people, pressure groups trying to make you not have the villain a banker or lawyer or anything else—as far as I can find out, no villains. The third censorship, which I fought, really was comedy censorship. The comedy department, the continuity acceptance department of NBC, twenty-odd years ago in 1935 and 1936 would read meaning into lines that, believe me, made me wonder. I used to go and say, "That's not dirty." That was essentially my role.

WALLACE: But Mr. Weaver, the fact remains, doesn't it, that there is censorship that prohibits full discussion of certain moral issues, certain religious issues, certain racial issues within dramas and in discussion programs on television?

WEAVER: But I would say really that is not one of the key issues of television, not today. For instance, *The Open Mind* has covered some absolutely wide-open subjects—I wouldn't even mention them on this program.

WALLACE: Let me ask you about this kind of censorship. Last year on CBS-TV, television interviewed Nikita Khrushchev and Marshal Tito. There was immediate reaction from Congress. One Congressman offered a resolution that TV reporters who wanted to interview Communists should first submit their questions to the Secretary of State for approval. Another said that Ed Murrow "should explain to the American people" why he did not put what the Congressman considered tougher questions to Marshal Tito. What's your opinion of this kind of thing?

WEAVER: We certainly have a division, I am afraid, in the country between those who believe in the open society, who want a freedom in the marketplace of ideas, who are perfectly secure in their belief that the people, when they are given information of all kinds, have enough judgment, even common sense and intelligence, to be right. Nikita Khrushchev's broadcast was, I thought, a splendid thing for which CBS deserves tremendous credit and they should do it much more often.

WALLACE: Who, in the final analysis, has got to fight the people who do want to impose censorship upon television? That is—I was about to say sensible censorship but who's to say what is sensible censorship? Who is to fight it? Does it again become the function solely of network management?

WEAVER: Network management is the only group that has the central position. While they must get support from their advertisers and their agencies—and I am sure they are for any sensible kind of arrangement; after all, they all really want a much better television service than they are getting—certainly it has to be network management.

WALLACE: Up to now you have said in various ways that television, perhaps, is not fully answering the needs of people of the country. Do you expect the major figures in television and the networks to change their ways as you would like to see them do?

WEAVER: Well, as a rule major figures change only under pressure, not from persuasion, and so I really think that they probably will not change their ways but that their ways will have to be changed for them through pressure of competition. That is the American way of life—competition—to give the people now throughout the country who have had an exposure to the magnificent, wonderful, marvelous kind of new miracle society that we're moving toward through television. As it shrinks its coverage, as it reduces its service, as it no longer brings the excitement and glamour of the Broadway theatre and the coverage of the whole world into the home and into the towns and big cities, together with the great cultural things, I think the people will reach out and want that material. They have gotten a taste of it. They know it is there and they'll want it. There are ways they can get it. If the network locks up the television set and says "O.K., so many Westerns, so many crime shows, so many game shows, news in station option time, and we'll throw them some bones on Sunday afternoon and that's good enough for them," first of all, I think, in time, those managements will be changed. But in the meantime the competition, the pressure on the networks will come from other ways of distributing this kind of programming material—and there are other ways of doing it outside of national television. Pay television is one way. Another way is through the theatres themselves. There are ways of doing that. It is too technical to go into here. But looking into the vacuum created as the networks abdicate certain areas of culture and information that they have been in, some new enterprise will move in and take over.

WALLACE: Pat Weaver, I surely thank you for coming and spending this half-hour and unburdening yourself. I sincerely hope that your good friends whom you have assaulted in number if not by name will look at you with even renewed respect for your speaking your mind in this fashion.

34. Television News: Reporting or Performing?

by John F. Day

Mr. Day was director of news for CBS at the time of writing this article. As a former newspaper reporter, correspondent, and managing editor, he records here the impressions of his first year in broadcasting. He is currently Editor and Publisher of the Exmouth Journal, Ltd., *in London, England. This article was a talk at the University of Minnesota School of Journalism and Mass Communications in February, 1956. It is reprinted with the author's permission.*

I have noticed recently a tendency of certain television newsmen to endorse products and to permit their names and their faces to be associated with the promotion and sale of these products. Personally, I feel embarrassed when I see these ads. And that is true despite the fact these persons do not work for CBS News and thus are actually no responsibility of mine. But I feel very strongly that newsmen should be newsmen, not pitchmen. And these men carry not only the nomenclature of newsmen, but of television newsmen.

Is television news going to destroy itself in commercialism before it gets out of its swaddling clothes?

Television as it exists today is show business. For that reason at least one branch of journalism is trying to adapt itself to a strange new world. In so doing it has taken on some of the elements of show business. Not all of these elements are new, of course, for TV news is in part an extension of radio news, and radio too is in show business. But television is show business with a capital S.

Television news has tried to fit into this glittering show world by a series of compromises which affect both the persons working in TV news and the product they put on the screen.

First, let's consider the effect on the individual.

Journalism has never known such a disparity in pay as exists at the network level in radio and television. The office boy, the news writer, and the news editor make more than the average for their jobs on metropolitan newspapers. Enough more to make recruiting from newspapers practicable, but surely not enough more to make the jobs utopian. The reporter who works with a camera team and develops pictorial stories requires new skill, and his job cannot be compared exactly with a particular newspaper job. But his pay is comparable to what a news writer or editor makes.

It is when a newsman becomes a performer, a director, or a producer that things get wild. Here it's what a man can command and what the traffic will bear. It's what he and his union and his agent can get.

To one who worked in newspapers as long as I did, it seems strange to have a news correspondent's *agent* making formal calls upon him to discuss better assignments and fatter contracts. But that's part of the element of show business, and that's the way the system works.

I should explain here that correspondents (men who are reporters but who also appear on mike and camera), as well as directors and producers, have basic contracts calling for staff salaries that are NOT in the high brackets. Thus persons who do not earn commercial fees are paid well but not fabulously. However, on top of the basic contract there is an elaborate fee system which makes it possible for some to soar into the upper five- and even six-figure annual pay brackets.

To state it mildly, this is not always fair. Too much depends on luck, agencies, and the whim of the public instead of the competence and depth of the newsmen. With certain reservations—such as the fact that unions set some of the fees—it is indeed free enterprise. But the system sometimes results in a man making three or four times as much as another who is as good or better and who works just as hard in the same field. It results in some men making many times more than the persons to whom they are responsible. This doesn't make for the best in responsibility, loyalty, and morale. There has been talk, of course, of abolishing the fee system or of pooling fees, but nothing has come of it. And I am not hopeful that anything will.

The wide-ranging pay; the fact a newsman may belong to the American Federation of Radio and Television Artists (note the word artists); the fact he may have one or more business agents; that he may even have a publicity agent; that he may get fan mail of the type no newspaper or magazine writer could get—all these are elements that identify television newsmen with show business.

Another element is the fact that television newsmen (and, to be sure, radio newsmen as well) are called upon to deliver commercials. Ever since it has had a news service, CBS, thank heavens, has tried to keep the news and the advertising separate by having persons other than the reporters deliver the commercials. We are not simon pure in this respect. The pressures from advertising agencies, and sometimes from persons within the company, are strong and unrelenting. In the matter of "lead-ins"—introductions to commercials that may in themselves endorse the product—a couple of sponsors have gotten a foot in the door by pressing our newsmen into reading them.

This may lead to the question of whether one can be half a virgin. I'm not positive about the answer to that, but I do believe one can do some flirting around without becoming a prostitute.

Seriously, CBS News strives with continuing vigilance to protect its newsmen from the chore of doing commercials.

Please understand, I have nothing against advertising. I am in fact one of its strongest advocates, because I believe its effect has been immeasurably great in the expanding of economy. But in this field I believe in segregation. Let someone not associated with the news deliver the sales pitch. It is worse than unfair to ask a man to try to deliver an explanation of some world crisis in one breath and an appeal to his listeners to buy a certain remedy for aches and pains in the next.

Now that I have pointed out some elements of commercialism as they affect individuals in television and radio news, I think it is important that I make this observation: Big as is the money that floats around in electronic journalism, I personally know of not one single case where a newsman has deliberately slanted a story, omitted a story, or added a story because his sponsor asked for it. I have no doubt there have been such cases. But they must be few, for talk of things like that travels fast. I'm happy to say that although electronic newsmen may become prima donnas, they don't become crooks.

I said earlier that because television today was show business, both the individuals and the product have made compromises to adjust that business. I have talked so far about the individuals. Now to the product.

Television news has been cut to a pattern that says, "Television is an entertainment medium; therefore television news must entertain."

Who says television HAS to be an entertainment medium? Some broadcasting officials act as though they have been handed holy writs saying, "This is your air and your electronic gadget; you're to use them solely for the purpose of making money and amusing the morons."

It's fortunate for all of us that the real leaders of the industry don't assume such an attitude and that the Federal Communications Commission keeps an eye cocked on the industry's obligation to operate in the public interest.

The men who created this modern miracle don't look upon it as merely an entertainment medium. For instance, there is Dr. Vladimir Zworykin, who is an honorary vice-president of RCA and who invented or developed many of the electronic instruments that make today's TV set possible. He, I read, has always thought of television's chief service not as entertainment but as extending human sight to places where not every man can go.

And FCC Chairman George McConnaughey had this to say not long ago: "Television has got to be more than merely a means of entertainment. It is an unequalled medium for enlightenment and education. There is no significant difference of opinion on the validity of the proposition. The mechanics for implementation are a different story."

Still and all, virtually everything in television today is geared to the thought that TV has to entertain. On behalf of the program planners it must be said they are trying to raise the public's taste as well as cater to it.

But primarily they're trying to raise standards of taste in the realm of entertainment. And this whole business of trying to uplift just a little while aiming mostly at a low common denominator is a bit like handing your wife a rose while whaling the daylights out of her.

Meanwhile electronic journalism must hew a path through the jungle if it is to achieve a goal of adding something to human knowledge. While recognizing that most people today regard television as an entertainment medium, electronic journalism must not forget this is a convenient assumption, not an unalterable law of the universe, or even a proved fact. While conceding that news must be presented interestingly if it is to compete and hold an audience, newsmen must not succumb to sensationalism. Just as it is true that a good newspaper doesn't have to be dull, television news can inform with liveliness and vividness. The point is that it must not forget that its primary purpose is to INFORM.

Radio news went through its formative stages in the 1930's; came of age during the 1940's. I am not particularly happy about the trend in radio toward a vast number of five-minute summaries, but the 15-minute programs such as our *World News Roundup*, *News of America*, and the Ed Murrow news have held their own, and I believe we will soon see the addition of some half-hour news-in-depth shows. So, basically, radio news has achieved stature; it knows what it can do and where it stands.

Television news, on the other hand, is in ferment. It is growing and it is improving—let there be no doubt about that. But it also is groping for answers.

No one on the outside has made criticisms of television news that we who are working in it haven't made. I have never known of any craft, trade, or profession so thoroughly self-examining and so unceasingly self-critical; so willing to try to find ways to improve itself.

We at CBS News carry on a continuing study that amounts to intense introspection. We keep asking ourselves such questions as: "What IS television news?"; "What should it be?"; "How can we best present the news each day?"

We have not resolved the basic question of what the television news program should do. The argument revolves around the matter of whether television should use only that news the medium can do best, or whether it should undertake to report ALL the news that is of importance or interest. The news television can do best, of course, is (1) the event "live" as it actually happens, and (2) the strictly pictorial story on film. But a full report—one designed to make the viewer reasonably well informed by the TV medium alone—requires use of the "ideal" story that is difficult or close to impossible to illustrate well.

I am a strong advocate of the complete news report. I believe that television is a basic medium for conveying information and adding to human knowledge. If we don't use it in this way, we abdicate all claims to its being a basic news medium. I think we MUST not become obsessed with the pictorial and the merely entertaining at the expense of the meaningful. We must find ways to present difficult stories more effectively. None of us who work in news at CBS is satisfied that we are making the best possible use of the visual medium. Too much of what we present is superficial and unimaginative. But we are trying, and trying hard, for better technical and editorial quality. We try incessantly to find people who can THINK in visual terms, who can plan the best possible pictorial coverage of the story which lies in the realm of ideas. We carry on continuing experimentation with and discussion of new types of cameras, film sound recording devices, and other equipment. We discuss the relative merits of sound film and silent film; of animation techniques; of showmanship—roughly comparable to "readability" in newspapering; of the length and type of film clips; of the balance between voice reporting and picture reporting.

And at the same time we must keep an eye on costs. Television news is an expensive product, no matter how you figure it. The viewer hears the newsman say, "Now we take you to San Francisco . . ." and presto: there is a film or a live picture of a California flood. But the picture doesn't just happen. It takes planning and it takes M-O-N-E-Y. In all probability we have had to install loops and connections in the San Francisco station. (When I first made my acquaintance with such figures as, "Installing loops and connections . . . $1,800," I thought those loops must be made of platinum, studded with diamonds. But they're just cables.) Then there is the line charge for the cross-country switch. That runs to another $1,500 or so. And these of course are more or less incidental costs to the covering of the news and the producing of the show. We had to have cameramen and reporters on that as well as many other stories. For the same show that used the transcontinental switch, we may have sent a cameraman and his excess-baggage equipment from Tokyo to Hong Kong to cover a story that runs a minute or minute and a half. Is the show worth all these costs? Well, circulation must be counted here. Some 11,000,000 persons or more see it.

Mention of audience size brings up that bugaboo "ratings." This is a bugaboo the broadcasting industry very largely brought on itself. Certainly it is important to have as accurate a picture as possible of how many people are watching or listening to a given program. But I am appalled by the importance placed on decimal-point changes by the broadcasting industry, by advertising agencies, and by companies who buy time. Rating systems are valuable as guides. Enslavement by them is tragic. Currently because it's all part of the show-business system, electronic news can't ignore the rating obsession. If the news doesn't do well against certain competition, it is liable to be moved to another time period or to lose its sponsor. Those are the facts of life. Even so, electronic journalism must keep its balance and remember that a high rating is not the end to which all means must be subjected. There IS such a thing as a quality audience, and there IS such a thing as the growing sophistication of the American television viewer. The late Mr. Mencken notwithstanding, the American public is not uniformly the "booboisie."

And here I must say that while I am critical of some practices in the broadcasting industry, I have no sympathy whatever for the intellectual, pseudo-intellectual, or would-be intellectual who declaims: "All television stinks; I wouldn't have a set in the house." In the first place, how does he know it stinks if he doesn't watch its development? He is mighty contemptuous of others who make judgments on so little evidence. It is more than irritating to have someone say, "Television news is nothing but poor-quality newsreel," and to find upon questioning that this critic has seen exactly two television news shows in his lifetime.

It is the duty of the responsible intellectual not only to "have a television set in his house" but to be concerned with it. Television is one of the greatest social phenomena ever to hit any civilization. No other single factor has more impact on the political, social, economic, cultural and even moral life of this country. Is television then to be pooh-poohed and scoffed at by persons allegedly interested in values? On the contrary, those who are genuinely concerned with the quality of American life should be making their weight felt in applauding the good and making sharp but constructive criticism of the bad fare served up on television.

Earlier I asked whether television news was going to allow commercialism to destroy it before it really gets out of its swaddling clothes. I have pointed out some of the dangers that make such a fate possible, but I am optimistic that the pitfalls will be circumvented and that television news will, in the not too distant future, achieve the stature of the good newspaper and the good radio program.

To do this it must improve both its editorial and its technical resources. It must overcome or at least lessen its logistical problem. Mobile units are a long way from being mobile. Films can't move with the freedom of words. The biggest newspaper goes to cover a national political convention with perhaps 15 persons. Television must go to cover a convention with a veritable army of men and machines. In fact, TV requires considerably more men behind scenes to get a man on camera than the army takes to get a man on the firing line. It must incorporate into its product the same sort of diligent, intelligent, meaningful, and objective reporting and writing and editing which has characterized the best in American newspapers and in radio and which has given American journalism its high place in the world. And then it must present this product with imagination and taste—yes, and, where suitable, with showmanship.

Certainly a large order; no doubt of that. But so unlimited is the horizon of television that I am reasonably confident these things will be achieved. Recently, in making an intramural estimate for future plans of CBS News, I hazarded a guess that our operation five years from now will have much greater scope and much higher quality in programming and will employ approximately twice as many people.

The technical people tell us that the following developments are here or on the way:

1. Color films, and live events in actual color.

2. Magnetic stripe on film for audio recording. Current sound-on-film quality is, as you know, generally pretty poor. The new process should approach high fidelity.

3. Magnetic tape for recording pictures. This is still some distance in the future, apparently; but eventually there will be a process that will permit recording and instant playback.

4. Closed-circuit systems that will nearly eliminate moving film about the country by plane. An editor in New York will monitor film on video tape in Los Angeles, choose what he wants, record it in New York, edit it, and use it on his next news show.

5. Facsimile machines instead of teletypes for the rapid movement of raw news copy and scripts.

6. Trans-Atlantic, and eventually worldwide, television networks through the scatter system of bouncing signals off the ionosphere.

All these, of course, are merely the tools. But what tools! Given brains, moral fibre, and clearly-sighted goals (no small order!) where else can television news go except forward?

35. Broadcasting and the Journalistic Function

by Lee Loevinger

Judge Lee Loevinger, former law professor, Minnesota Supreme Court Justice, and head of the Justice Department's antitrust division, was appointed to the Federal Communications Commission in 1963. This selection is taken from his first speech as a Commissioner. It was originally delivered before the Association for Education in Journalism at the 1963 national convention in Lincoln, Nebraska. It is reprinted with the author's permission.

> The time, it is to be hoped, is gone by, when any defence would be necessary of the "liberty of the press" as one of the securities against corrupt or tyrannical government.

Thus, in 1859, John Stuart Mill began his classic statement on the liberty of thought and discussion. With eloquence and cogency seldom equalled and never surpassed he argued that complete liberty of thought and discussion is necessary for learning and demonstrating truth. It is, he said, a pleasant falsehood that truth always triumphs over persecution, and it is a piece of idle sentimentality that truth has any inherent power denied to error which permits it to prevail. In words that are often forgotten but never irrelevant he declared:

> If all mankind minus one were of one opinion, and only one person were of the contrary opinion, mankind would be no more justified in silencing that one person, than he, if he had the power, would be justified in silencing mankind. Were an opinion a personal possession of no value except to the owner; if to be obstructed in the enjoyment of it were simply a private injury, it would make some difference whether injury was inflicted on only a few persons or on many. But the peculiar evil of silencing the expression of an opinion is, that it is robbing the human race: posterity as well as the existing generation; those who dissent from the opinion still more than those who hold it. If the opinion is right, they are deprived of the opportunity of exchanging error for truth: if wrong, they lose, what is almost as great a benefit, the clearer perception and livelier impression of truth, produced by its collision with error. . . .
>
> Complete liberty of contradicting and disproving our opinion is the very condition which justifies us in assuming its truth for purposes of action; and on no other terms can a being with human faculties have any rational assurance of being right. . . .
>
> The beliefs which we have most warrant for have no safeguard to rest on but a standing invitation to the whole world to prove them unfounded.[1]

Recognition of the importance of the function of furnishing information and ideas to the public is the foundation of the First Amendment to the Constitution which withdraws from the federal government all power to abridge the freedom of the press. But it is the journalistic function of disseminating information and ideas which is given a constitutional protection, not simply the publisher's privilege of making a profit. Mill took some pains to point out that the principle of individual liberty, or freedom of discussion, is not involved in government regulation of trade as such.[2] Many businesses have had their profit-making opportunities curtailed by legal limitations thought to be in the public interest. There is nothing inherently unconstitutional or improper in this. If publishing or broadcasting is to enjoy an immunity from legal interference greater than that of other commercial enterprises, this must rest upon the performance of some public function not performed by other business enterprises. Neither entertainment nor commercial advertising will, by itself, justify such an immunity. The journalistic function of disseminating news and ideas will.

The point was well stated by Chafee, reporting for the Commission on Freedom of the Press, when he said:

> Because of the great value of this affirmative task to society, the press is granted by the Constitution an immunity from law which is not accorded to any other profit-making group or indeed to anything else except religion. The press is to be free so that it can give the community the service needed from the press. This essential principle has been too much overlooked of late. Leaders of newspapers and radio have frequently been inclined to glory in the fact of protection without bothering about what is protected. Many of them have behaved as if the First Amendment were a high board fence behind which they could hide and do whatever questionable acts they pleased—pay substandard wages, refuse to bargain collectively with their employees, break solemn promises about the quality of future radio programs, form monopolies. Meanwhile, . . . they have not given enough thought to improving the performance of the press as a whole.
>
> Another point often forgotten is that the First Amendment was not adopted to protect vehicles of advertising and entertainment. They are legitimate and beneficial activities, but so are stock-broking and circuses, which receive no constitutional immunity. The more newspapers and radio allow advertising and miscellany to swamp news and ideas, the greater the risk of losing some of their privileged position. Freedom of the press exists to enable the press to perform its essential task of dispensing news and ideas.[3]

Another aspect of what has been aptly termed "the first freedom"[4] is that it requires a diversity of voices. As Judge Learned Hand has observed,

> . . . neither exclusively, nor even primarily, are the interests of the newspaper industry conclusive; for that industry serves one of the most

vital of all general interests: the dissemination of news from as many different sources, and with as many different facets and colors as is possible. That interest is closely akin to, if indeed it is not the same as, the interest protected by the First Amendment; it presupposes that right conclusions are more likely to be gathered out of a multitude of tongues, than through any kind of authoritative selection. To many this is, and always will be, folly; but we have staked upon it our all.[5]

Similarly, Justice Black, speaking for the Supreme Court, has said that the First Amendment

. . . rests on the assumption that the widest possible dissemination of information from diverse and antagonistic sources is essential to the welfare of the public, that a free press is a condition of a free society.[6]

It thus becomes appropriate to inquire how diverse the sources are from which the public receives its news. While newspaper circulation has been increasing fairly steadily during the last half century, in both absolute and per capita figures, the number of daily newspapers has been declining. These are the data:

Year	Population [7]	Daily Newspaper Circulation [8]	Daily Newspapers [9]
1910	92 million	22 million	2,202
1920	106 million	28 million	2,042
1930	123 million	40 million	1,942
1940	132 million	41 million	1,878
1945			1,749
1950	151 million	54 million	1,772
1960	179 million	59 million	1,763
1962	187 million	60 million	1,760

However, the decline in the number of newspaper news sources is even greater than indicated by these figures. For the number of daily newspapers under common, or chain, ownership has increased from 62 in 1910 to 560 in 1960, while the number of cities with competitive newspapers has decreased from 689 in 1910 to 73 in 1962. These are the data:

Year	Newspaper Chains [10]	Chain Dailies [10]	Owners Separate Daily [11]	Cities With Dailies [12]	Non-Competitive Cities [12]	Competitive Cities [12]
1910	13	62	2,153	1207	518	689
1920				1295	743	552
1930	55	311	1,686	1402	1,114	288
1940	77	364	1,591	1426	1,241	185
1950	70	386	1,456	1422	1,305	117
1960	109	560	1,312	1461	1,382	79
1962				1476	1,403	73

The dramatic decline in the diversity of newspaper sources that is indicated by these data raises the question whether we may look to broadcasting to provide the diversity which we are losing in newspaper publishing and to perform the journalistic function. The formulation of an answer to this question requires some consideration of the scientific and engineering parameters of broadcasting.

Broadcasting is a method of transmitting messages by electromagnetic radiation. While energy is radiated in many apparently different forms, they are all electromagnetic in nature, obey the same basic laws, and travel through space at the speed of light, about 186,000 miles per second. Under some conditions electromagnetic radiations exhibit wave properties, while in other phenomena they behave as a continuous flow of small discrete quantities of energy. For communications purposes, radiant energy may be regarded as a wave phenomenon having frequency and wavelength. . . .

There are about 600 television stations operating commercially and about 70 noncommercial or "educational" television stations. Although there are substantial geographical areas which have no satisfactory television reception, it is believed that television is available to over 90 per cent of the population.[13] The barrier to more television stations has been, until recently, the limited number of channels available, the high power required, and the extreme sensitivity to interference. The availability of UHF channels together with the requirement that all receivers sold in interstate commerce be equipped to receive all channels will certainly result in the establishment of many more TV stations. However the relatively high cost of constructing and operating a TV station and the paucity of program material probably means that for the foreseeable future there will be substantially fewer TV than radio stations.

It is apparent that as the number and diversity of newspapers has decreased, the number of broadcasting stations has increased. Indeed, the increase in broadcasting, with its competition for advertising revenue, has undoubtedly contributed to the decline in the number of newspapers. Has broadcasting provided the diversity which we have largely lost in newspaper publishing?

The question is much easier to ask than to answer. To begin with, there is substantial cross-ownership between publishing and broadcasting. Of the total number of AM stations more than 10 per cent are affiliated with newspapers.[14] This percentage has been steadily declining since 1940, when more than 30 per cent of a much smaller number of stations had newspaper affiliations.[15] However, the trend simply reflects a changing profit picture, as newspaper investments have been shifted from AM to TV, and more than 30 per cent of the TV stations are now affiliated with newspapers.[16] As previously noted, only about one-fourth of the commercial FM stations represent truly independent enterprises, the rest being owned by AM broadcasters.

In broadcasting, as in publishing, there are chain or multiple owners. Unfortunately there are not complete and exact data available regarding the extent of multiple ownership in broadcasting. The FCC is now installing electronic equipment which is expected to enable it to secure such data. Using present records and sources, the staff has estimated that about 45 per cent of the total number of commercial stations are owned by one of approximately one thousand multiple station owners. Two countervailing observations must be added to this conclusion. First, the number of multiple owners of broadcasting stations is relatively large, and the number of stations owned, on the average is rather small, being about 2.5. Second, these over-all figures are not really a fair indication of the extent of concentration of control in this field. That can be ascertained only in the light of a variety of other factors, including the size of the communities involved, the location and geographical concentration of multiple-owned stations, the power, frequency, and service areas of the stations involved, and a collation of multiple broadcasting ownership with inter-media affiliation. Thus, in the 25 TV markets over 70 per cent of the TV stations are controlled by multiple owners, and less than 30 per cent are independently owned.[17] Further, virtually all of the multiple-owned stations in such markets are affiliated with AM or AM-FM stations, and about a third of all TV stations in these markets are affiliated with newspapers. These data acquire greater significance when it is realized that these 25 markets have a population exceeding 70 million, or about 40 per cent of our total national population; these markets contain over 30 million TV homes, which is 60 per cent of the total number of TV homes in the country, and the stations in these markets reach about 50 per cent of the households in the country. Every indication is that within the next decade the concentration of population within these metropolitan areas will increase.

The affiliation of TV stations with networks, and their dependence upon networks for program material, may be even more significant. As is well known, there are just three TV networks in the United States. These three networks provide the major source of program material for all but 35 TV stations in this country.[18] In most communities the choice of TV programs is even more limited, by reason of the distribution of stations. There are 311 TV markets in the nation.[19] More than half of these—167—have just one TV station. Another 60 markets have just two stations. So well over two-thirds of all the TV markets in the country have a choice of programs from not more than two stations. Another 66 markets have three stations. Just 18 communities in the entire United States have more than 3 commercial TV stations. Of this 18, there are 13 with 4 stations, two with 5 stations, one with 6 stations, New York has 7 stations, and Los Angeles has 7 VHF plus 2 UHF stations.

Unfortunately there is no simple, convenient, and valid method of summarizing, expressing, or appraising the significance of such concentration data. It is apparent that the total number of broadcasting stations has increased in recent years and is continuing to increase. This is simply a result of the fact that broadcasting is a new medium. Still, if we count broadcasting stations with newspapers as sources of news, the total number of independent enterprises furnishing news to the public has increased despite the decrease in the number of independent and competitive newspapers. However, with an increasing population, a growing economy, more efficient means of travel and communication, the proliferation of problems and ever-expanding intellectual horizons, we need and should expect an increase in the number and variety of news sources. From this viewpoint the growing concentration of control of broadcasting stations and among all the mass media is cause for grave concern.

However, those who are concerned with the social significance of broadcasting must not lose sight of the fact that it is a *mass* medium. Indeed broadcasting is much more of a mass medium than even newspapers. There are about 56 million households in the United States, more than 90 per cent of which have TV sets. Total morning and evening daily newspaper circulation is about 60 million.[20] I have not found statistics as to the degree of duplication in newspaper circulation, but assuming a duplication ratio as low as one-third—which is below the estimates I have been given—it is apparent that more homes have television than have daily newspapers. Less noticed and discussed but more pervasive even than television is contemporary radio. There are three times as many radio sets in use as television sets, and the public is buying four times as many radios annually as television sets.[21] Americans own more than 184 million radio sets, which is one for every man, woman, and child in the country. The number of AM stations has quadrupled since the end of World War II. Radio has become a locally oriented medium and competes with newspapers for local advertising more than TV does.[22] Americans watch television in the living room or the recreation room several hours every day, but they listen to radio in the kitchen and the bedroom, in the automobile, on the beach, while walking down the street, and in almost every other conceivable place during almost every waking hour.

It is sometimes overlooked that the press requires its public to be at least literate. It cannot communicate to the illiterate and has small appeal to the much more numerous group of semi-literate who read with difficulty or reluctance. In any event, reading requires some slight effort, more than radio listening or television watching. Broadcasting speaks to any who will listen and it informs the illiterate equally with the educated. It is clearly the most popular—in every sense of the word—of all the communications media.

Since broadcasting by its very nature requires the protection of an exclusive government license to operate in a part of the electronic spectrum, it seems proper to ask what public ends are served by reserving this large segment of the spectrum in the face of the clamorous competing demands for spectrum space from other services.

It seems clear that the most significant public interest served by broadcasting, and the end that most clearly justifies the spectrum allocations made to it—is the performance of the journalistic function and its contribution to the maintenance of a political democracy and a free society. This conclusion provides a guide to the proper role of government in relation to broadcasting.

Complaints have often been made that broadcasting—and particularly television—has mainly programs that are banal, boring, and bad, that there are excessive and offensive commercials, and that there is a woeful lack of public service programming. It should be noted that these complaints have come principally from critics, intellectuals, and the educated elite. Although the commercial rating services appear to have been exposed as ranging from the fraudulent to the unscientific, and as being almost entirely worthless, there have been a few proper and apparently useful studies of public attitudes toward television.[23] These show beyond much doubt that the public at large likes and is generally satisfied with television, although it does not consider it excellent or beyond improvement. The average viewer regards television primarily as entertainment and finds it relaxing and pleasant. He objects to the commercials because they are too interruptive, too repetitive, and irritating; and he thinks there is too much violence in the shows for children. He would like television to be more informative, but he does not watch the educational or information programs in any significant numbers. He does watch the news programs regularly.

The intellectuals and the educated elite are less pleased with television for several reasons: they are more discriminating and have different tastes, they have a wider range of interests and less need for television, and television generally is designed for the mass taste rather than for an appeal to the intellectuals and the educated. However, the educated elite, along with their brethren of the masses, watch entertainment in preference to information programs in overwhelming numbers (9 to 1) when given the choice. Other than entertainment, only the news programs are popular with all segments of the public.

These attitudes and habits are not confined to this country. They are also found in Britain, which has the largest and best-developed broadcasting system outside the United States. In Britain, as in the United States, the intellectuals resist and disdain television; and they use the term "idiots' lantern" for the device which sophisticated teenagers in this country call the "boob tube."[24] However, the famous "third programme" of BBC, which presents entirely informative and educational programs and cultured entertainment attracts only about 2 per cent of the audience.

On the other hand, all who are concerned with the mass media—a term which I use in a descriptive, not a derogatory sense—must take some account of the fact that the intellectual leaders, the critics, and college graduates generally do criticize and disdain most television programming. I think it would be a mistake to attribute this merely to intellectual snobbishness, rather than to a genuine judgment that television is falling far short of achieving its best potential quality, that much television programming is actually objectionable, and that television is failing to perform its most important functions adequately.

In all these circumstances and given these differing views and evaluations of the medium, what should be the role of government in relation to broadcasting? The issue is not a simple one and the answer is not easy. But it must be answered by those who exercise the power of government in the communications field; and it is answered by them, either explicitly or implicitly, whether they act or fail to act in the field. I think the answers are more likely to be sound if they are explicitly articulated and related to my own views of this problem, noting, however, that they are still tentative and subject to change upon further study and reflection.

To begin with, it seems to me that the intellectuals' disdain for television programming is somewhat unrealistic. The rate of consumption of program material in television makes it inevitable that most programs will be trivial at best. By way of comparison, consider the situation in the motion picture industry. Movie theaters use one or, at most, two programs a week or perhaps 100 a year. Yet movie exhibitors have great difficulty in securing enough pictures worth showing to keep their theaters operating continuously. Even in the heyday of motion picture production—before television hit the fans—there were several hundred pictures a year being produced but the great majority of them were exactly the same kind of trash that television is showing today. But television uses upwards of half a dozen programs a day, or more than 2,000 a year. This is at least twenty times the rate of consumption in the movie industry, and, because of the wide exposure of television, the possibility of repetition of programs is less. To expect the mass production of intellectual quality is simply foolish. If we insist upon the right to watch television day after day and night after night, and if we demand variety in the programs shown, it is inevitable that the majority of programs will have little content and no value except to amuse the idle, the bored, and the ignorant.

The inevitability of mediocrity in mass production is as true of all other fields of intellectual creation as it is of television. As television is the *bete noir* of intellectuals, book publishing is their beau ideal. Yet James T. Farrell, one of our leading authors, has recently complained of the "vast amount of junk printed."[25] He wails that "There is a conspiracy of malignant mediocrity surrounding me," and charges that "the publishing system" is encouraging too many writers and too much writing.

The truth is that every golden age is one that is past. In the present we are always overwhelmed by the chaff and the dross; it is only after the winnowing of time that we discern the grain and the gold and count them as the glory of the time of their production.

Beyond these considerations, I do not deem it the proper role of government in a democracy to establish standards of taste or to dictate the intellectual or cultural level of expression of the mass media. Much television programming is trash by my standards. But I would not ban all trash from the air if I could. That which I disdain is esteemed by others. One man's trash is another man's treasure; one man's vast wasteland is another's verdant vineyard. For my own taste, I prefer radio to television and much prefer reading to the intellectual indolence of surrender to any form of broadcasting. I still regard the book as the greatest visual aid to education yet devised. But I do not know of any practical method of imposing my preference upon the public, and, were it possible, I do not know of any legal or ethical warrant for attempting to establish my taste as a standard for others. I think there is, or should be, a corollary to Voltaire's famous declaration in favor of free speech. Although I abhor that which you choose to read or hear or see, I shall defend to the death your right to make your own choice in these matters.

Applied to broadcasting, this view implies that the public must actually be offered a choice. The physical parameters of the broadcasting process which limit the opportunities to broadcast and require some regulation by government suggest that government should exert its power to insure that there is a range of choice adequate to provide for all tastes.

How is this to be done?

In the same year that this country first set up the Federal Radio Commission, Great Britain sought to insure quality and variety in radio programming by setting up the British Broadcasting Company as an enterprise separate from, but subsidized and controlled by the government.[26] Despite the generally acknowledged quality of performance by BBC in both radio and television, Britain was not wholly satisfied with this system, and in 1954 Parliament established the Independent Television Authority to provide supplementary TV broadcasting service supported by advertising.[27] At least one objective and well-qualified American observer has concluded that the presence of two competing broadcasting organizations has improved British broadcasting and provided a wider range of choice for the audience.[28] While the results of Britain's laboratory experiment in the effects of competition in broadcasting are instructive for this country, some aspects of the British system seem quite uncongenial to America. Both BBC and ITA are subject to governmental control through appointment of their directors by the government.

Hours of broadcasting are limited by the government, and the broadcast day is much shorter than in this country.[29] The amount of advertising is limited, and commercial spot announcements are not permitted to exceed 10 per cent of the day's program time.[30] Editorializing and local political broadcasts are forbidden,[31] although news coverage generally is good. In addition to the two government-chartered broadcasting companies (BBC and ITA), there are some ten contractors or program companies, which produce television programs, and an Independent Television News service, which is supported by the program companies.[32] On the whole the programming of television in Britain has been more independent and diverse than one would expect from its economic structure. But the system is based upon a degree of government participation and supervision that would certainly not be acceptable in this country, and it seems significant that recent developments have been in the direction of the American system.

Another possible method for securing variety in broadcast programming is to undertake an appraisal of the programs presented and to refuse to renew licenses of those who do not present an adequate variety—or quality of program, if that is what is sought. There's been much discussion of this technique by the FCC, the Commissioners, the broadcasters, critics of the Commission, partisans of the Commission, and many others. Ostensibly this is the approach which the Commission follows. In fact I believe that the Commission does not, and cannot, use this technique for one crucial reason: it has no operational criteria which might enable it to apply this technique. The Commission does get a good deal of information from applicants and licensees concerning their proposed or past programming. These reports classify programs according to categories roughly corresponding to those set forth in the 1960 statement of policy on programming.[33] These comprise fourteen categories ranging from "opportunity for local self-expression," through children's, religious, educational, and news programs to, "entertainment programming." I have no particular quarrel with these categories as representing what the Commission originally stated them to be—"major elements usually necessary to meet the public interest, needs and desires. . . ."

However, there are two insuperable difficulties to the use of any categories of programs as a basis for judging the performance of a broadcasting station. First, there are no standards whatever as to the quantitative proportions that are desirable, or even permissible, among such categories of program classification. Second, the classification of programs by descriptive category gives no information at all about the most important aspect of programming, which is its quality. Entertainment—which is by far the largest category for most stations—covers everything from rock and roll and Westerns to opera and Shakespeare.

The Commission does employ two rubrics in talking about programming standards: the broadcaster must make a positive and diligent effort to determine the tastes, needs, and desires of the public in his community and to meet them; and the broadcaster must comply with the representations he makes to the Commission as to programming. But neither of these ideas advances us very far toward operational criteria.

As to community needs, the Commission does not know and has no means of ascertaining whether the diligent efforts of applicants and licensees are effective or not. Inevitably any but the most stupid applicant or licensee will design his programs to meet the reported desires of the community—or vice versa (which is about equally probable). Any sizable community will contain groups representing every variant of taste, desire, and need. In any event, the desires of the community are never ascertainable or expressible in sufficiently specific or definite terms to provide much of a standard for judging performance in any but the most extreme case. Finally, the injunction to go to the community and see what it wants simply encourages the tendency to seek the largest mass audience without making any provision for the needs or desires of minority groups—such as college graduates and music lovers.

Even the standard of compliance with representations as to programming is not very useful until we are prepared to say which deviations may be an improvement and which a degradation of the promised service. As for the quality of programming—if we desired to do so we would not know how to begin defining a standard, what instructions to give the staff to suggest an applicable standard, or what questions we should ask to begin formulation of such a standard.

In short, I believe that under our system of government any effort at direct control of programming is not only wrong but futile.

In saying this, I do not suggest that the Commission should not and does not have a duty to prevent the broadcasting of positively objectionable matter. Clearly it does.[34] Lotteries, frauds, obscenities, and incitements to riot or violence are improper for broadcasting and may be grounds for denying or revoking a license. As to such matters, broadcasting may be subject to stricter standards than other media.[35] For example, the law may properly permit sale of the unexpurgated version of "Lady Chatterley's Lover" as a book, but its most intimate scenes would hardly do for dramatic representation on television during the dinner hour. However, my concern in this discussion is not with these rather obvious, aberrant, and rare extremes, but with the issues involved in developing a rational, consistent, and practical approach to the problem of insuring that ordinary broadcasting does serve the public interest.

I think that there are practical operational techniques and criteria which avoid the difficulties and objections I have mentioned, are consistent with democratic principles, and which offer reasonable assurance that broadcasting will operate in the public interest. I would have the Commission recognize these as cardinal principles in this field:

First, the Commission should seek the maximum attainable dispersion and diversity of station ownership and control.

Second, the Commission should require adequate performance of the journalistic function by all broadcasters.

Third, the Commission should encourage enterprise, experimentation, and innovation in broadcasting.

Of these principles, the first is surely the most important. We can insure a variety of viewpoints and of programs only by having a diversity of broadcasters with differing viewpoints and program policies. The Commission has given some recognition to this principle in its rules providing that no person or enterprise may be licensed for more than 7 AM stations, and 7 FM stations, and 5 VHF-TV stations plus 2 UHF-TV stations.[36] But I think the Commission rules are too liberal and their construction and application has been too lax. An applicant who owns 20 other broadcasting stations may get the same consideration in contention for his 21st as an applicant with no other broadcast interests. Indeed, experience gained at other stations may be an advantage. It seems to me that if the fully competitive nature of broadcasting is to be maintained and if we are to have diversity in viewpoints and programming, then there must be a strong presumption in favor of awarding licenses to those with no or the fewest other interests in the field.

In considering multiple ownerships, the Commission should not count by arbitrary categories but should consider all other interests and affiliations in the communications field. It makes no sense to say that a man with 7 small AM stations cannot acquire an eighth, although that same station may be acquired by a corporation with 7 TV stations, 7 FM stations, and 6 large AM stations plus a string of newspapers. Generally I would consider newspaper ownership or affiliation a substantial negative factor in determining qualification for a broadcast license. Needless to say this is not because of any prejudice against publishers but solely because I believe in promoting the widest diversity in the operation of the mass media.

It has been argued that multiple owners and newspaper publishers frequently make the best broadcasters and produce the best programs. That may be true. But if we must choose between program quality and diversity I would choose diversity without hesitation. Dictation of the broadcasting programs of the nation by a private monopolist, or by a few oligopolists, is not discernibly preferable to government control. The basic objection to government censorship is not that the power of control is exercised by government officials—who really are just as decent and well-intentioned as broadcasters or publishers. The objection to government censorship is that the power to control broadcasting should not be held by any single person or agency. The power should not exist except as the dispersed responsibility of numerous individual broadcasters. Every increase in multiple ownership is a step away from dispersed and individual responsibility and a step toward centralization of control and monopoly. In an area as sensitive and vital as mass communications I think we should take no steps in the wrong direction.

The matter is complicated by introducing conventional anti-trust concepts and seeking to define the relevant market. The Commission has assumed that each broadcasting station serves a particular community which constitutes its market. Consequently the Commission has followed a "duopoly rule" by which it has refused to permit one licensee to operate two facilities of the same category in one community. Perhaps there was a time when this rule was appropriate. Today it seems to me to be plainly inadequate. It is just as undesirable for one person to control the local newspaper and TV station or radio station as to control two radio stations in the same locality—and for the same reason.

Furthermore the local market for audience, advertising, and attractions is by no means the only significant market. For purposes of news and ideas there is a national community that is at least as important as any local community. As the journalistic function is one of the most important functions of broadcasting, any reduction in the number of independent enterprises serving that function within the national community is an injury to the public interest. While there are practical and legal difficulties with divestiture on these grounds of present broadcast holdings, I would oppose any significant increase in the concentration of control of the mass media as incompatible with the public interest, the legal mandate of the FCC, and the political principles of a free society.

Consideration of program quality also leads to the same conclusion. In the long run, diversity of control and competition in broadcasting is far more likely than any other course to produce programming that will please such minority groups as intellectuals, college graduates, drama and music lovers, and others who seek literate entertainment. Neither taste nor intellectual quality is a function of cost or expense. On the contrary, the pursuit of the mass audience necessary to sustain and secure maximum profits from large operations seems to preclude taste or intelligence and force programs to the lowest common cultural denominator. The existence of a number of competitive stations in the same community will almost certainly cause some of them to seek the smaller specialized markets of such minority groups by offering programs designed for their discriminating tastes. In the absence of the economic spur of competition there is no reason to believe that these groups will be served by broadcasters. Surely no amount of oratory or exhortation can be nearly as effective as the influence of market forces resulting from a competitive economic structure.

This has proved to be the case in radio. It is quite significant that nearly all the current criticism of broadcast programming is directed at television. In television competition has been very limited, for a number of reasons. Contemporary radio, on the other hand, is rather highly competitive. (There are nearly 9 times as many AM and FM stations as TV stations.) Radio does provide programming for a wide diversity of tastes from very lowbrow to very high brow, and even—if that be more elevated—egghead.

The FCC is now expanding TV into the UHF band and many more stations will be licensed. If we can adhere to the principles I have suggested and disperse the new licenses among a large number of independent and diverse licensees, I believe that most of the present complaints about television programming will be met.

It is not as clear that increasing competition will necessarily insure adequate performance of the journalistic function by broadcasters. Presumably it will insure that some broadcasters offer the news for those who desire to hear it, but this will not suffice either to satisfy the public service obligation of all broadcasters or to inform the mass public. Since it is the journalistic function which gives the principal social value to broadcasting, I would measure broadcasting performance principally by the degree to which it performs this function. Specifically, I would require as a minimum that each station devote at least as much broadcasting time to news as it does to commercial advertising. There is no transcendental virtue in this particular quantitative relationship and it is, at best, a crude measure. But it is clear and definite and provides operational criterion. In effect, it makes each broadcaster pay for the time he takes from the public domain for his own commercial use by devoting an equal amount of time to public service.

It may also, quite incidentally but effectively, provide some check on over-commercialization. We should surely be concerned that broadcasting not be—in Herbert Hoover's trenchant phrase—"drowned in advertising chatter." [37] Yet we must be somewhat sympathetic with the broadcasters' pleas that it is the revenue from commercials which enable them to perform public service. We should also take cognizance of the fact that newspapers today devote well over 50 per cent of their space to advertising, although this source supplies only about two-thirds of their revenue.[38] In contrast, broadcasting derives substantially all of its revenue from advertising, but, on the average, devotes less than half as much of its "space" to commercials as most newspapers. It is an indication of the inescapable difference between broadcasting and other media that commercials become objectionable, and sometimes intolerable, even when they are in much smaller proportion to total content than newspaper advertising.

There are several reasons for this. To begin with, broadcast programming is strictly linear; one thing is presented at a time and the audience has no choice about watching and hearing what is presented, while newspaper advertising is easily and often ignored. Further, broadcasting time—and therefore content—is limited. Every moment devoted to commercials is necessarily taken away from substantive program features. Newspapers, in contrast, can add space almost without limit to accommodate whatever advertising, news, and feature volume the day affords. Differences in size of newspapers from day to day and the large Sunday edition are taken for granted. Broadcasters have the same amount of time available every day. While these problems are inherent in the nature of the medium, a third factor is the creation of the broadcasters themselves. This is the interruptive nature of commercials. While other media attempt to sustain attention and mood, broadcasting seems to make an effort to lose attention and shatter mood by frequent and ill-timed interjection of commercials into all types of programs. This is the aspect of broadcasting most frequently and most vehemently criticized by all segments of the audience. Thus both public duty and self-interest combine to require that commercials not be excessive in number or time.[39]

Still, establishing a standard of excess for commercials that will be applicable to all stations is no simple matter. The principle that at least as much time should be devoted to the journalistic function as to the commercial function provides some sort of an automatic balance between commercials and public service. It avoids a purely arbitrary limitation on commercials and guarantees some reasonable amount of attention to broadcasting's most important social use. It seems to me to be the best and most practical method of judging "program balance."

Finally, I would have the Commission encourage enterprise, experimentation, and innovation in broadcasting. This means many things and the burden of action in this area will necessarily be on the industry rather than on the government. The industry must remember that enterprise means more than the private holding of title and reception of profits. Enterprise connotes initiative, risk-taking, and action. Profits are the stimulus, not the social objective. Likewise, government intervention or regulation is a means to an end and certainly not an end in itself. The economic structure, the market system, and the multitudinous forms of government regulation and intervention all exist for the satisfaction of various human needs and wants and will ultimately be judged by the degree to which they contribute to the satisfaction of the public needs and desires.

In the broadcasting field, as I have indicated, the most important function performed is that which I have termed the journalistic function. This is what raises broadcasting from a business to profession—when it is. But so far broadcasting has largely relied on newspaper resources and newspaper techniques for its operation in this field. This is, presumably, the reason why newscasters are not yet generally recognized as journalists, as evidenced by their recent exclusion from the International Press Institute.[40] Broadcasting must develop its own news sources and news presentation techniques. The electronic miracle of television is surely not employed to its fullest advantage when it is used to present the picture of a balding newscaster reading the evening news from a script. The integrity of journalism is surely compromised when a newscaster, without pause or change of pace, moves from the telling of a news story to the statement of a commercial plug. It is time that broadcasting developed some new formats and techniques for the presentation of news. I cannot and do not presume to tell the industry what they should be. Many possibilities are obvious and worth exploration, not the least of which may be the use of the television screen to augment the flow of information by combining a projected text, possibly of headline significance, with a verbal statement.

More important, broadcasting should establish its own independent news resources. At the present time, broadcasting relies on the AP and UPI and on the networks for news. There are a few notable exceptions among individual stations which have set up their own news-gathering staffs; but there is still no general or widespread news-gathering agency devoted exclusively to serving the journalistic function of broadcasting. There has been widespread and vociferous complaint among broadcasters regarding both AP and UPI. I do not now attempt to judge whether or not these complaints are well founded. It is clear that both organizations are, must, and should be, devoted primarily to serving newspapers. They do not and should not secure and disseminate news in a form that is adapted primarily to broadcasting. Obviously this is not going to be done until there is an organization established for the purpose of serving broadcasters.

The networks are, of course, dedicated to broadcasting and do have news-gathering facilities. However, the networks are commercial enterprises basically engaged in providing entertainment and selling advertising. Networks are not available to all stations, are not controlled by the stations, and, indeed, themselves have proprietary interests that are competitive with many stations. As commercial enterprises operating in a national market, networks are principally interested in national, rather than regional or local news. In all, the networks are not news-gathering agencies of the kind that newspapers have had for years, and that, I think, broadcasting must have if it is to function adequately. What is needed is a Broadcast News Association, an independent organization controlled by and responsible to broadcasters, to serve the broadcasting industry as the Associated Press serves newspapers. It should gather news and present and disseminate it with a view to its broadcast presentation by television and radio. How much the techniques of broadcast news-gathering and presentation may differ from those of newspapers I do not know. However, I hazard the guess that television has not yet begun to develop its potential techniques to present news in an informative and interesting manner. It will not do so until it has established its own independent news sources. It is high time that this effort be initiated. Perhaps the Pottstown decision,[41] and the suggestions for action from within the industry,[42] may help to stimulate some action. It is interesting to note that there is an independent organization in the television field devoted to covering news of sporting events.[43] The broadcasting business, with gross revenue of $1.6 billion from television and nearly $700 million from radio should be able to establish and support its own independent organization to gather hard news, to become truly competitive with newspapers, and to utilize the magnificent modern techniques of electronic communications to expand the intellectual horizons and extend the intelligence sources of the public it is supposed to serve.

How far the FCC can go in stimulating and encouraging such enterprise and innovation I am not prepared to say just now. It is obvious that the power to grant, withhold, and renew broadcasting licenses, and the mandate to take such actions in the public interest, enables the Commission to exert a powerful influence, if not actually to require action. However, it is surely better for all if the initiative in such matters is taken by broadcasters and if the Commission is not forced to take action to compel broadcasters to fulfill their responsibilities to the public.

In summary, I suggest three principles as practical operational criteria for insuring the public responsibility of broadcasting in America today.

First, the FCC should secure variety in programming through maximum diversity and dispersion of station ownership.

Second, the broadcasters should provide, and the FCC should require, adequate performance of the journalistic function. This means at least as much time for news as for commercials.

Third, broadcasters should engage in, and the FCC should encourage enterprise, experimentation, and innovation. An independent Broadcast News Association appears to be one of the things that broadcasting enterprise should now be prepared to provide the public.

Undoubtedly there will be some who object that these suggestions involve undue interference with the business of broadcasting, just as there will be others who will say that this approach does not do enough to insure high quality, education, and culture in broadcast programming. To both sets of critics there is a single answer. There is something more important than either profits or culture—and that is democracy. That is the value that I seek to serve above all. If we lose the democratic foundations of the Republic, then nothing else matters. If we preserve democracy in this country, then we will be able to achieve anything else to which we aspire. That, in any event, is my faith.

NOTES

[1] John Stuart Mill, *On Liberty* (1st ed., 1859), chap. II.

[2] *Ibid.*, chap. IV.

[3] Zechariah Chafee, Jr., *Government and Mass Communications,* Report from the Commission on the Freedom of the Press (University of Chicago Press, 1947), pps. 794–795, 797.

[4] Morris L. Ernst, *The First Freedom* (1946).

[5] *United States* v. *Associated Press,* 52 FSupp 362, 372, quoted with approval by J. Frankfurter, concurring in *Associated Press* v. *United States,* 326 US 1 (1945).

[6] *Associated Press* v. *United States,* 326 US 1, 20 (1945).

[7] *Statistical Abstract of the United States;* and U.S. Department of Commerce Bureau of the Census, *Current Population Report.*

[8] Raymond B. Nixon and Jean Ward, "Trends in Newspaper Ownership and Inter-Media Competition," *Journalism Quarterly* 38:3 (Winter 1961); *Statistical Abstract of the United States* (1962 ed.), p. 524.

[9] Nixon and Ward, *op. cit.*, supra note 8; *Statistical Abstract of the United States* (1962 ed.), p. 524; *Historical Statistics of the United States, Colonial Times to 1957,* p. 500.

[10] Data stated for 1940 and 1950 are for 1939 and 1949 respectively. Nixon and Ward, *op. cit.*, supra note 8; Agee, "Cross-Channel Ownership of Communications Media," *Journalism Quarterly* (Dec. 1949), p. 415.

[11] Computed from preceding data.

[12] Data for 1950 are interpolated from 1944–1945 and 1953–1954 data. Nixon and Ward, *op. cit.*, supra note 8; 1962 *Editor and Publisher International Yearbook;* ANPA General Management Bulletin No. 6, Jan. 30, 1963, p. 16. Competitive cities include cities where newspapers have joint printing plants. The number of such cities increased from 4 in 1940 to 21 in 1962.

[13] *Television Magazine,* June 1963, p. 99.

[14] 1963 Broadcasting Yearbook, p. 14.

[15] Levin, *Broadcast Regulation and Joint Ownership of Media* (New York University Press, 1960), p. 5; also see Nixon and Ward, *op. cit.*, supra note 8.

[16] See references in notes 14 and 15.

[17] FCC staff study as of July 1961.

[18] *Television Magazine,* June 1963, p. 60.

[19] *Television Magazine,* June 1963, p. 99.

[20] *Editor and Publisher Yearbook*—1963, p. 215.

[21] *Advertising Age,* Jan. 15, 1963.

[22] *Ibid.*

[23] See Gary A. Steiner, *The People Look at Television* (1963). This is the most elaborate and sophisticated study that has been reported. Also see Burton Paulu, *British Broadcasting* (1956), and Burton Paulu, *British Broadcasting in Transition* (1961). I am also grateful to KSTP, of St. Paul-Minneapolis, Minnesota, for graciously showing me a copy of an unpublished research report on "Public Opinion About Television." This summarizes the results of a survey made at the Minnesota State Fair in the fall of

1961 in which 16,000 respondents answered an extensive questionnaire regarding television. The conclusions of all these studies are not only consistent but closely similar.

[24] Burton Paulu, *British Broadcasting in Transition* (1961), p. 178.

[25] *Washington Post*, July 30, 1963, p. A 3.

[26] Burton Paulu, *British Broadcasting* (1956).

[27] Burton Paulu, *British Broadcasting in Transition* (1961).

[28] *Ibid.*, pps. 191–220.

[29] *Ibid.*, pps. 78–79.

[30] *Ibid.*, pps. 46–47.

[31] *Ibid.*, p. 16.

[32] *Ibid.*, p. 61.

[33] FCC No. 60–970, Report and Statement of Policy, July 29, 1960, *Federal Register*, Aug. 3, 1960, p. 7291.

[34] See *Trinity Methodist Church* v. *Federal Radio Commission*, 62 F2d 850, 61 App DC 311 (1932), cited with approval in *Noe* v. *F. C. C.*, 260 F2d 739, 104 App DC 221 (1959).

[35] *Cf. Burstyn* v. *Wilson*, 343 US 495 (1952).

[36] 47 CFR secs. 3.35, 3.240 and 3.636.

[37] Speech as Secretary of Commerce to the First Annual Radio Conference in 1922.

[38] *Advertising Age*, Jan. 15, 1963, p. 56.

[39] See "Advertising: Chatter Irks Buyers of TV Time," *New York Times*, May 21, 1963, p. 56.

[40] *Washington Post*, June 8, 1963, p. A 13. [Editor's note: A resolution passed by the IPI Assembly, meeting in June 1967, amended this provision and now allows the participation of television and radio journalists.]

[41] *Pottstown Daily News* v. *Pottstown Broadcasting Co.*, Sup. Ct. of Pa., No. 155, Jan. 1963; also see *Associated Press* v. *United States*, 326 US 1 (1945).

[42] See "No Creative Contributions from TV Stations?" *Broadcasting*, July 8, 1963, p. 25. In this article, the author suggests that TV stations should form and support a cooperative to finance experimental programming.

[43] *New York Times*, July 7, 1963.

36. The Role of the Commentator as Censor of the News

by Gunnar Back

Mr. Back is associated with the Triangle Stations in Philadelphia, Pennsylvania. His remarks were written for the formal opening of the State Historical Society's National Mass Communications History Center at the University of Wisconsin, January 25, 1958. They are reprinted here with the author's permission.

We who broadcast the news take for granted, I hope, that the radio and television news commentator censors the news, either deliberately or without conscious intent, generally the latter. Even the impartial or objective broadcaster doesn't really exist and cannot, though his listeners often want to believe he does. He can be impartial, or a non-censor, only in varying degrees.

As our chairman, Quincy Howe, has shown in a recent issue of the *Saturday Review*, the news commentator as opinion-giver and interpreter of the news has been on the wane since the war. He remains in network radio, and there are a number—of which Mr. Howe is one—who continue to fill the assignment with integrity and influence. On the other hand, I can think offhand of a network news commentator with a considerable national following in radio who used to rely for interpretation of Washington events on one or two sources which were completely right-wing conservative, for the newscaster himself knew little of the Capitol from first hand and seemed to regard it at best as a sinister place. If he spoke of Americans for Democratic Action at all, this commentator spoke in derision; or he left out news of ADA because he felt it was an organization dangerous to his America. His broadcasts are labeled "news," and I would presume that a large part of his vast audience get much of what they know about what's going on anywhere from what he says. His conservative approach has continued to be appealing to the business people who buy time or who determine who stays on the air. I can think of another commentator of some influence who gives short shrift to the NAM and the U.S. Chamber of Commerce when their version of what's good for the country makes news.

It would at least help to designate these two broadcasters as "commentators," to properly free them to select what they want to talk about. But this would still not free them of the obligation to try to look at the other points of view, and comment on them with argument or fact, rather than emotion. Yet the presentations of which I've just spoken are designated as "news," with only a glancing suggestion that they also contain "opinion." Of course, there are other well-known network radio broadcasts that are plainly labeled "comment" or "analysis," or "news and views," which is as it should be.

Now, the local radio news commentator, as such, has all but vanished. On network television, I know of no news commentator who appears with regularity, as is the case with network radio. What we get on network television is the so-called documentary, irregularly broadcast. With the exception of such regular documentary comment as that of Edward R. Murrow, these telecasts are often sardine-canned into times when nothing like an *I Love Lucy* audience is available. In non-network television, the opinion broadcaster is a rarity, and the local documentary news commentary appears infrequently for reasons of lack of staff, money, or energy in the burdened newsman who'd like to undertake it.

Last summer I had a look at a survey made by Trendex, an audience research group by which many network newsmen may live or die, and one as reliable as any, I suppose, among the sampling techniques the broadcasting industry bows to. Trendex asked a nation-wide cross-section of people whether they depended on newspapers, television, or radio for local, national, and international news. Almost 32 per cent depended solely on radio and television for local news. Almost 39 per cent counted on radio and television for national and international news. A surprisingly small number, under 7 per cent, said they counted on newspapers *and* radio or television. They made one or the other their sole source of news, so it seems.

National surveys of local news programs in television have shown that the local television news program invariably has a larger, and usually a much larger, audience than the network newscaster.

I think we need, therefore, to look now at how local radio and television are carrying out the trust so generously placed in our hands, a trust granted us, true enough, because we can supply the news in the easy, quick, capsule form twentieth-century Americans seem to enjoy.

Local newscasters in radio and television are usually regarded by their public as news commentators, an impression many of these newscasters don't try to discourage. But for every local newsman with some claim to the title—either as writer or broadcaster, or both—there are a half dozen news readers on the air. They read what is written for them in a radio or television newsroom, or, far more often, they tear off prepared news from a teletype machine. These newscasters have a far greater impact on the 39 per cent of the public who depend solely on radio and television for news, than do the diminishing band of network news commentators and analysts.

There are well over three thousand AM radio stations in the United States. The competition of television has driven most of them into the news and music category. News on the hour every hour is the boast of most of these three thousand stations. More than 20 of these radio stations are available to us in what we call the greater Philadelphia area, and this does not count the FM stations. The newscasts of this score of stations every hour on the hour give us a veritable downpour of news meant to bring awareness, enlightenment, and understanding. What is true of Philadelphia is true across our land in varying density.

What's the origin of this downpour, assailing the ears of the 39 per cent who count on radio and television for all of the news? Most of it comes from the radio news roundups, condensed by United Press and Associated Press staffers from their major wire to fit the requirements of the 5-minute newscast, or the 15-minute one, when there is time for that. On any night a car driver can move across the dial and hear the same news, in the same words. There are exceptions, of course. Many radio and television stations have newsrooms, news writers, and news reporters. But it appears news doesn't have to be prepared professionally in local news rooms in order to sell.

Members of this panel are familiar with these 5- and 15-minute prepared news roundups provided by the principal news services. They have been devised and are turned out to fill a need developed by radio and television stations. The wire services compete strongly to sell more of these news reports teletyped for direct air use.

The censorship of time in the news which a good part of America hears is well illustrated by the tendency to make the 5-minute newscast the major format in the whole broadcasting day. The United Press recently made a survey of the some 1800 stations which buy its radio wire. The 664 stations answering the questionnaire revealed that some put on the air up to 48 five-minute newscasts a day, including local and network. The average was 12 a day plus. The average 15-minute-long broadcasts of news in these 664 stations was 2 a day plus. The majority of the UP clients were in favor of reducing the seven daily 15-minute prepared newscasts (on national and international news) offered them so that 5-minuters could come more frequently on the wire.

The short newscast is easier to sell. It can be bought cheaply; and too many radio and television programmers and policy-makers are convinced that news can be handled adequately by a combination of a radio wire and a personable, intelligent announcer who sometimes reads the newspapers, although that's not required.

My experience for many years with these 5-minute newscasts has been this: the style is made popular—difficult concepts in the news are reduced to workaday language—and the need to be sparing of words gives you something like this, which I picked up entirely at random: Quote: "John Gates, editor of the *Daily Worker,* resigned today from the Communist

Party. He said he quit because the party was of no more importance in the country." End of news story.

Millions of listeners—who will hear no more of the John Gates story, and read no more on the subject—presumably will never really know why John Gates resigned, or what has happened and is happening to the Communist Party in the United States. Many Americans had cringed before this party so long, had been encouraged to do so, and had supported stamping out this thing of which nothing was known except that it must be feared.

I don't want to quarrel with the writer of those quick lines on John Gates. He works fast under pressure, he's obliged to get everything important in, and the whole must be compressed within a 5-minute period. The writer, and his colleagues, were obliged to treat the Rockefeller Brothers Fund report on the state of our readiness with the same brevity.

Several years ago, when a client was found for one of my free-lance newscasts, on a news and music station with a large and faithful audience, the personnel were considerably surprised when I reported in at 2 in the afternoon for a 6 o'clock evening broadcast. They didn't know that it takes time to read a whole news wire, evaluate what must and can be done with it, and then it takes time to write it, ever reducing to the minimum for the time the sponsor will buy.

The radio station manager, considered one of the astute ones in the industry, told me that he wished he could go back to having his station's newscasts written, wished he could use me for longer newscasts and commentary, but he had found that advertisers were waiting for a chance to sponsor his short news periods, and raised no complaint at announcers cutting the 5-minute summaries from the machines and putting them directly on the air, more often than not without reading them first. He sensed that his newscasts ought to sound different from the others heard in the city, that something as important as news ought to have more direct supervision, and certainly far more air time at any given point on his schedule.

Another characteristic of these 5-minute news roundups which can be heard across the land in precisely the same terminology is that each 5-minuter, provided fresh each hour, must be topped by the newest story within that hour, the most important within the hour if that's possible, but the newest in any case. This comes out of the necessity of providing something new, of being on "top" of the latest developments. I have seen again and again instances of obscure members of Congress with utterly preposterous ideas getting the lead story in these roundups coast to coast. A Midwest member, a politician obsessed with economy, conditioned by the provincialism of his Congressional district, says he will introduce a bill to abolish the Central Intelligence Agency as wasteful of money, too secret, and impotent. In a dull hour, that could get top play in literally thousands of these 5-minute news roundups over the whole country. Senator Joseph McCarthy benefited greatly by this kind of top play.

The Associated Press and United Press make no requirement that any station broadcast the roundups as they come in on the machine. It is entirely the station's responsibility to use them as they are, expand them, or use them only as a supplement to main news wires. How the material is used is up to the station itself, which generally uses it as it comes.

It is a case of the rule rather than the exception that station announcers—or announcers called newsmen—and with local reputations as news commentators, quickly edit the 5 minutes of national and overseas news provided by the teletype to make room for the local news which also goes into the 5 minutes. This editing is guided by time consideration, haste, and ignorance, or at best a fumbling at the importance of the news and its emphasis, and usually under instructions to sacrifice national and international for local. This censorship by brevity and ignorance is practiced by many local news commentators with established claims to the title on the local scene. The total audience involved is far greater than that reached by the responsible network newsmen, or local newsmen who have enough skill to defeat these hazards.

In Philadelphia, with an audience potential of millions, two local TV news programs show daily ratings (for what they're worth) at least twice as high as those of two network news commentators also heard from the two stations. The local newscasters of whom I speak are on the air twice nightly.

The network news commentators have 15 minutes. The two of us locally have 10 minutes. The network commentator, of course, does not have to give any of his time to state of Pennsylvania news, the local news in the Philadelphia region, to say nothing of news of Southern New Jersey.

The viewers who give their allegiance to me, who assume that I will keep them informed and perhaps give them guidance in today's complexity of events and ideas, expect me to talk about London and Washington and the Mid-east, as well as Philadelphia and Camden, New Jersey, and Philadelphia's Main Line.

This imposes on me as the commentator a daily discipline of selection, emphasis, use of film—national and local—which each new day says anew to me: What you want to say, what you need to say just can't be done in 10 minutes. The network news commentator has a reporter or two at Little Rock, the day's film is flown out of Little Rock at expense and effort, in time for his news. He can devote 4 minutes to covering the story. I can devote less than half that much time, and my film, if I have any at all, is a day old. Yet I seem to have more audience. The necessity of tight yet meaningful writing presses on me—nagging at all my knowledge and experience.

Day by day I must make decisions. Do I note simply that Dag Hammarsjkold has gone to London, or can I afford another line or two to suggest why he's going there? The first is the easy way; it helps save time for the minute and a half sound interviews I have with Mayor Richardson Dilworth of Philadelphia on a municipal problem. Or, is it best, since it's meaningless after all to say Mr. Hammarsjkold has gone to London, to leave the matter out until the next day, when perhaps there'll be hard news on the visit and a little more time to deal with it. The members of the Bagdad pact are meeting again in Ankara. Have I the time to remind the viewers again what countries belong to the pact, the degree to which the United States is joined to it, why Secretary of State Dulles is attending the meeting?

Some news commentators on the local scene content themselves merely with noting that Hammarsjkold is travelling and the Ankara Bagdad pact meeting is to begin.

Recently, in reporting the appearance in Washington before Lyndon Johnson's Senate investigating subcommittee of the head of a large aircraft company, I noted that he too made the uniform complaint of other defense contractors: the Pentagon is confounded by red-tape, too little money is available, there is a lack of central authority. This seems to approximate the truth—if the uniformity of the testimony before the subcommittee means anything. Now, this witness had also added that he did *not* take the gloomy view that we have lost the missiles race, nor does he favor a "czar" in the Pentagon. This was somewhat buried in the first wire service account I used. In those final desperate 15 minutes when you're trying to cut your copy to your air time, when the director is waiting for his film cues, and you have but a minute or so to comb your hair and check your tie (no newspaper reporter, or radio reporter need be concerned about that)—in this desperate time—because I needed 10 seconds—I cut out this part of the report as a lesser part of what he had to say.

Listening to my radio as I drove home that night, I heard this part of the testimony as it deserved to be, far up in the front of the wire service news roundups.

A few minutes before air time that same night, General Nathan Twining, the Chairman of the Joint Chiefs of Staff, issued a sharp statement against what he felt was the unfair impression the Senate investigation was giving. General Twining said that the armed forces of this country were ready, today or any day, to go to war, if necessary, as a team, on the basis of plans carefully laid and fully unified.

Now I suspect that news item did not make some of the local television news programs that were about to go on the air in a few minutes, as mine was. In my case, it required a minimum of 25 seconds for adequate telling. It meant throwing something else out of the straitjacket which is a television newscast, a newscast of items so inter-dependent because news items may also be film cues, and film usually cannot easily be thrown out on a moment's notice. Including the 25 seconds of the Twining statement meant eliminating something else in an already spare presentation—as likely as not a local story. We non-network broadcasters all seem to get prodded, lightly or otherwise, from the upstairs for not emphasizing local news, or in the judgment upstairs for seeming not to do so. I know of a number of TV newscasters who take great pride in oiled performances, who will not run the risk of such intrusions, unless the last-minute story is a matter of major disaster or revelation, or ordered in by the performer's writer. Some of these news performers are artists in the field of performance, they are devotedly followed by large segments of viewers who get no news anywhere else, except by accidental listening or newspaper headline scanning. So General Twining's anger was probably lost on quite a few people.

I don't think it will be anything new to members of this panel when I say that we local broadcasters, all of us, keep hearing from within our organizations: emphasize the local, cut the national and international, reduce the length of your treatment of each story, be newsy, give the folks a fast-paced summary of news events, right down to the last stabbing and knifing in South Philadelphia.

Harold E. Stassen comes to Philadelphia, and you film an interview with him. He has some ideas on how to approach the Kremlin with less rigidity. A four-minute interview. You edit it down to two. You try again, and Mr. Stassen is down to a minute and a half. But it takes 25 seconds to introduce him. In that one story alone, you've used up a fourth of your total news time.

These things are the daily lot of the news commentator who tries to meet his obligations and the responsibility he has of condensing the news as meaningful as possible to the narrow precincts of his medium. It's a grappling with the raw stuff of a day in the world, a problem common to the network and the local newscaster, who, in the end, is also regarded as a news "commentator" to so many Americans. It's an experience painfully calling on awareness, experience, and training.

This local newscaster is almost the total source of news to thousands, even millions of listeners and viewers who hear few other newscasts or documentaries on the air, or read the newspapers with any assiduity.

Since the war, we have seen a decline of the influence of this kind of news broadcaster, whether we call him that, or "commentator." Increasingly, so it seems to me, those who are left on the radio networks, are dropped by the affiliates unless they are sponsored. Able network newscasters in television are having poor luck with sponsors. No new commentators appear on network radio. None at all have found a place in network television.

All across the land, managers of programming and policy of radio and television stations have been aware of the rekindled interest in news since the Soviet satellites, and to some degree they have felt the responsibility falling on them. But I think they have met that with an increasing number of 5-minute newscasts, originating locally (when the network doesn't supply the 5 minutes). This is supposed to be the logical answer to the desire or need for more news. Not any more information at any one time, but news at many more times in the day. I know of only one major radio station which has proudly, and justifiably so, announced that since the Sputniks, the news on the hour every hour will be increased from 5-minute duration to 15. I'm sure there are others, but probably not many. And occasionally, only occasionally, a major radio station reveals, with a justifiable pride, that it maintains a news staff equal in size to the principal newspaper in the city.

At this moment, in our country's well over three thousand radio stations and more than five hundred television stations, a belt-line of the latest news is being prepared or broadcast—which is, as someone has put it, a glut of occurrences.

It's a flow of news which certainly tells us *what's* happening—fast and generally accurately—it mentions the ideas of our times, without really going into them. It's a flow of news on the air which makes us seem the most informed people of the world. But, except for the contribution of the careful workmen in our craft, it's a flow of news censored by well-meaning ignorance, the inexperience of those who deliver it, and it suffers from time limitations and conformity to industry policy practices.

We've got the "who, what, when, and where," but these days there ought to be more of the "why" in the news.

37. Reporting or Distorting: Broadcast Network News Treatment of a Speech by John F. Kennedy

by Thomas H. Guback

Dr. Guback is Research Assistant Professor in the Institute of Communications Research at the University of Illinois, where he teaches courses in popular culture, public opinion, and mass communications. This 1962 study is published here with his permission.

The American public forms opinions of foreign policy (and domestic policy) matters almost entirely on the basis of what it reads, sees, and hears in the media. To the extent that the media faithfully depict and make known background details, current developments, and informed views—in other words, the full story—the public will either have adequate or inadequate information on which to understand events, place them in context, and formulate opinions. One function of the news media in a free, democratic society as we understand the term is to provide the public with such sufficient material as will enable it to reach decisions on matters of policy, whether that policy pertains to the European Economic Community, Algeria, cultural exchange, or the Cold War.

The ways in which we view the world, and the events going on in it, are largely products of what is reported to us by the media. The media select what is presented to the public, they select the versions of reality, and the portions of the truth, the public uses to formulate opinions. It has been pointed out that the media function in ways to preserve the status quo, to reinforce prevailing attitudes. And at least in the context of the Cold War, this prevailing attitude has largely taken the form of a good-bad stereotype—"we are good, they are bad." The act of the media feeding back to us what is already a prevailing attitude, in this case "we are good," can be as much a distortion of reality as it is a gross failure on the part of the media to perform their functions in a democratic society. By neglecting to inform us about the times we have not been "good," the media create an illusionary world for us to live in where we eat of the opiate of smugness and self-righteousness.

Because local news media, in this study, radio and television, do not have the resources and personnel nor inclination to cover national and international events, the public has come to rely upon the networks for coverage and reporting in these fields. One looks to three television and four radio networks for detailed treatments of such news events. The extent to which the networks fulfill their responsibility to inform can be determined by comparing the actual event with what is reported about that event. With such a fidelity comparison, one can determine the emphasis given certain aspects of the event: the elements selected, inflated, and stressed beyond their original size, and the elements suppressed and neglected.

This case study compares an event, a speech by John F. Kennedy, with what was reported about that event on radio and television network news programs.*

During the height of the 1960 presidential campaign, the then-Senator Kennedy delivered a major speech at a Democratic fund-raising dinner in Cincinnati on October 6. It had been previously announced as a major statement on our Cuban policy. The nature of the speech, the nature of the speaker, were hardly of the type which could be overlooked as not being newsworthy. In fact, the news media were quite aware of the importance of the speech and did a fine job of alerting the public. John Daly told his ABC television viewers on October 6 that Kennedy had been busy that day "preparing what is billed as a major pronouncement tonight on U.S. relations with Cuba." Taylor Grant told his Mutual Broadcasting System audience that Kennedy's "staff members have been sending out word all day that this will be an important speech tonight, in their opinion, and only once before during the entire campaign have they released an indication of the subject matter in advance in this manner." On NBC radio, Peter Hackes noted that Kennedy was "working on what his aides call an important speech on Cuba, which he'll present tonight in Cincinnati." Lowell Thomas, broadcasting on CBS radio, told his listeners that "Kennedy's standpoint on Cuba is being given top billing by his aides. They believe his remarks on Cuba are just what most Americans want to hear."

However, as the record demonstrates, Americans never heard everything Kennedy had to say, or for that matter, a balanced summary of it. In his speech, the future President of the United States charged that the trouble in Cuba was due, in large measure, to a lack of "imagination and compas-

* The material for this analysis is drawn from published government documents. The Kennedy speech appears on pages 510–516 of: *Freedom of Communications, Part I.* The Speeches, Remarks, Press Conferences, and Statements of Senator John F. Kennedy, August 1 through November 7, 1960. Report of a subcommittee of the Committee on Interstate and Foreign Commerce, United States Senate, 87th Congress, 1st Session, January 6, 1961.

The scripts of radio and television network news broadcasts for October 6 and October 7 appear on pages 308–380 of: *Freedom of Communications, Part IV.* The 15-Minute Radio and Television Network Newscasts from the Period September 26 through November 7, 1960. Final Report of the Committee on Commerce, United States Senate, prepared by its Subcommittee of the Sub-Committee of Communications, 87th Congress, 1st Session, December 12, 1961.

sion to understand the needs of the Cuban people." He pointed out that the United States had "refused to help Cuba meet its desperate need for economic progress" by supplying "nearly all our aid . . . in the form of weapons assistance." He told his audience that the American government had frequently brought pressure on the Batista government to act in specific ways, and that another "and perhaps most disastrous of our failures was the decision to give stature and support to one of the most bloody and repressive dictatorships in the long history of Latin American repression." Kennedy stated that "it was our own policies—not Castro's—that first began to turn our former good neighbors against us."

Going into greater detail on Kennedy's speech, we find that he asked: "What did we do wrong? How did we permit the Communists to establish this foothold 90 miles away?" He gave four answers. He noted, first, that we refused to give the desperately needed economic aid:

> In 1953 the average Cuban family had an income of $6 a week. Fifteen to twenty per cent of the labor force was chronically unemployed.
>
> Only a third of the homes in the island even had running water, and in the years which preceded the Castro revolution this abysmal standard of living was driven still lower as population expansion out-distanced economic growth.
>
> Only 90 miles away stood the United States—their good neighbor—the richest Nation on earth. . . .
>
> But instead of holding out a helping hand of friendship . . . , nearly all our aid was in the form of weapons assistance—assistance which merely strengthened the Batista dictatorship—assistance which completely failed to advance the economic welfare of the Cuban people—assistance which enabled Castro and the Communists to encourage the growing belief that America was indifferent to Cuban aspirations for a decent life.

Senator Kennedy then pointed to certain acts of our government which he felt enflamed the feelings of the Cuban people. This was his second answer:

> Secondly, in a manner certain to antagonize the Cuban people, we used the influence of our Government to advance the interests of and increase the profits of the private American companies, which dominated the island's economy. . . .
>
> Of course, our private investment did much to help Cuba. But our action too often gave the impression that this country was more interested in taking money from the Cuban people than in helping them build a strong and diversified economy of their own.
>
> The symbol of this short-sighted attitude is now on display in a Havana

museum. It is a solid gold telephone presented to Batista by the American-owned Cuban telephone company. It is an expression of gratitude for the excessive telephone rate increase which the Cuban dictator had granted at the urging of our Government. But visitors to the museum are reminded that America made no expression at all over the other events which occurred on the same day this burdensome rate increase was granted, when 40 Cubans lost their lives in an assault on Batista's palace.

In giving another answer to his question of "What did we do wrong?", Kennedy noted that we continued to support a dictator who we knew was murdering his own people:

The third, and perhaps most disastrous of our failures, was the decision to give stature and support to one of the most bloody and repressive dictatorships in the long history of Latin American repression. Fulgencio Batista murdered 20,000 Cubans in 7 years—a greater proportion of the Cuban population than the proportion of Americans who died in both World Wars, and he turned democratic Cuba into a complete police state—destroying every individual liberty.

Yet our aid to his regime, and the ineptness of our policies, enabled Batista to invoke the name of the United States in support of his reign of terror.

Administration spokesmen publicly praised Batista—hailed him as a stanch ally and a good friend—at a time when Batista was murdering thousands, destroying the last vestiges of freedom, and stealing hundreds of millions of dollars from the Cuban people, and we failed to press for free elections.

We stepped up a constant stream of weapons and munitions to Batista—justified in the name of hemispheric defense, when, in fact, their only real use was to crush the dictator's opposition, and even when the Cuban civil war was raging—until March of 1958—the administration continued to send arms to Batista which were turned against the rebels—increasing anti-American feeling and helping to strengthen the influence of the Communists. For example, in Santa Clara, Cuba, today there is an exhibit commemorating the devastation of that city by Batista's planes in December of 1958. The star item in that exhibit is a collection of bomb fragments inscribed with a handshake and the words: "Mutual Defense—Made in U.S.A."

Even when our Government had finally stopped sending arms, our military missions stayed to train Batista's soldiers for the fight against the revolution—refusing to leave until Castro's forces were actually in the streets of Havana.

Kennedy's fourth answer was that we failed to convince the Cuban people, and the people of other Latin American nations, that we really were in favor of freedom and independence:

> Finally, while we were allowing Batista to place us on the side of tyranny, we did nothing to persuade the people of Cuba and Latin America that we wanted to be on the side of freedom.

Continuing with his remarks, Kennedy stated that:

> It is no wonder, in short, that during these years of American indifference the Cuban people began to doubt the sincerity of our dedication to democracy. They began to feel that we were more interested in maintaining Batista than we were in maintaining freedom—that we were more interested in protecting our investments than we were in protecting their liberty—that we wanted to lead a crusade against communism abroad but not against tyranny at home. Thus, it was our own policies—not Castro's—that first began to turn our former good neighbors against us. And Fidel Castro seized on this rising anti-American feeling, and exploited it, to persuade the Cuban people that America was the enemy of democracy. . . .

Kennedy went on to say that the "great tragedy today" is that "we are repeating many of these same mistakes throughout Latin America."

> For we have not only supported a dictatorship in Cuba—we have propped up dictators in Venezuela, Argentina, Colombia, Paraguay, and in the Dominican Republic. We not only ignored poverty and distress in Cuba—we have failed in the past 8 years to relieve poverty and distress throughout the hemisphere. For despite the bleak poverty that grips nearly all of Latin America . . . our aid programs have continued to concentrate on wasteful military assistance until we made a sudden recognition of their needs for development capital practically at the point of Mr. Castro's gun.
>
> Today time is running out in Latin America. Our once firm friends are drifting away. Our historic ties are straining under our failure to understand their aspirations. And although the cold war will not be won in Latin America, it could very well be lost there.

In short, Kennedy's speech advocated a foreign policy based on plowshares rather than the sword. In only two paragraphs, 175 words of his 2,700-word address, did Kennedy attack Richard Nixon, who had visited Cuba "five years ago," and who "could not see then what should have been obvious." Kennedy charged that "if this is the kind of experience Mr. Nixon claims entitles him to be President, then I would say that the American people cannot afford many more such experiences."

The manner and extent to which the future President's speech was reported to the public stands as a shocking condemnation of the performance of network news. When it is remembered that the importance of the speech was known before time, that the speech was delivered in a major metropolitan area equipped for the transmission of news, that the speech was made by a person who could (and did) become the next President, that the speech dealt with a newsworthy subject of national and international import, there can be no excuse for the lack of adequate balanced reporting.

When Kennedy's remarks were reported, the network news treatments were general, lacked detail, missed the main substance, and concentrated chiefly on Kennedy's two paragraphs devoted to Nixon. Kennedy's aides, in calling the media's attention to the speech, apparently included enough about the theme of the address to permit some scanty news stories to be written about it *before* it was actually delivered. However, by October 7, the day *after* the speech was made, there was no excuse for not knowing the full content of the speech and treating it in the way a "major pronouncement" should be treated.

The speech was mentioned in broadcasts on the same day it was delivered. *Freedom of Communications* lists fourteen radio and television network newscasts for that day. Of these fourteen, five failed to mention a single word about the speech, and failed to alert the public that such a speech would be made that night. Of the remaining nine, four simply indicated that Kennedy would be speaking on Cuba. What follows is the full, complete text of what was reported to the public.

John Daly, ABC-TV:

Senator Kennedy was in Cincinnati, where he spent much of the day preparing what is billed as a major pronouncement tonight on U.S. relations with Cuba.

Peter Hackes, NBC:

Kennedy will spend most of today working on what his aides call an important speech on Cuba, which he'll present tonight in Cincinnati, after which he'll fly here to Washington. . . .

Taylor Grant, MBS:

Democratic presidential nominee, John Kennedy, and his aides are hoping to attract attention tonight with a speech the Senator will make on the subject of Cuba. His staff members have been sending out word all day that this will be an important speech tonight, in their opinion, and only once before during the entire campaign have they released an indication of the subject matter in advance in this manner.

Kennedy's previous approaches to this topic have included criticism of past administration policies which he claims brought about the current situation in Cuba, but he has never been critical of the caution exercised by the Eisenhower government since Castro came into power.

Fulton Lewis, Jr., MBS:

> The Kennedy forces have charged that under the Eisenhower administration the Communist frontier has been moved from Europe and Asia to 90 miles from the Florida coast and have let it go at that. Whether Senator Kennedy does more tonight, remains to be seen, but it can be confidently expected that he will, before he gets through.

The remaining five network newscasts went into slightly more detail, apparently because the reporters were able to draw upon advance copies of the speech or other material provided by Kennedy's aides.

David Brinkley, NBC-TV:

> Senator Kennedy said tonight that Cuba is gone, and that there is not a great deal his or any other administration could do about it.
>
> And he said the Eisenhower administration is responsible. He said Vice President Nixon was there 5 years ago, praised the Batista dictatorship, and that now these same mistakes are being repeated in other Latin American countries.
>
> Kennedy concluded the job for the next President will be to avoid having the same thing happen in other Latin American countries, and he said in another speech, coming later in the campaign, he'll tell how he would try to do that.

Douglas Edwards, CBS-TV:

> In Cincinnati, Kennedy is charging the administration is responsible for Cuba becoming an armed Communist camp; that Nixon should have seen the need for helping the Cuban people during a visit 5 years ago.

John Daly, ABC:

> Senator Kennedy, in a speech tonight in Cincinnati, declared that—the Eisenhower administration "must accept full responsibility" for Cuba's becoming an armed Communist camp only 90 miles from U.S. shores. The Democratic presidential candidate, in one of his most scorching attacks on the administration, also cited the Cuban crisis in hurling a new challenge at Vice President Nixon's claim of experience qualifying him for the Presidency.

Edward P. Morgan, ABC:

> Kennedy, in a speech prepared for delivery in Cincinnati tonight, dwells on the Cuban crisis—notes Nixon's visit there and later to South America—contends the American people cannot afford many more such experiences. Without proposing any specific solution of his own, Ken-

nedy says the administration must accept complete responsibility for extension of the Iron Curtain "almost to our front yard." He blames the Castro takeover on Republican preoccupation with American business profits, the support lent ex-dictator Batista. "The greatest tragedy," says Kennedy, "is that we are repeating many of the same mistakes throughout Latin America."

Lowell Thomas, CBS:

Kennedy. Tonight Senator Kennedy is concentrating on the problem of Cuba. The Democratic candidate, addressing a fund-raising dinner in Cincinnati, again attacking the administration's handling of Castro. Kennedy, who has repeatedly called Castro a serious threat to this hemisphere, promises to move quickly to counteract Castro, if elected. Kennedy's standpoint on Cuba is being given top billing by his aides. They believe his remarks on Cuba are just what most Americans want to hear.

This skimpy reporting may be justified and rationalized on the ground that the speech had not yet been made, that the broadcast remarks were preliminary and simply alerting the public to the event. However, on October 7, the day *after* Kennedy made the speech, there can be no excuse for skimpy reporting and inadequate coverage. *Freedom of Communications* lists thirteen network newscasts for October 7. Of these, eight failed to mention the speech. Of particular note is one major network news program, *Douglas Edwards with the News,* on CBS-TV, which devoted only 34 words to the speech on October 6, and completely ignored it on October 7.

Of the five network broadcasts mentioning the speech, each offered an abbreviated and perfunctory report of what Kennedy said.

Bob Siegrist, MBS:

In Havana, meanwhile, Castro's propaganda outlets were blasting a newly discovered target—Senator Kennedy—whom the Castro propagandists denounced as a "millionaire illiterate" and "international gangster" for having described Castro-Cuba as a Communist satellite and a threat to Western Hemisphere security in what Kennedy billed, last night, as a major address on the Cuban situation—an address in which, to no one's surprise, Kennedy blamed the Eisenhower administration—and Dick Nixon in particular—for the Cuban mess.

John Daly, ABC-TV:

On the general battlefront, Republican National Chairman Senator Thruston Morton charged that Senator Kennedy is now attacking the same administration Cuban policies he once supported. Senator Morton labeled the Kennedy charge of U.S. bungling in Cuba as just "another vivid example of Monday morning quarterbacking."

Daly never explained, however, on either October 6 or 7, the extent of U.S. bungling as described by Kennedy. Lowell Thomas, CBS radio, who on October 6 told his listeners that Kennedy's "remarks on Cuba are just what most Americans want to hear," devoted only five lines of his 15-minute broadcast on October 7 to the address:

> The Democratic candidate for President is making headlines in Cuba. You can guess the reason. Last night Senator Kennedy called Cuba a Soviet satellite and asked why President Eisenhower allowed Communist influence so close to our shores. Well, Castro's puppets have something to say about that. They're terming Kennedy "a millionaire illiterate, an international gangster."

The reporting of the address on the Huntley-Brinkley NBC-TV news program concentrated entirely on Kennedy's charges against Nixon:

> Kennedy last night was in Cincinnati, where he accused Nixon of visiting Cuba, 5 years ago, and failing to see what was happening there.
> [Videotape of Kennedy speaking]
> (Kennedy tape ends: ". . . the American people cannot afford it.")

The fifth and most extensive comment on the Kennedy speech was made by Fulton Lewis, Jr., on the Mutual Broadcasting System:

> Kennedy and Cuba: Senator Kennedy took off after Mr. Nixon on the Cuban question in Cincinnati last night, and drew a considerable hand from his audience; but his political onlookers here have a feeling that he may well have stuck his neck into a controversial noose, because he apparently was not prepared as well as he could have been on his facts and he not only missed an opportunity to land some blows on the chin of the Eisenhower administration, and thus, indirectly, on Mr. Nixon, but he got himself out on several limbs on which he is very likely to be sawed off, disastrously.
>
> He built up a big to-do about the evils of the Castro regime, and what a monstrous situation it is to have a Communist satellite 90 miles off our front doorstep at the tip of Florida, and then he attempted to make Vice President Nixon responsible for this by saying that Nixon was in Cuba just 5 years ago, and should have seen the impending disaster that was then in the making, the dissatisfaction of the people with their conditions, their poverty, et cetera, et cetera. Of course, in the first place 5 years is a long time, and the conditions that prevailed in Cuba 5 years ago might not reflect any of the impending developments and events that were to take place 3 years later, and in fact probably did not.
>
> The disastrous fact is, for Senator Kennedy, that there is a highly

> authentic, sworn-testimony record of all the events of the Cuban evolution into Castroism, taken at great length by the Senate Internal Security Subcommittee in late August of this year. . . .

At this point, Lewis went into great detail, quoted from the record, and tried to show that Nixon could not possibly have seen anything wrong in Cuba because, according to the record, the economy was booming and Cuba's relations with the United States were in good condition. Lewis promised his audience that in future broadcasts he would go into still more detail about the

> shameful, shocking story of intrigue and infiltration of the State Department at the working levels, to the extent that the Department deliberately set about to do everything it could to undermine the Batista government, even though Batista had promised to hold free, open elections . . . but he never got around to it because all the while, the State Department was building up Castro at the instigation of the left-wingers and ultra-liberals who had infiltrated the picture.
>
> But this testimony, I think you'll agree, makes the statement by Senator Kennedy in Cincinnati last night, that Mr. Nixon should have seen these dire conditions in Cuba in 1955 and should have taken steps to have the Government in Washington do something drastic toward helping the Cuban economy, look rather stupid and poorly informed.

Not only did Lewis fail to report the substance of what Kennedy said, he then proceeded to tell his listeners the *reverse* of what Kennedy said:

> The fact is, the Cuban economy was in a boom status at the time. There was no poverty and unemployment and disaster to see on the part of the population because actually, the country had a labor shortage, and anybody who wanted to work could get a job at good wages for the asking.

Lewis' statement on this point is refuted by Kennedy when he said in his speech that "in 1953 the average Cuban family had an income of $6 a week. Fifteen to twenty per cent of the labor force was chronically unemployed . . . this abysmal standard of living was driven still lower as population expansion out-distanced economic growth." Unfortuntaely, Kennedy's statement was not broadcast, but Lewis' was and it became part of that material the public draws upon in formulating opinions.

The conclusion to be drawn from this case study is that Kennedy's major speech on our Cuban policy was largely ignored by network newscasts. When the speech was reported by radio and television, the description never extended beyond the political attack technique of the campaign, never probed beyond the slogan-like "Communist camp 90 miles from our doorstep," never reached beyond Kennedy's charges against Nixon. There was no mention of the four answers Kennedy gave to his own question: "What did we do wrong?" The public, if it wanted to know what we did wrong, could never have found out by listening to the news reports. No one was told that Batista, whom we supported, had murdered 20,000 Cubans, that Cuba desperately needed economic—not military—aid, that our government advanced the private interests of our businesses in Cuba, or that "it was our own policies—not Castro's—that first began to turn our former good neighbors against us." The harsh truths, the points which showed that we were not "good," were not reported to the public. The status quo and prevailing attitudes were permitted to go unchallenged while the "they are bad" stereotype was reinforced by reference to the notion that it was not our fault that Cuba was lost—it was the activities of "left-wingers and ultra-liberals who had infiltrated" the State Department.

Another important point to be observed is this: the speech was viewed by the media, and presented to the public in the guise of political campaign gamesmanship. That is, the 175 words Kennedy devoted to Nixon's visit to Cuba were magnified out of proportion and inflated to make them appear as if they had been the most significant part of the speech. The address was presented to the public in the form of campaign charge-countercharge—one candidate shouting to the public that his rival had struck out or committed an error.

This analysis of the coverage of the Kennedy address must be viewed with the following in mind. It was known before time that Kennedy would be making an important speech on our Cuban relations. In fact, several broadcasters mentioned the significance of the speech and alerted their audiences. Yet, with the advance notice, how can one justify the inadequate coverage of the speech on network news programs? When one considers that this was a major address, on a major subject, by a major speaker, and notes the quality and quantity of the treatment rendered, one is forced to question the fidelity of accounts of other events that may not be considered as significant or as newsworthy by the broadcast media.

This case study also supports the hypothesis that the public is not getting the whole truth of an event, but rather a very select and simplified version of the event, processed, packaged, and merchandised by giant industrial organizations. In place of faithful accounts, the public is getting a few fragments of truth fastened together by paste-pot journalism and slipped into a context that does not represent reality. It would seem that the gamesmanship aspects of the political campaign, and the oneupmanship of Cold War politics, have been used as pretexts for selecting these truth fragments, neglecting some and inflating others until they are distorted beyond recognition. A modern, democratic society can exist only as long as its media make faithful representations available to the public.

When this process is consciously or unconsciously disrupted, the consequences are grave, for the public can only formulate opinions on the basis of what is communicated to it.

38. News Coverage in 1959

by Chet Huntley

Mr. Huntley started radio work in 1934 and has since worked for all three major networks, covering assignments ranging from the founding of the United Nations to overseas reporting. Joining with David Brinkley in coverage of the 1956 national political conventions, he established a unique and efficient format for presenting the day's significant news. His honors include the Peabody award, an Overseas Press Club award, Nieman Fellows award, and many others. This article is the text of the annual A. L. Stone Address, delivered at the University of Montana in 1959, and later published in the Spring, 1959 issue of the University's Journalism Review.

It is published here with Mr. Huntley's permission and that of Journalism Dean Nathan Blumberg of the University of Montana.

It's exciting and thrilling to be home again!

It occurred to me that the last time I was in Missoula was in the summer of 1931. I was working a hay crop on a ranch over in Ross's Hole and a horse nipped me on the arm, with the consequence that I wound up here in the Northern Pacific Hospital with a thriving infection. I recall that a friend came by and I rode with him from Missoula to my home in Bozeman in his old Model T Ford. Just this side of Butte, a connecting rod burned out and we put the tongue of my shoe in between the connecting rod and the crank-shaft. To my knowledge that Ford went the rest of its life on my shoe-tongue. I suppose the only moral of that story is that the Model T invited, commanded, and permitted a good deal more ingenuity than the contemporary models.

But the airplane is not the ideal vehicle in which to come back to a place you have known so well almost 30 years before. All that time is made meaningless by the rush of miles beneath the wings. I have the feeling I should have come back here on the back of a good horse . . . certainly with nothing faster than a Model T. It is in situations like this that one of my generation realizes that he is not entirely adjusted to this age of supersonic flight and space probings.

Nor is journalism wedded to this age of speed with completely harmonious and meritorious results. Journalism has sacrificed too much of the "what" to the "when." A thousand and one AM's, PM's, Sunday papers, radio and television stations are demanding of the wire services that the bulletins be rushed to them within seconds of the event. The consequence is that very few editors are concerned with what really happened, and less with why.

You will observe . . . or you are aware . . . that much of our important news, particularly that dealing with foreign affairs and with Washington developments, is not the crisp, bulletin-type material. The action on an important bill in the Congress is spread over many days and there may have been no one critical point in the whole debate. An international conference begins and draws to an end and its decisions come filtering out, piece by piece, over a period of time. But here again, our editors run the calculated risk of distorting the basic story by insisting that the reporters or correspondents on the scene file some sort of crisis angle which will serve as a "peg" for the account of what was really going on. And the correspondent must come up with two angles per day: one fresh one for the AM's and another for the PM's. I need not expound on the danger, I should think, of permitting the search for angles to overcome the search for genuine substance.

It is the custom these days to deplore from the lecture platform. I suppose it is an old custom. Even in those halcyon days of no Soviet threats, the Bob LaFollettes, the Bob Ingersolls, and the Teddy Roosevelts were deploring something.

However, I am sure that those immortals of the American lecture circuits were able to deplore with a somewhat better chance of seeing something done about what they were deploring. America was pretty well in command of her own destiny in those days and it was reflected in the way those giants behaved on the lecture platform. Roosevelt would assume that immovable stance on his thick, short legs, and his neck, bulging over the back of his collar, would grow crimson as he pounded home his convictions. Ingersoll would shout his salutation, "Good evening, Ladies and Gentlemen," from way off stage, creating the effect that his ideas simply could not wait another moment to explode. And LaFollette would remove his coat, then his vest, then his tie in a sort of erudite and intellectual striptease. There was no doubt in the minds of those men about their ability to make America into the shape of her promises and her potential. There was little to get in the way of it.

LESS THAN FULL COMMAND

Today, unfortunately, we do not have the complete command over our own destiny which the nation had then. A decision in Peiping, or Cairo, or Moscow can very genuinely affect our well-being.

All of which has a bearing on American journalism. No longer can we write or report with the complete assurance which graced the style of journalists then. Today, we are forced to write in the face of, or in the shadow of, dilemmas and imponderables. It is not always easy.

But true to the tradition of the lecture platform, permit me to deplore for a few moments.

I wish I could say to you that American journalism today provides little justification for deploring. But such is not the case.

In a few hours from now, the foreign ministers of East and West will engage in a round of negotiations which will contain the very real and the very breathless possibilities of war or peace . . . or continuation of the Cold War.

My wife and I came to Missoula by way of New York to Fort Wayne, Indiana, to Chicago, to Billings, and thence to here. So we had an opportunity to glance at a few newspapers enroute. In rare instances was the Geneva story given the space or position commensurate with its importance. Local murders, local thievery and assorted cussedness, even baseball, competed with the Geneva story. The number of newspapers in this country giving the Geneva story the treatment it deserves constitutes an exception to a general practice. The over-all effect, for how many million readers, is one of a strong implication that all is well and that we are justified in an indulgence of pastimes and unrewarding entertainment.

I fear American journalism must bear some of the responsibility for a traditional American unpreparedness in the area of foreign affairs. Foreign affairs have a bad habit of involving the American citizen as nothing else ever involves him . . . in real life-and-death situations. Three times in one generation foreign affairs have gone awry and the American citizen has either marched off to war or has seen his sons and daughters do it.

I have a friend in New York, an experienced correspondent for Associated Press, who was brought home several years ago to be foreign news editor. Each time I see him he is lamenting the lack of interest in foreign news by the editors of the country, who are constantly demanding less and less of that commodity on the AP wire.

I have little first-hand information of the problems of newspapers, since I cut my teeth on radio journalism. Indeed, there are problems . . . the high costs of publishing, the increasing problems of circulation. I can sympathize with these problems and difficulties. But our sympathy need not be excessive for the publisher who obviously regards the newspaper as only so much bait for his advertising, or a decoy for his play for local or regional political and social power. There is something revolting about the too-frequent combination of an exceedingly wealthy publisher and a newspaper which is a journalistic fraud. His fortune is his to be resented by none of us, unless he has made it by peddling to us a bad or irresponsible newspaper. In that event, he invites the scorn and the wrath of all of us.

In any survey of American journalism, I should think we cannot be too critical of the two major American wire services: Associated Press and United Press International. They are the product of the editors of the country and will be as responsible as editors demand.

Putting aside the newspapers and periodicals, suppose we consider radio journalism for a moment. The chief thing wrong with radio journalism today is that not enough people are listening to it to satisfy most of us. Still, we at NBC have discovered that the radio news audience is still one which any newspaper in the world would like to claim as its circulation. We in the news department at NBC are pleased and flattered with the repeated declaration of the president of the company that news has placed the network in the financial black and is keeping it there.

During World War II, radio news demonstrated what it could do. It kept a nation reasonably well informed . . . quickly, accurately, and without editorial axes to grind. A tremendous public confidence in radio news was established. As for the radio commentators or news analysts, while the public tended to split on the question of their accuracy or degree of responsibility, still the public found them interesting. I would argue that radio still provides a fine medium through which to obtain a pretty good idea of what is going on. Generally, a radio station or a network has no sacred cows in its stable which have to be fed some special diet of editorial bias.

LIMITATIONS OF TELEVISION

Television has cut heavily into radio listening and while radio news has suffered less than the entertainment function, it, too, is not what it used to be. Television journalism has not filled the gap. I think this is not anyone's fault; it does not imply an absence of public responsibility. Rather, television simply has a built-in set of limitations which make it something less than the ideal medium for the dissemination of news.

The primary limitation is one which involves time. I think we in television journalism bear the responsibility to remind our audiences occasionally that they should not depend upon us for a run-down of the day's entire news budget. It cannot be done in 15 minutes; it cannot be done in 30 minutes. Film moves at a specified rate of 36 or 90 feet per minute through a projector, depending on whether you are using 16 or 35 mm. I can create the verbal image of a Sir Winston Churchill deplaning in Washington in three or four seconds. It will take film perhaps 30 or 40 seconds to create that image.

There is, however, in television one tremendous compensation: given a news story with dynamics or impact it is fully transmitted with the loss of none of its power.

Therefore, our responsibility in television journalism becomes clear: in a 15-minute program do not attempt to cover the day's news budget. Rather, if we can shed some degree of illumination on four or five stories, we have turned in a rather good performance.

I think I need not expatiate on television's remarkable power to bring into home or office an important event such as a political convention, a sports contest, or whatever. That is television at its finest.

Parenthetically, it might interest you to know that television news is not a great employer of young journalism graduates or hopefuls. Thus far, television appears to have enjoyed a great attraction among all the other media. Therefore, television has tended to employ the best from other media. My advice, to anyone looking for employment in television news, is to try to establish his reputation with a newspaper or wire service or in radio and then shift over to television, or prepare to start in television at a very low salary in some very small station far away from the big metropolitan centers.

Let us talk for a moment about journalism generally . . . whether it be newspaper, radio, television, or periodical journalism.

The city of Washington has grown rather suddenly from a rather quiet and somnambulent little Southern city into a big metropolis. Simultaneously, our federal government has grown into a tremendous, sprawling organization. And with these growths of Washington and the federal government, the news corps in the city of Washington has grown proportionately. Today, the larger newspapers, the television-radio networks, and the wire services have departmentalized their Washington bureaus. One man is assigned to the Supreme Court, another to the White House, another to the Pentagon, another to the Senate, another to the House, and so on. In so far as space will permit, or as time will allow, the work of these specialists is splendid. Very little goes on which does not come to their attention.

But the result of this specialization is a kind of fragmentary reporting. There is not time nor opportunity for a good reporter to stand back and look at Washington and take in all of it. The pieces are rarely put together to give the people of the country a kind of composite view of what is going on in their name.

Fortunately, there are two institutions which tend to compensate for this weakness. One is the columnist. Occasionally, the columnist will try to put the pieces together and catch the general policies and attitudes of the federal government. The other is the institution known as the *New York*

Times. It not only provides news, yards and yards of it, pertaining to all the branches and departments and agencies of the government, but it also, particularly in the Sunday edition, attempts to take in the whole structure.

THE DISEASE OF CYNICISM

There is a disease now common in this country and it assails not only journalism but every profession and it threatens every individual. It is the disease of cynicism. I fear we are losing our old wonderful knack of having heroes, our old habit of hating the villains, our old ability to be awe-struck, bemused, fascinated, interested, and absorbed. There is a blandness about everything. We tend to believe that no man acts out of unselfish motive. We inquire as to the price of every man.

For example, the Premier of Cuba, Fidel Castro, is having a horrible time in his attempt to prove to us cynics that he is not a dictator, not a Communist, not a sadist, nor something else unpleasant. He could be any one or all of these things, of course, but we have tended to pin them on him prior to any evidence. The cynics simply could not get it through their heads that a man today could be a genuine, self-sacrificing patriot. This cynicism does not become us and it is a debilitating ailment for a great nation. We have seen it spread throughout France to the point where the finest institutions of that great country are endangered. And particularly is cynicism a fatal disease when it besets the profession of journalism. It may be quite true that great men and great ideas are not in as great supply as we might wish. It may be true that we have seen national heroes revealed with feet of clay. It is true, indeed, that in spite of the great national outpouring of selflessness and bravery, blood and treasure fifteen years ago, our ideals were scoffed at by Communism. All these things are true; but they still do not justify the conclusion that wickedness is paramount over virtue or that every soul, every mind, and every ideal can be bought in the market place.

But in spite of this disparagement, in spite of the relatively unhappy state of American journalism, I always take great comfort in the knowledge and the conviction that it is within the means of every citizen of this country to be well informed if he will. He may have to carry quite a subscription list, and he may have to thumb through quite a list of publications, and he may have to do a bit of dial twisting; but it is all there. It is a comfort to know that there are newspapers such as the *New York Times,* the *Christian Science Monitor,* the *Washington Post and Times Herald,* the *St. Louis Post-Dispatch,* and others. There are periodicals such as *Foreign Affairs, The New Leader,* and *The Reporter.* The airplane has brought us the rich and rewarding air editions of such splendid British publications as the *Manchester Guardian,* the London *Economist,* and a host of others. As for radio and television, I shall leave you on your own, lest I be either overly modest or overly vain.

There is something else still at work in American journalism which prevents the outlook from being completely sordid. The American journalist is not, generally, overpaid. I am certain that he vies with the American school teacher and university instructor for the lowest professional salaries.

Then what is it that attracts so many and so much talent to journalism? It can only be that the search for basic fundamental truth goes on, undiminished. I have no idea how many hundred journalists I know, many of them rather well. Their income is a never-ending source of lament, but few of them would ever contemplate leaving the profession. And with all of them, no matter how crass or irresponsible the policy of the publisher for whom they work, there is that drive, that search, that respect for truth. It is this quality among our journalists which precludes some of our publications being much worse than they are.

39. The Television News Commentator

by Eric Sevareid, Martin Agronsky, Louis Lyons

This is a reprint from a series of television programs originally produced by Station WGBH-TV in Boston, Massachusetts, under a grant from the Fund for the Republic. The programs, moderated by Curator of the Nieman Fellowships at Harvard, Louis M. Lyons, were designed to bring television audiences informed discussion of the problems and performance of the American press in reporting the leading questions of the day. Mr. Sevareid and Mr. Agronsky are well-known national news correspondents for the Columbia Broadcasting System. The interview is reprinted from The Press and the People, *published by the Fund for the Republic in 1959.*

VOICE: Today and every day the American people must make decisions on which their whole survival may depend. To make sound decisions the people must be informed. For this they depend on the nation's free press. How well is the nation's press doing its essential job? The people have a right to know the truth. They have a responsibility to ask.

ECHO: The right to question.

VOICE: The Press and the People.

As Moderator from Harvard University, the winner of the Peabody

Award for television and radio journalism and the Lauterbach Award for outstanding contributions in the field of civil liberties, Mr. Louis Lyons.

LYONS: Our guests are two of the most perceptive commentators on radio and television.

VOICE: One guest is a veteran of twenty years with CBS, a war correspondent on three continents, winner of a dozen awards, and chief of CBS's Washington Bureau—Mr. Eric Sevareid. Mr. Sevareid says, and I quote: "The bigger our information media, the less courage and freedom of expression they allow. Bigness means weakness."

Our other guest was also a distinguished war correspondent, recipient of many journalism awards, a Washington correspondent for sixteen years, now with NBC—Mr. Martin Agronsky. Mr. Agronsky says, quote: "The special story, the report in depth, becomes more and more difficult to keep on the air because of the tremendous rising costs of TV production."

LYONS: Our guests are a very special breed of radio and television journalist. They are the broadcasting counterpart of our nationally syndicated columnists, our Lippmanns, our Alsops, our Childs's. They don't just read news headlines in front of a camera, they do their own reporting. They make their own news judgments. The role of an Agronsky or a Sevareid is to get to the core of the big news and give us their digest of its meaning.

Mr. Agronsky, we've heard you quoted as saying that the special report that goes into news in depth is disappearing from television because of the very high cost. Won't you explain?

AGRONSKY: It really doesn't need an elaborate explanation. It means exactly what I said. It costs so much for anyone to take an hour, let us say, of television time on a network basis that only the greatest corporations can afford it. And the greatest corporations, faced with that kind of cost, are inclined to compare the costs that they put out for a news program to the costs that they would put out for an entertainment program—for a shoot-'em-up Western. Finding that the ratings and, in their opinion, the audience are infinitely smaller for the news program, they don't care to spend their money that way. The inclination is more and more not to do it but to spend that kind of money for what they regard as something that would get them a large audience.

LYONS: Mr. Sevareid, you say that the larger a news medium, the less chance there is in it for courage and freedom of expression. Won't you discuss that?

SEVAREID: I think it's axiomatic. Courage of controversy, courage of innovation in the realm of ideas goes in inverse ratio to size of the establishment. The investment in any given item that is produced—whether it is a TV program or a Hollywood movie or a big mass magazine—is so enormous that they must find a great denominator in terms of audience, whether it is the lowest common denominator or not. The risk is too great. This is not true of the small-capital media, like the stage or book publishing or small magazines.

LYONS: The bigger the stake, the less you can afford to hazard the loss of a customer.

SEVAREID: On the whole, of course, I must say for bigness that you can do things in terms of extensiveness. The scope is enormous. Small corporations could not possibly do some of the tremendous shows around the world that the great networks can do. That is, it's not all loss; I don't mean that.

LYONS: Mr. Agronsky, what Mr. Sevareid has just said makes me wonder whether cost is the only reason that the networks with their sponsors don't go into some of these special programs. To what extent is it because of a concern about controversy that sponsors prefer Westerns to information?

AGRONSKY: I think to a considerable extent. I accept Mr. Sevareid's observations completely in that area. I think both of us have had any number of experiences personally which tend to substantiate what he has said. I can cite one to you. Last year I was doing a half-hour interview program on NBC with the Reverend Martin Luther King down in Montgomery, Alabama. As a demonstration of the kind of pressures that networks face, which is rather difficult for them to withstand, when I did the show in Montgomery, Alabama, local segregationists permitted it to go to the network but disconnected the cable that led to the local transmission. So it was cut off in Montgomery.

If you get into a segregation story, for example, the network must reckon not only with sponsors—and there I would expand Mr. Sevareid's observation—it must also reckon with regional susceptibilities. It must reckon with the sensitivity of local stations on the network which are unwilling to participate in controversial issues that directly affect their area.

SEVAREID: Martin, I was going to take slight issue with you on the first thing you said—about these programs declining. I'm not quite sure they are. There is one matter: the thing that is so costly is film. CBS has eliminated a number of special news programs because they were based on film. When you get into that realm, the laboratory and other costs are perfectly fantastic. I am sure that's one of the basic reasons why *See It Now* went off as a regular or weekly or monthly program. CBS cancelled the one I did on Sundays for three years that required chiefly film. It was very expensive. Now CBS has another show, also a half hour, that Howard Smith is doing, called *Behind the News,* that doesn't use anything but stock film or some stills or interviews. It is far cheaper to do. But I'm not sure that it isn't quite as good in terms of bringing understanding; it might even be better. I don't think there is a rule that operates here, except the rule of the cost of film.

AGRONSKY: Put it this way: let us say that a large corporation that had been sponsoring an hour news special at a prohibitive cost, or what we might regard as prohibitive cost, had the opportunity to spend twice as much money for an entertainment program with Bob Hope, or what have you, on it. I'm inclined to think that the corporation and their advertising agency would not hesitate for a moment to spend twice the amount of money. It is not that they are priced out by the initial cost of the program, they are priced out by what they regard as the value received; in other words, the audience and the ratings that they can get.

SEVAREID: That's right. The total cost of network broadcasting for a sponsor—this universal cross-country thing—is so fantastic, the necessity to get the greatest mass audience, that industries that can afford it are not great in number and are chiefly consumer-item companies.

AGRONSKY: And they are guided entirely by the return on the investment.

SEVAREID: Well, they are businessmen.

LYONS: We're told that they feel the public is more interested in the entertainment—the whodunit or the Western—than in good news reporting. Mr. Sevareid, do you feel they underestimate the public's interest in information?

SEVAREID: I always have thought so.

AGRONSKY: Yes.

SEVAREID: Reading of newspapers and magazines encompasses millions and millions of people in this country, but there is an extra dimension, an extra difficulty, when you come to television. This is partly show business. It is partly entertainment. So are newspapers, of course, but not quite to the same extent. People normally do not sit down in front of the television set in the same frame of mind in which they pick up the newspaper or magazine. That is why television commercials irritate, and newspaper commercials or advertising do not.

LYONS: What about radio? I think of radio as costing less and having consequently more time, as it certainly used to, for discussion. What has happened, or has the more dramatic television squeezed it out? How is the news doing on radio?

SEVAREID: Not doing well.

AGRONSKY: On NBC—Eric is CBS and I'm NBC—on NBC, news is doing exceptionally well on radio.

SEVAREID: In terms of finances, you mean.

AGRONSKY: Well, in terms of getting sponsors, and in terms of a very considerable coverage that I'm inclined to feel—and this is not just a house plug—NBC has every right to be proud of. They do have the news-on-the-hour programs, which mean five minutes of news at every hour on the hour.

SEVAREID: Is that really news?

AGRONSKY: No, it isn't. I would agree with you on that. You haven't the talent.

LYONS: Your colleague, Edward R. Murrow, says a five-minute news program is not news.

AGRONSKY: All right. You haven't the time, I'll grant you. Nevertheless, in the time that you are allowed on a five-minute program—which I shudder to say is two minutes and fifteen seconds—you can take an angle and say something, not the way you would like to, not with the freedom you would have in five minutes, and certainly not with the freedom you would have in fifteen minutes. Beyond these, NBC also does a great spread of feature stuff. For example, we started a thing called *Image of Russia* on radio. Also, I suggested I would like to do something with Justice Douglas on a comparison of the Russian and American judicial systems, and I did a half-hour radio interview with Justice Douglas. They are running the whole thing intact, a good solid half-hour. Now, you can cover a subject. Generally they are finding that news has paid on NBC and they are widening its scope all the time.

SEVAREID: It's almost the only thing that pays on radio at all.

AGRONSKY: That's right.

LYONS: Mr. Agronsky, you cover Washington with its complex of news of great importance to us out of the national government. How much total air time on the three television networks do we get in a day on the news from Washington?

SEVAREID: On television?

LYONS: Yes.

AGRONSKY: Shockingly little, shockingly little.

SEVAREID: You get very little. I don't think that the normal daily fifteen-minute network television news programs are successful in terms of content or in giving people understanding. I don't think they ever can be. You'd be better off from fifteen minutes of radio, listening to somebody really giving it to you.

LYONS: I think the only complaint we have about the television news and information is whether we get enough of it.

AGRONSKY: Well, you don't. Let's be specific about it. CBS carries a fifteen-minute show and NBC carries a fifteen-minute show, and ABC carries a fifteen-minute show. Now, in that fifteen minutes, actually how much does Washington get? At the outside, Washington gets two or three minutes. I defy anyone to cover Washington in two or three minutes.

SEVAREID: Yes, but you see television did not carry over from radio the old-time radio commentator, who was palatable because you didn't have to look at him, possibly, but you could listen to his voice, his reasonably palatable voice for as long as fifteen minutes explaining news. This has not been done on television. The closest you come to it is the documentary approach, where you do a long take-out with film and interviews and some commentary, and those are rather infrequent.

LYONS: What about the editorial? We expect the newspaper not only to give us news but editorial comment. Is there any reason why we should feel that the station should be limited just to the facts? What is the problem of getting editorials on the air?

SEVAREID: No problem at all, except the people who run stations. They can do it if they wish.

AGRONSKY: Certainly they can do it.

LYONS: Would you say that the public takes a rather different attitude, that it judges television by a rather different standard from the newspaper, and perhaps the government does too? Does that make it more difficult to get into comment?

AGRONSKY: It is a different standard, because in television you reach a tremendous number of people, and I suppose they feel that when you hit the mass audience there are certain limitations, certain criteria that they feel forced to apply to television that do not apply to newspapers.

SEVAREID: There's a feeling with television, and also radio, that it is a kind of public trust that belongs to everybody. This is partly because of the way it started with a government hold on it, parceling out wave lengths. It has become a national whipping boy. Newspapers are not. This atmosphere, coupled with endless monitoring by Congressional committees and executive people—this relationship with government which the press does not have—has created a situation where it takes a fairly brave station owner to go out and editorialize as a company, though I think he should.

LYONS: I am thinking particularly of a CBS interview with Khrushchev. You remember how much criticism arose—even the President in a press conference tended to be critical—and yet nobody criticized a newspaper correspondent for getting an interview with Khrushchev.

AGRONSKY: Exactly. James Reston of the *New York Times* went over and interviewed him and nobody would say beans about it.

LYONS: Is it because of this potential regulation by the government?

AGRONSKY: No. They give a greater weight to the influence. They feel that the audience is so tremendous that there should be limitations and restrictions applied to the TV medium that are not applied to the newspapers. And I think wrongly.

SEVAREID: Then a legitimate thing is legitimate so long as it is not too big.

AGRONSKY: Yes.

LYONS: Let me take up another aspect of that. We've all heard of equal time. We know under some conditions somebody is entitled to time to reply. To what extent is the correspondent, the commentator, held back if he is inclined to critical comment, by feeling that this may involve his station in having to give somebody fifteen minutes tomorrow night? Does that get in the way?

AGRONSKY: Not really, because it would be rather ludicrous. Even a politician with a rather poor sense of perspective and a poor sense of humor would feel that he was demeaning himself if he were to say, "Well, now, Sevareid or Agronsky criticized me for five minutes. I want five minutes equal time." No, it wouldn't operate that way.

LYONS: Mr. Agronsky, you have handled some tough assignments, including assignment to Senator McCarthy when he was active. As you think of it, what would be the toughest one you have handled?

AGRONSKY: I would say McCarthy. He proved very expensive for me on radio, very expensive indeed. When I worked for ABC, I had a co-op program, which meant individual sponsors in different cities. Mr. McCarthy was not above bringing pressure to bear on the individual sponsors, wherever the opportunity presented itself. I was able in specific cases—I won't name stations or cities—to determine that on such-and-such a station a sponsor dropped me because of influence brought to bear by McCarthy supporters.

SEVAREID: I think Martin once lost about twenty-four stations from one broadcast. Is that correct?

AGRONSKY: Yes, I did, but that was not McCarthy, that was MacArthur. I did a broadcast on the return of General MacArthur, when he addressed the joint session of Congress, and I made an observation, that I don't intend to repeat, about the General. In my opinion it was not in bad taste, nor was it inaccurate, but it demonstrated that I didn't believe in, oh, I suppose, in the elevated stature that had been given to General MacArthur. There was the most incredible response. I lost about—I think it was twenty-two stations as a result. Not stations, sponsors.

SEVAREID: I didn't have that kind of trouble, but I did have a Senator in the Midwest—over the same kind of broadcast about the same General—take a punch at me.

LYONS: We know that a reporter needs only his pad and pencil—indeed, he doesn't always need that—to get into all sorts of places, meetings, groups, discussions. Are there places where the television reporter is barred but where the reporter can get in? Does that limit the range of news that television can cover?

AGRONSKY: Physically, mechanically, it is infinitely more difficult to cover the news. One of the problems in TV—I think you'd agree, Eric—is that you almost have to anticipate where the news is going to break.

SEVAREID: You're talking about bringing cameras in.

AGRONSKY: Yes.

LYONS: For instance, you can't, I believe, do any televising in the Congress. You can in some hearings and not in others. Am I right?

AGRONSKY: That's quite right.

LYONS: I'm wondering to what extent television reporting is limited by this feeling about the camera.

SEVAREID: It is limited on the House side of the Capitol because of Speaker Rayburn. He doesn't like the clutter and the fuss of cameras, though we've reduced that a great deal. On the Senate side it is up to the individual committee.

AGRONSKY: It's entirely up to Rayburn. I had a rather interesting experience with him just last week as a matter of fact. Mr. Sam has always been very kind to me in terms of permitting me to do interviews with him, and we did a Congressional show last Sunday. He gave me an interview, and I wanted him to sit on the rostrum. I thought it would be rather dramatic in the empty House chamber. He wouldn't permit it, nor would he permit us to do it in the House cloakroom or in the reading room or any place like that. He insisted on doing it in his office. He said flatly that, were he to do it, all the precedents that he had established in this area would be broken.

LYONS: Mr. Sevareid, what would you think of as the most important stories that have not been well or sufficiently dealt with in television—current stories?

SEVAREID: Some of them are the same kind of stories that have not been dealt with sufficiently in the press. The China story is one. We had an immense break a few weeks ago—with some initiative of our own involved, I must say—with film that got out of China from a German photographer. Here was the first real pictorial story of the Chinese communes. I thought it was one of the greatest documentaries that television has ever done.

AGRONSKY: Well, China represents a restriction applied by our government to all news media.

LYONS: One of your Washington colleagues, Ed Lahey, of the Knight papers, said on our program a little while ago that inflation was the hardest story to write and the worst-covered story. Later, on another program, the Harvard economist, Dr. Galbraith, said the same thing. What do you say about that?

SEVAREID: All economic problems are hard to do with pictures.

AGRONSKY: It is tremendously difficult to take a thing like that and portray it. I mean, how do you picture a deflated dollar?

SEVAREID: After all, Mr. Lyons, what are the differences in news by ear and eye, television and press? You can't use statistics very well, either on radio or television, through the ear alone. The press can.

LYONS: I wanted to ask you what you think of as the chief differences between newspapers and broadcasting. Are they just competitors or do they to some extent supplement each other because of different limitations?

AGRONSKY: I think that we can add a quality to news coverage dimension that newspapers simply cannot. I have in mind an obvious example—the conventions. There is an immediacy, a feeling of the sweat and the confusion—there is a spirit that emerges from the floor of the convention that you can go right into. You can see the people worrying and fretting and sweating and fighting and being confused and being pushed around. You get that quality on to the screen that no reporter can possibly capture in writing about it.

Or take the McCarthy hearings, for example. Certainly the best reporter in the world could not produce the quality of Joe's voice. He couldn't do it. No matter how well you write it, you've got to hear it, you've got to see that face, you've got to see the contempt and the scorn and the caustic way with which he dealt with witnesses. You get a feeling from watching that you can't reproduce on the printed page.

SEVAREID: There are great limitations, of course, on both. But there's one great difference, to put it very briefly, in radio and TV as compared to the press. In broadcast journalism you have what I would call Page 1—a sort of front page with a sports section and an opposite-editorial page with a few columnists or their counterparts. You don't have Pages 2, 3, 4, 5, and 6 and so on.

LYONS: You have to hit the highlights.

SEVAREID: This is one reason that discourages a radio or TV reporter from pursuing certain exclusive stories. Unless they turn out to be Page 1 quality, he really can't use them much. This is one great limitation. But what Martin is talking about—this extra dimension of feeling—is tremendously important. I recommend to anybody the speech that Archibald MacLeish gave at Minnesota on poetry and journalism. He said that understanding without feeling is not understanding. Facts are nothing by themselves.

AGRONSKY: I think of the segregation story, for example. You remember that white girl who wanted to go to school, despite the fact that Negroes were going to go to the school, and who was finally talked out of it, apparently by her parents or by the officials. It struck me as a perfect example of the thing that I have in mind. That girl, appearing on television, had a quality that was evident in the way she spoke and in the way she looked which was something that could not be captured by a reporter writing it down.

LYONS: Mr. Sevareid, as a foreign correspondent for many years, how would you rate the performance of the newspapers in foreign coverage? Take the *New York Times* with its great foreign service. How would you rate it as compared to what you would like to do as a television correspondent? I'm thinking in terms of what Mr. Agronsky said about the importance of seeing the faces and the confusion at the conventions.

SEVAREID: I don't know how you can compare these two. They are different media. I could not exist in terms of understanding foreign news without papers like the *Times*. But then you ask yourself, thinking again of MacLeish's distinctions about facts and real understanding, except for this massive reporting of facts, is even the great *New York Times* really a great medium in that sense? I sometimes wonder. The great divorcement of factual knowledge from human understanding of things is what MacLeish called a danger to our whole civilization. I think you see it even in that paper. There was a time when there were great literary journalists—the artist at work. You don't see much of that in the press now.

LYONS: CBS has a great foreign correspondent corps, NBC has a great foreign correspondent corps. But to what extent do they get through to us? A minute and a half from Paris by Schoenbrun, and two minutes somewhere else—is there any chance we can get more from these men?

SEVAREID: I would love to see it.

AGRONSKY: Certainly.

SEVAREID: We were better on radio some years ago in that respect.

AGRONSKY: Certainly there's a better chance if the networks want to do it. That's your answer.

LYONS: Thank you very much, gentlemen. The old newspaper cliché, that you give people what they want, has a special twist in television. Here we see the individual advertiser, the sponsor, deciding whether people want news or entertainment, controversy or comics. And this has seriously complicated the problem of getting news and information through what is chiefly a medium of entertainment. As we have seen, our television journalists are concerned about this problem, and clearly it is our problem too.

VIII. Advertising and Entertainment

David Brinkley once pointed to the mixed parentage of broadcast news: show business, sales and advertising emphasis, and corporate ethics. These constitute the essential framework in which not only news but all programming must be presented and considered.

How complex are the advertising pressures involved? How compatible are the entertainment and education objectives which are expected to co-exist in broadcasting? What dangers are seen in the present arrangement by writers, advertisers, psychiatrists, and critics?

Rod Serling, one of television's bright writing talents, considers here some of the continuing problems confronting television writers in the mixed context of pressures from sponsors set against the realities of television programming. In "The Solid Gold Egg," Leonard Lavin presents a challenging and articulate view of the problems involved within the dynamics of an electronic medium. Jack Gould's "Control by Advertisers" and Dr. Fredric Wertham's "Can Advertising be Harmful?" offer refreshing points of view set against the context of an address by E. William Henry, former FCC Chairman.

40. About Writing for Television

by Rod Serling

Mr. Serling is a distinguished writer. By 1957, when Patterns *was published, he had had over a hundred television plays produced. He had won two Emmys for best original teleplay writing; two Sylvania awards; the first Peabody award ever given a television writer; and many others. He is still most active both as a writer and as a spokesman on behalf of better television.*

The present selection is taken from Mr. Serling's introductory essay, preceding the scripts for his four television plays: Patterns; The Rack; Old MacDonald Had a Curve; *and* Requiem for a Heavyweight, *as published by Simon and Schuster in 1957 under the title:* Patterns: Four Television Plays with the Author's Personal Commentaries.

We are indebted to both Mr. Serling and Simon and Schuster for permission to reprint here. Readers are referred to the above-noted volume for Mr. Serling's individual commentaries preceding each play.

There is probably no single "absolute" anyone can use as a yardstick to describe the nature of the television writer, his background, his fortes, or the nature of his advent into the realm of television writing—save for the simple statement that there are no absolutes.

The TV writer is never trained to be a TV writer. There are no courses, however specialized and applied, that will catapult him into the profession. And it was especially true back in the twilight days of radio that coincided with the primitive beginnings of television that the television playwrights evolved—and were never born. In my case the decision to become a television writer arose from no professional master plan. I was on the writing staff of a radio station in the Midwest. Staff writing is a particularly dreamless occupation characterized by assembly-line writing almost around the clock. It is a highly variable occupation—everything from commercials and fifteen-second public-service announcements to half-hour documentary dramas. In a writing sense, it serves its purpose. It teaches a writer discipline, a time sense for any kind of mass-media writing, and a technique. But it also dries up his creativity, frustrates him, and tires him out.

It's axiomatic that the beginning free-lance writer must have some sort of economic base from which he operates. Usually it is a job with at least a subsistence wage to give him rent money and three square meals a day while he begins the treacherous and highly unsure first months of writing on his own. The most desirable situation encompasses an undemanding job that draws little out of the writer's mind during the working day so that his nocturnal writing will be fresh, inspired, and undiverted. In my case this was a wish but never a reality.

I used to come home at seven o'clock in the evening, gulp down a dinner, and set up my antique portable typewriter on the kitchen table. The first hour would then be spent closing all the mental gates and blacking out all the impressions of a previous eight hours of writing. You have to have a pretty selective brain for this sort of operation. There has to be the innate ability to single-track the creative processes. And after a year or so of this kind of problem, you have rent receipts, fuel for the furnace, and a record of regular eating; but you have also denied yourself, as I did, a basic "must" for every writer. And this is simple solitude—physical and mental.

The process of writing cannot be juggled with another occupation. The job of creating cannot be compartmentalized with certain hours devoted to one kind of creation and other hours set aside for still another. Writing is a demanding profession and a selfish one. And because it is selfish and demanding, because it is compulsive and exacting, I didn't embrace it. I succumbed to it.

I can pinpoint the day and almost the hour that it happened for me. After two years of double-shift writing, I had made approximately six sales to network television programs. These weren't bad scripts. There was usually a kind of strength to them that showed in dialogue and a sense of character. But they were stamped with the lack of professional polish. They showed in many ways that they were done on a kitchen table during the eleventh and twelfth and thirteenth hours of a working day. They were always sharpened, but never to their finest points.

So, on a midwinter day, I gave in to free-lance writing. This was not the overtly courageous plunge that some writers make. In my case I had just finished a three-week assignment as a staff radio writer, planning a documentary series designed to honor certain towns and cities in the listening area. (In regional radio, adjoining localities are forever being honored. This is designed to make for excellent public relations, but it is only on rare occasions that it makes for even a modicum of good listening. In most cases, the towns I was assigned to honor had little to distinguish them save antiquity. Any dramatization beyond the fact that they existed physically, usually had one major industry, a population, and a founding date was more fabrication than documentation.)

I had just turned in a sample script to the program director that was essentially above and beyond the call of duty, and well beyond the call of truth. My script called for a narrator and a 30-piece live orchestra, and contained the kind of prose that made Green Hills, Ohio, look like the Alamo!

When I was called into the P.D.'s office my script was lying face down on his desk, like a thumb in a Roman arena. He leaned back in his swivel chair and studied me pensively, as if searching for some velvet-glove language that could be utilized to castigate me without breaking my spirit.

"Serling," he said, "it's this way. Your stuff's too stilted. You seem to be missing the common touch. We're looking for grass roots here. We want to be close to the people. We're obliged to use the 'folksy' approach. In short, we want our people to get their teeth into the soil."

As he was talking I knew exactly what he meant. The "folksy approach" did not include a 30-piece orchestra, or prose out of Norman Corwin's *On a Note of Triumph.* It needed only two elements: a hayseed M.C. who strummed a guitar and said, "Shucks, friends"; and a girl yodeler whose falsetto could break a beer mug at twenty paces. This was getting the teeth into the soil. And the little thought journeyed through my brain that what these guys wanted was not a writer but a plow!

During the next couple of hours two things occurred to form and then cement a resolve of mine. The staff writer, in addition to writing, acts also as a kind of roving "idea man" for several current and varied types of programs. One of my duties was to supply "gimmicks" for an afternoon ladies' show. That afternoon I stopped by the studio to watch the tail end of one of its performances. The master of ceremonies was a semi-literate, ex-tent revivalist with curly hair and an absolutely devastating smile. He was winding up his show with a three-minute sermon on the boys in Korea and how we should pray for them. The ladies in the audience were totally captivated. There wasn't a dry eye in the studio. The program went off the air and the M.C., his eyes half closed, walked softly out of the studio, past the sighs and fluttering eyelashes of the good ladies who had come to see him and who now stared worshipfully and respectfully up at him. He nodded to me and we both got on the elevator. It went down one floor and a girl got off. The wavy head went up, the look of soulful ecstasy left the broad and dimpled face. He winked at me, nudged my elbow and said, "I wonder if she lays?"

(This same fat-faced, sanctimonious slob had told me earlier that he used to travel with his father, who was also an itinerant evangelist. But at the time, he said, his father did the preaching and he was the one who made the money: he had the "Bible concession.")

Two hours later I got my next assignment: to dream up an audition show for a patent medicine currently the rage. It had about 12 per cent alcohol by volume and, if the testimonials were to be believed, could cure everything from arthritis to a fractured pelvis. I spent two minutes studying the agency's work sheet, which stated the general purpose of the program. I read as far as the second paragraph: "This will be a program for the people. We'd like to see a real grass-roots approach that is popular and close to the soil." The pattern of whatever future I had was very much in evidence. I was either going to write dramatic shows for television, even at the risk of economics and common sense, or I was going to succumb to the double-faced sanctimony of commercial radio, rotating words as if they were crops, and utilizing one of the approaches so characteristic to radio—writing and thinking downward at the lowest possible common denominator of an audience. That afternoon I quit the radio station.

I sat that night with my wife, Carol, at a Howard Johnson's restaurant and after a few false starts—"You know, honey, a man could make a lot of money free-lancing"—I talked out my hope. Free-lance writing would no longer be a kind of errant hope to augment our economy, to be done around the midnight hour on a kitchen table. Free-lance writing would now be our bread, our butter, and the now-or-never of our whole existence. My wife was twenty-one, three months pregnant, and a most adept reader of the score. She knew all about free-lance writing. She'd lived with it with me through college and the two years afterward. She knew that in my best year I had netted exactly $790. She was well aware that it was a hit-or-miss profession where the lush days are followed by the lean. She knew it was seasonal, and there was no definition of the seasons. She knew that it was a frustrating, insecure, bleeding business at best, and the guy she was married to could get his pride, his composure, and his confidence eaten away with the acid of disappointment. All this she knew sitting at a table in Howard Johnson's in 1951. And as it turned out, this was a scene with no dialogue at all. All she did was to take my hand. Then she winked at me and picked up a menu and studied it. And at that given moment, the vision of medicine bottles, girl yodelers, and guitar-strumming M.C.'s faded away into happy obscurity. For lush or lean, good or bad, Sardi's or malnutrition, I'd launched a career. I'll grant you the perhaps inordinate amount of sentiment attached to all the above, but if this were a novel, patent medicines, Howard Johnson, and my wife, Carol, would all be part of an obligatory first chapter.

This was the nature of television in 1951. The medium had progressed somewhat past the primitive stage. There was still a sense of bewilderment on the part of everyone connected with the shows. And it was still more the rule than the exception to find the opening camera shot of almost every television play trained on the behind of one of the cameramen. But by this time there were six half-hour "live" shows that came out of New York, and two or three one-hour shows. On the Coast there were a dozen or more filmed half-hour anthologies. The television writer's claim to the title "playwright" had been made, but as yet was not universally accepted. The TV play, once called by Paddy Chayefsky "the most perishable item known to man," enjoyed no longevity through the good offices of the legitimate stage and the motion pictures. The motion-picture industry looked down at its newborn cousin somewhat as the president of a gourmet club might examine an aborigine gnawing a slab of raw meat. The movie people had no way of knowing at the time that this bumbling, inexpert baby medium would one day compete with them and come dangerously close to destroying them. For at that time the television play went on and off the air with few cheers and with no one to mourn its passing. The video diet was a lean mixture of wrestling and occasional football. These were the days of the 10-inch screen, the 1931 movies, and Gorgeous George. The television dramas extant were still in the process of feeling their way around, trying to find some kind of level of performance, some reason for being, and some set of techniques. At the time there existed no species referred to exclusively as "television writers." There were radio writers who were extending themselves a bit, realists who knew that the golden days of radio drama were dimming into twilight. There were screen writers doing television films as a stop-gap between picture assignments. There were also some embryonic playwrights who used the new medium as a kind of finger exercise for what they hoped would turn into legitimate writing later on. But neither the industry nor the public was prone to make any association between writing of real quality and the sort of thing done for television.

My first television script had been sold in early 1950 to an NBC film series on the Coast called *Stars Over Hollywood.* It was brave and adventuresome. Beyond this, the production and conception of the program were symptomatic of absolutely the worst features of Class-B moviemaking. The plots were an ABC mishmash, with the depth and levels of an adobe hut. The performances were rarely ever able to overcome the scripts. The piece I sold them was called "Grady Everett for the People." It starred Burt Freed, who turned in a pretty fair performance, considering everything. I don't recall too clearly the essence of the story or the way it was done, but I have a vivid recollection of the payment involved. It was exactly $100 for all television rights. I never met anyone connected with this production, nor did I set foot in the studio. But as of this writing it has been on at least twenty-four times at odd hours and odd channels. I will claim immodestly that it surpassed wrestling; beyond that, I'll make no value judgment whatsoever.

The singularly distinguishing feature of television drama in those early days was a paucity of payment, sets, and theme. And to go along with this was a bleak desert which represented the area of identity of the television writer. He was practically anonymous; he had an ill-defined respect for his talents and no protection at all for his work. He had few prerogatives in terms of its production and only the barest of recognition for his contribution.

The *Kraft Television Theatre,* the oldest of the one-hour dramatic shows, wouldn't permit a writer at rehearsal until the day of the show. His presence at that late hour was probably a guarantee against intrusion. For by then the lines had already been changed, the interpretations made, the blocking and camera arranged. The writer could protest, but only as a gesture. The show was a *fait accompli* prior to his arrival. In the kindred areas of rewriting, casting, music, et al., the writer had even less to say. I cite the *Kraft Theatre* as an example not to single it out for a necessarily unique mistreatment of authors but because it was one of the few dramatic programs existent at the time. It aimed for quality and often achieved it. This was the show that did things like Molière's *A Doctor in Spite of Himself;* Ibsen's *A Doll's House;* Galsworthy's *Justice* and *Loyalties.* It also produced plays like *Valley Forge, Berkeley Square, Comedy of Errors,* and *Macbeth*—and some of these shows were produced as early as 1947. The reader can pretty much gather what the policies of the lesser programs were with regard to their conception and treatment of the men and women who wrote the material.

These first four and five years of television were the cradle days of a baby whose birth may not have been accidental but whose process of maturing was far from being planned. But thanks to programs like *Kraft* and some others, television was expanding its technique and coverage. And along with it came an expansion in quality. Besides *Kraft* there was *Studio One.* Tony Miner, a pioneer without coonskin, was producing the plays on *Studio One,* whose expanse was becoming as much horizontal as vertical. One of his productions took place on a submarine. He used actual water and a mock-up submarine, and he did it on a nickel-and-dime budget that today wouldn't pay for a cast on a half-hour show. It evolved as a striking and powerfully realistic illusion, and it pointed the way to a new horizon in live television.

Celanese Theatre went on the air and did things like Maxwell Anderson's *Winterset,* and it did them well and effectively. *Celanese* was directed by a man named Alex Segal, who became one of the early "names" in television. He was later to direct some television plays by Rod Serling, who at the time of Alex's arrival on the television scene was still writing prayer messages for an ex-tent revivalist in Ohio. But at this moment in the evolution of television drama there were even a few intellectual diehards who began to see the potential of it, and began to realize that a television play could come close to the legitimate theater, and even surpass it sometimes in terms of flexibility. Along with television's expansion and progress came the birth of a new school of television actors and actresses, men and women associated with the medium and known because of it. Like Hollywood and Broadway before it, television began to produce its own stars, and also like Hollywood and Broadway the writer was the last of the company to achieve an identity. To his everlasting credit, he did it on his own. The networks financed no campaign to make Chayefsky a known and associable quantity. Several million viewers began to make that association on their own. Plays by Paddy Chayefsky, Horton Foote, Bob Arthur, David Swift, and David Shaw were stamped with that particular quality that forced recognition. The programs and networks helped, of course. The medium was improving to a point where it allowed them to help. They began to supply the financial and technical aid to enlarge the scope of the television drama. Now a writer could conceive of a story that played on more than two sets with more than four actors. He could write with an eye toward the fluidity of movement that came with three cameras. His sets and costumes were no longer slapped together as incidental accouterments to a one-shot performance. They were given thought, preparation, and time.

But the major advance in the television play was a thematic one. The medium began to show a cognizance of its own particular fortes. It had the immediacy of the living theater, some of the flexibility of the motion picture, and the coverage of radio. It utilized all three in developing and improving what was actually a new art form. As indicated previously by the plays on *Kraft*, *Studio One*, and *Celanese Theatre*, one could see that the television play was beginning to show depth and a preoccupation with character. Its plots and its people were becoming meaningful. Its stories had something to say. There was a flavor to it well beyond the early Hollywood half-hour film which shoved a product out that was obviously molded at an early age and became moldy at a late one. This product was sprinkled with a kiss, a gunshot, a dab of sex, a final curtain clinch, and it was called drama. Parenthetically it might be stated here that Hollywood did little to help in the evolution and improvement of television as a medium, at least in terms of drama. What accolades are deserved here should go to Chicago and New York.

In terms of technique, the "close-up" that had served as such a boon to the motion pictures was further refined and used to even greater advantage in television. The key to TV drama was intimacy, and the facial study on a small screen carried with it a meaning and power far beyond its usage in the motion pictures. I can't forget, for example, the endearing passage in Paddy Chayefsky's "Marty" between Marty and the girl in the little all-night beanery. This scene was a close-up of the two through the entire playing. And the wonderfully fabulous thing of two lonely people finding each other was played on the two faces. I am also reminded of one of my own things—a totally different piece from "Marty" but one which utilized the same kind of television technique that was so uniquely television. This was "The Strike," produced on *Studio One* in June 1954. There was a moment in the play when Major Gaylord (extremely well played by James Daly) was recounting an experience during World War II when he was obliged to fire on an American soldier in the dead of night on a Pacific island. The camera stayed tight on his face for almost three solid minutes, and we had a moving, poignant, and almost heartrending picture of the fatigue and fear that go hand in hand in the province of wartime combat.

The physical and the fates conspired to force the maturing of the television drama. It was no longer a novelty; it had become a fixture. As such it competed with every other kind of entertainment; consequently it was forced to become better, to become different, and to aim higher. It was a medium that in a one-hour time period could play to an audience greater than a Broadway play reached in one solid year of SRO crowds. With this kind of potential and with this kind of impetus, however young, however groping, television was something to be reckoned with.

Television today remains a study in imperfection. Some of its basic weaknesses and mediocrity are still with us. There is still wrestling, soap opera, overlong commercials, and some incredibly bad writing. There is really no defense for any of this, but there is an explanation. You need only look at a calendar to remember that only seven or eight years have gone by and the medium remains a young one and a groping one. There still remain new techniques to learn, new fields to examine, and a myriad set of roadblocks to progress that still have to be breached. But there is still time and there are still ways. Radio was around for twenty-odd years before it really found its niche and ultimately wrote out a finis to its potential. Television hasn't exhausted its potential or altogether found its niche. And in the area of drama it has already far surpassed that of its sister medium.

Like any mass medium, it might still die from internal strangulation. But for those of us who professionally cast our lot with it in its early days, we haven't given up. For us the heartening thing is that there are still things to strive for.

A FEW RECOLLECTIONS

A *Writer*—at least this writer—measures his career not so much in terms of years as in individual moments. They are the good moments: the big sale, the well-received show, the award at the end of the year. The television playwright must savor his success and his good moments very hurriedly, because they're temporal at best. But the bad moments—his failures, the script rejections, the incisively bad reviews—cling to him with much more tenacity and for longer periods than the moderate successes.

Between late 1951 and 1954 I lived in Ohio, commuting back and forth to New York to take part in story conferences and the rehearsals of my shows. This was expensive and time-consuming, but was a concession to my own peculiar hesitancy about all things big, massive, and imposing. New York television and its people were such an entity. For some totally unexplainable reason, every time I walked into a network or agency office I had the strange and persistent feeling that I was wearing overalls and Li'l Abner shoes.

I remember one incident during those early days when I had flown into New York to discuss a rewrite on a script called "You Be the Bad Guy," which starred Macdonald Carey. The script editor, Dick McDonagh, asked if I'd like to meet the star of the show. I was ushered into a small office where the cast was assembled for the reading, and there was introduced to Mr. Carey, who turned out to be an extremely pleasant, affable guy, who stood up and shook my hand and complimented me on the script. I remember standing in the center of the room wondering what the hell I could do next, and deciding that I had outworn my welcome and my purpose and should at this time beat a retreat. I looked busily and professionally at my watch, nodded tersely to all assembled, mumbled something about it being a pleasure to see them all but that I had to catch a plane going west, and then turned and crashed into the wall, missing the door by two feet. Then, in backing out of the room, I ran into an oncoming secretary and dropped my briefcase, exposing not only scripts and writing material, but a couple of pairs of socks, some handkerchiefs, and some underwear. (I traveled light in those days.) My exit from the J. Walter Thompson offices that day could not have been more pointed and obvious had it been staged by Max Leibman. But as a postscript to the story, I remember Dick McDonagh gripping my hand before I left the building and saying, "Look, little friend, these people don't give a good healthy damn what you carry in your briefcase, or how you leave a room. All they care about is what's in there!" He pointed at my head. Then he slapped me on the back and wished me well, and I headed back to the airport and Ohio.

On a writer's way up, he meets and does business with a lot of people. And in some rare cases there's a person along the way who happens to be around just when he's needed—perhaps just a moment of professional advice, a brief compliment to boost the ego when it's been bent, cracked, and pushed into the ground, a pat on the back and eight words of encouragement, when a writer's self-doubts are so persistent, so deep-rooted, and so destructive that they affect his writing. Dick McDonagh gave me many moments and several words of encouragement and enough pats on the back to keep me propelled forward. He once told me that there might be a day when he'd be reading some of my plays in a book anthology. He may know very well by now how prophetic were his words.

But I wonder if he also realizes how instrumental he was in having it happen.

The writer in any field, and particularly the television writer, runs into "dry periods"—weeks or months when it seems that everything he writes goes the round and ultimately gets nowhere. This is not only a bad moment but an endless one. I remember a five-month period late in 1952 when my diet consisted chiefly of black coffee and fingernails. I'd written six half-hour television plays and each one had been rejected at least five times. What this kind of thing does to a family budget is obvious; and what it does to the personality of the writer is even worse. The typewriter on my desk was no longer a helpmate; it took on the guise of an opponent. The keys seemed stiff and unyielding. The carriage seemed bulky and sluggish, and the wastepaper basket would get crammed by the hour with discarded pages—a testimonial to my unsureness as to what to write and how to write it. Toward the end of this, I got a letter from Mr. Worthington Miner. I've mentioned Tony Miner earlier. Then as now he was a major-league, top-drawer television producer. And to get a letter from him, particularly a letter asking to see scripts, was like a third-string pony league pitcher getting a telegram from John McGraw telling him to come up and pitch for the Giants. I flew into New York to see him, my briefcase bulging with manuscripts. (There wasn't even room for socks.) Tony read them, and during our second meeting informed me that he'd like to buy at least six of them. He was putting together a new show to be sponsored by an auto company, and my work impressed him. The feeling I got in that given moment was something akin to what a person feels when he is notified that he's just won the Irish sweepstakes. The knees begin to give out and there's a roar that begins some place down deep in the gut and starts to travel toward the throat. Fifteen minutes later I was on the telephone calling my wife and guzzling a Scotch on the rocks I ordered from room service (tipping the bellboy a whole buck), and adding up in my mind how much are six times six or seven hundred dollars. One week later, back in Ohio, I got another letter from Tony Miner apologizing and explaining that the show he was putting together had been shunted off to another agency and he would not be producing it. The guy who had won the Irish sweepstakes couldn't find his ticket stub. It was that kind of feeling. For some perverse reason I saved Tony's second letter; my wife put it into a scrapbook. And sometimes I take a look at it as a piece of memorabilia to document a bad moment that on the scale of a career's ups and downs represents the bottom of the barrel. A writer's career is studded with the near sales, the close hits, the almost-but-never-wases. And afterward, when he becomes accustomed to eating a little higher off the hawg, the bad moments get remembered. And no matter what you eat, it tastes like pheasant under glass.

♣

Besides the good and bad moments in a career of writing, there is also an indefinable hard-to-peg turning point, a crossing of the Rubicon when suddenly you find your name somewhat known in the agencies and on the networks. You announce it at the reception desk and the girl nods knowingly and doesn't ask you to repeat it or query you as to its spelling. Exactly when this happens and how, you're never quite sure. But it does happen. Afterward the process of writing is never any simpler, the ideas are never easier to come by, and your craft and technique don't seem appreciably altered. But there is a difference, as if the long grind upward levels out a little bit and the going becomes a little easier. In my case it happened because of a single show that emanated live out of New York City. This was the *Lux Video Theatre.*

Over a two-year period they bought twelve of my shows and produced eleven of them. Since that time, *Lux* has gone the way of so many dramatic shows. They moved West, went into an hour form, and in this case began to use old movie properties instead of originals. But in its New York half-hour days, the *Lux Video Theatre* proved itself symptomatic of the basic difference between what was Hollywood television and what was then New York City television. It was a show that consistently aimed high. Its whole conception in terms of dialogue and production was adult, never hackneyed, and almost always honest. It touched upon themes like dope and marital infidelity. It did things like adaptations of short stories by Faulkner and Benét; it encouraged the submission of original scripts by any writer who knew how to write, regardless of what his credits were. The definitive characteristic of this show was that it never got rutted into a "type" program. It was never a till-death-us-do-part marriage between the policy of the program and the type of story and ending. The most meaningful and probably the most valuable thing that I can say about the *Lux Video Theatre* is applicable to all of television. On the basis of individual shows, it was as often unsuccessful as it was successful. But it always tried. And though its sights were sometimes aimed higher than its capabilities, it was rarely dull. If this could be said of the entire medium, flags could be raised on all the antennae.

THE PROBLEM AREAS

Defensiveness in a television writer is a kind of occupational disease. His newfound stature has been somewhat therapeutic in combating it, but it remains in varying degrees. A writer is still thought of in some circles as a hack, plain and simple. His work is still regarded by some as merely an appendage to a sales message. And the medium he writes for is still maligned as being principally a display case and not an art form.

The TV writer falls prey to some of his criticism because he deserves it. A sizable bulk of television writing still must be dismissed as inconsequential or simply bad stuff, but there also exist reasons for this. And if they don't stack up as reasons all the time, they are at least in a sense explanatory of a condition. The mass medium writer has two major problem areas in which he must write. These two areas represent roughly the nature of the medium and the writer's identity.

THE NATURE OF THE MEDIUM

Built into television drama are innate and homegrown problems that do not exist in any other art form. Television, while unique in its potentials, is further unique in its limitations. In play-writing this is particularly true. For example, in no other writing form is the author so fettered by the clock. The half-hour program will sustain a story for only 23-odd minutes. The hour program calls for a 48- to 50-minute play. It is unheard of that a legitimate playwright must write within so rigid and inflexible a time frame. But the TV writer must. It is further arbitrary that his play must "break" twice in a half-hour show and three times in an hour show to allow time for the commercial messages. Obviously, there are some plays that will not in any circumstances lend themselves to such an artificial stoppage. The "break" will hurt the flow, the continuity, and the build, but the "break" must come. And what do you do about it?

This time problem extends over into another area: production. The average hour television show rehearses for eight or nine days. This means a little over a week allotted to reading, staging, blocking, line learning, camera, dress rehearsals, and, finally, production. Contrast this with a Broadway play that rehearses on an average of one month to stage a production that runs only twice as long as its television counterpart.

Very recently, when *Playhouse 90* began telecasting on a weekly basis with a 90-minute play each week, it was thought that here at last was the time frame long and flexible enough to aid the writer in handling plot, character, and pacing. I for one gratefully accepted the assignments for the first two plays of this new series, thinking, as did most others, that with about 70 minutes allotted the play, it would be moving out of an igloo into a mansion. But once again television's own peculiar limitations cropped up and, instead of aiding the playwright, the new time frame within this program did nothing but hurt him. For instead of a regular three-act arrangement, *Playhouse 90* took a host of sponsors, each demanding at least two commercials. The result was that during the ninety minutes the show had to be divided into twelve- and thirteen-minute segments, each separated by a commercial, so that the over-all effect was that of a chopped-up collection of short dramatic segments torn apart and intruded upon by constantly recurring commercials. Scenes had to be automatically "curtained" at a high emotional pitch to accommodate the stoppage of action, the commercial, and then pick up the thread of story line. It is obvious that a succession of phony curtains or emotional high points will eventually dilute the effect of any play. An audience can get used to and almost oblivious to bomb blasts if they occur often enough.

The physical limitations of the television drama are part and parcel of the innate problems of the writer. Four or five basic sets represent the maximum stretching of both facilities and imagination. I might parenthetically state here that television's "intimacy," so often its strength, is an outgrowth of this weakness. We had to be intimate. We didn't have room to be anything else. In New York, the mecca of live television drama, the set problems are the greatest and show the least possibility of improvement. Most of the shows are berthed in old movie houses, buildings that are the victims of the young medium which now utilizes them. They are segmentized, over-extended, and asked to serve in a capacity they were not designed for. This lack of space is often reflected in the techniques of television playwriting. The author must often probe vertically because there just aren't enough inches to let him spread out horizontally.

But while time and space present hurdles, the basic, the most important limitation of the television dramatist is not totally physical. In a sense it is more philosophical. And this happens to be the simple and fundamental fact that our economy is geared to advertising. For good or for bad, the television play must ride piggy-back on the commercial product. It serves primarily as the sugar to sweeten the usually unpalatable sales pitch. It's the excuse to wangle and hold an audience. The play is forced to become a kissing cousin to an entity totally foreign to it. The audience, during a one-hour viewing of a drama, is forcibly deprived of that drama and in its place is exposed to three minutes of Madison Avenue dynamics. The audience must then make its own mental and emotional realignment to "get back with" the sole object of its intentions. That it can do it at all is a tribute to mass intelligence and selectivity.

I don't really believe there exists a "good" form of commercial. There are some that are less distasteful than others, but at best they're intrusive. And even in the most absolutely palatable form, they thrust a cleaver into the over-all effect of a television drama—and they do it three times during its all too brief playing, and even more during the 90-minute shows.

I make reference to this by way of pointing out a basic weakness of the medium. I do not presume to suggest any antidotes or alternatives. At the moment none seems possible. A sponsor invests heavily in television as an organ of dissemination. That organ would wither away without his capital and without his support. In many ways he hinders its development and its refinement, but by his presence he guarantees its survival.

Still, I don't think it is possible to generalize about the sponsor or the agency or the networks themselves. They vary as to the intensity of their dogmas, the legitimacy of their concerns, and the extent of their interference in a given television play. But, at their very worst, their interference is an often stultifying, often destructive, and inexcusable by-product of our mass-media system. It extends into an area of dramatic creation that should by rights lie well outside their bailiwick and well beyond their scope of prerogative. I think it is a basic truth that no dramatic art form should be dictated and controlled by men whose training, interest, and instincts are cut of entirely different cloth. The fact remains that these gentlemen sell consumer goods, not an art form.

A few years ago on a program called "Appointment with Adventure" I was called in to make alterations in some of the dialogue. I was asked not to use the words "American" or "lucky." Instead, the words were to be changed to "United States" and "fortunate." The explanation was that this particular program was sponsored by a cigarette company and that "American" and "lucky" connoted a rival brand of cigarettes. After establishing beyond any doubt that my leg wasn't being pulled and that this wasn't some cheap, overstated gag, I did the only thing a writer can do in television in the way of a protest. I asked that my name be withheld from the script. It was not that the alteration of the language in this case was of particular consequence or to any large degree changed the story. But in the matter of principle I felt that this was ludicrous interference, and I didn't want to be part of it. I'll never forget the man from Talent Associates, the outfit that produced the show, explaining to me that this was not the happiest state of affairs, but that writers, as well as any creative people connected with the show, should keep in mind that it's altogether proper for a sponsor to utilize certain prerogatives since he's paying for what goes on. Extending this kind of logic, we might assume that it is altogether proper for a beer-company executive to have a hand in managing a baseball club whose games are televised under his sponsorship.

Exactly where is the line of demarcation between the play and the commercial? No one seems to know. Ideally, the sponsor should have no more right of interference than an advertiser in a magazine. Theoretically, at least, this advertiser has no say over the policy of the magazine he buys space in, nor should he have even to a minute degree. But in television today, the writer is hamstrung and closeted in by myriad of taboos, regulations, and imposed dogma that dictate to him what he can write about and what he can't.

In the television seasons of 1952 and 1953, almost every television play I sold to the major networks was "non-controversial." This is to say that in terms of their themes they were socially inoffensive, and dealt with no current human problem in which battle lines might be drawn. After the production of "Patterns," when my things were considerably easier to sell, in a mad and impetuous moment I had the temerity to tackle a theme that was definitely two-sided in its implications. I think this story is worth repeating.

The script was called "Noon on Doomsday." It was produced by the Theatre Guild on the *United States Steel Hour* in April 1956. The play, in its original form, followed very closely the Till case in Mississippi, where a young Negro boy was kidnaped and killed by two white men who went to trial and were exonerated on both counts. The righteous and continuing wrath of the Northern press opened no eyes and touched no consciences in the little town in Mississippi where the two men were tried. It was like a cold wind that made them huddle together for protection against an outside force which they could equate with an adversary. It struck me at the time that the entire trial and its aftermath was simply "They're bastards, but they're our bastards." So I wrote a play in which my antagonist was not just a killer but a regional idea. It was the story of a little town banding together to protect its own against outside condemnation. At no point in the conception of my story was there a black-white issue. The victim was an old Jew who ran a pawnshop. The killer was a neurotic malcontent who lashed out at something or someone who might be materially and physically the scapegoat for his own unhappy, purposeless, miserable existence. Philosophically I felt that I was on sound ground. I felt that I was dealing with a sociological phenomenon—the need of human beings to have a scapegoat to rationalize their own shortcomings.

"Noon on Doomsday" finally went on the air several months later, but in a welter of publicity that came from some fifteen thousand letters and wires from White Citizens Councils and the like protesting the production of the play. In news stories, the play had been erroneously described as "The story of the Till case." At one point earlier, during an interview on the Coast, I told a reporter from one of the news services the story of "Noon on Doomsday." He said, "Sounds like the Till case." I shrugged it off, answering, "If the shoe fits. . . ." This is all it took. From that moment on "Noon on Doomsday" was the dramatization of the Till case. And no matter how the Theatre Guild or the agency representing U.S. Steel denied it, the impression persisted.

The offices of the Theatre Guild, on West 53rd Street in New York City, took on all the aspects of a football field ten seconds after the final whistle blew. Crowds converged, and if there had been a goal post to tear down, they would have done so. The White Citizens Councils threatened boycott and the agency people somberly told me that this was no idle threat. They had accomplished effective boycotts down South against the Ford Motor Company and the makers of Philip Morris cigarettes.

In the former case, it seemed that Negro workers had been permitted to work on assembly lines alongside whites; and in the case of Philip Morris, there had been a beauty contest in Chicago where one of the winners was a Negro girl. This was all it took for a wrathful wind to come up from the South. I asked the agency men at the time how the problem of boycott applied to the United States Steel Company. Did this mean that from then on that all construction from Tennessee on down would be done with aluminum? Their answer was that the concern of the sponsor was not so much an economic boycott as the resultant strain in public relations.

These, therefore, were the fears, and this was the antidote. The script was gone over with a fine-tooth comb by thirty different people, and I attended at least two meetings a day for over a week, taking down notes as to what had to be changed. My victim could no longer be anyone as specific as an old Jew. He was to be called an unnamed foreigner, and even this was a concession to me, since the agency felt that there should not really be a suggestion of a minority at all; this was too close to the Till case. Further, it was suggested that the killer in the case was not a psychopathic malcontent—just a good, decent, American boy momentarily gone wrong. It was a Pier 6 brawl to stop this alteration of character. The script was then dissected and combed so that every word of dialogue that might remotely be "Southern" in context could be deleted or altered. At no point in the script could the word "lynch" be used. No social event, institution, way of life, or simple diet could be indicated that might be "Southern" in origin. Later, on the set, bottles of Coca-Cola were taken away because this, according to the agency, had "Southern" connotations. Previously, I had always assumed that Coke was pretty much a national drink and could never, in the farthest stretch of the imagination, be equated with hominy grits and black-eyed peas, but I was shown the error of my thinking. And to carry the above step even further, a geographical change was made in the script so that instead of being a little town of undesignated location, it was shoved as far north as possible, making it a New England town. It is conceivable that the agency would have placed the action at the North Pole if it hadn't been for the necessary inclusion of Eskimos, which would prove still another minority problem. For it to open in New England, with the customary spires of a white church in the background of the set—so typically Yankee and Puritan—was somewhat ludicrous to behold. But this was to be a total surrender, and there would be no concessions made even to logic.

"Noon on Doomsday" was, in the final analysis, an overwritten play. It was often tractlike, much too direct, and had a habit of overstatement. What destroyed it as a piece of writing was the fact that when it was ultimately produced, its thesis had been diluted, and my characters had mounted a soap box to shout something that had become too vague to warrant any shouting. The incident of violence that the play talked about should have been representative and symbolic of a social evil. It should have been treated as if a specific incident was symptomatic of a more general problem. But by the time "Noon on Doomsday" went in front of a camera, the only problem recognizable was that of a TV writer having to succumb to the ritual of track covering so characteristic of the medium he wrote for. It was the impossible task of allegorically striking out at a social evil with a feather duster because the available symbols for allegory were too few, too far between, and too totally dissimilar to what was actually needed. In a way it was like trying to tell a Jewish joke with a cast of characters consisting of two leprechauns. This track covering takes many forms in television. It is rarely if ever successful, and carries with it an innate transparency that shows it up for what it is.

When Reginald Rose, in an exceptionally fine play, "Thunder on Sycamore Street," took an uncompromising swipe at a brand of lunacy in our country that recognizes equality as applying only to those whose roots are in the third-deck planking of the *Mayflower*, he had to couch his theme in a language acceptable on Madison Avenue. It was the story of a family in a residential street being bullied and pushed around by their neighbors because the guy happened to be an ex-convict. The story was originally written about a Negro family. The central conflict in every line of dialogue pointed to the Negro-White problem and the altogether basic premise that sooner or later human beings are going to have to live together side by side. Mr. Rose's enforced track covering was simply exchanging an ex-convict for a Negro. And this is a process a TV writer has to learn and to perfect. He must hunt and peck until he finds a more acceptable minority than the Negro—often the American Indian. This is, of course, somewhat limiting—since it is a difficult minority problem to play in New England—but television sponsors and agencies are prone to accept slight inconsistencies when it comes to skirting a sticky issue. I am afraid that eventually we TV writers may run out of substitutes. I suppose, then, because we are pretty inventive and imaginative guys, the standard minority scapegoat will turn out to be a robot, and this will step on no toes whatsoever. But in the meantime, a medium best suited to illumine and dramatize the issues of the times has its product pressed into a mold, painted lily-white, and has its dramatic teeth yanked one by one.

Sometimes television is faced with a problem where it is physically impossible to substitute an idea. Last year I was faced with such a problem when I wrote a script called "The Arena," which was done on *Studio One*. In this case, I was dealing with a political story where much of the physical action took place on the floor of the United States Senate. One of the edicts that comes down from the Mount Sinai of Advertisers Row is that at no time in a political drama must a speech or character be equated with an existing political party or current political problems. Some of these problems, however, are now so hoary with age and so meaningless in modern context that they are stamped as acceptable. Slavery, for example, can now be talked about without blushing. Suffrage is another issue that need make no one wince. The treatment of the lunatic in chains and dungeons can no longer be considered controversial. But "The Arena" took place in 1956, and no juggling of events can alter this fact. So, on the floor of the United States Senate (at least on *Studio One*), I was not permitted to have my Senators discuss any current or pressing problem. To talk of tariff was to align oneself with the Republicans; to talk of labor was to suggest control by the Democrats. To say a single thing germane to the current political scene was absolutely prohibited. So, on television in April of 1956, several million viewers got a definitive picture of television's concept of politics and the way the government is run. They were treated to an incredible display on the floor of the United States Senate of groups of Senators shouting, gesticulating, and talking in hieroglyphics about make-believe issues, using invented terminology, in a kind of prolonged, unbelievable doubletalk. There were long and impassioned defenses of the principles involved in Bill H. R. 107803906, but the salient features of the bill were conveniently shoved off into a corner of a side-of-the-mouth *sotto voce*, so that at no time could an audience have any idea what they were about. In retrospect, I probably would have had a much more adult play had I made it science fiction, put it in the year 2057, and peopled the Senate with robots. This would probably have been more reasonable and no less dramatically incisive.

The problem of censorship in television is not only a writer's problem. What narrows his frame of reference must of necessity narrow the area of television entertainment available to the audience. When the television drama is forced to go around Robin Hood's barn tying itself into verbal knots to evolve as stainlessly nonpartisan, whatever nonsense comes out as the replacement is the nonsense that an audience must live with on its television sets. Perhaps if some thoughtful people would write to sponsors, pleading for an adult airing of issues on a dramatic program, to counteract those cranks who hoist up the Stars and Bars whenever a play suggests a racial controversy, the sponsor or agency would realize that not to attack a controversial theme might be just as destructive as attacking it.

Television critics have tried to champion the writers' cause in the area of censorship but have done so obliquely and as a result, in many ways, have hurt instead of helped. For when these St. Georges of the Press tilt their lances at what is humdrum, ineffectual, and hackneyed, too often the victim they single out is the writer or the program's producers. They slay a dragon, but it's often the wrong one. The people who put on a television play, from the writer on through the entire staff, are not the people to make bleed for what is an innate weakness in the treatment or theme. Let the critics go on record as condemning the whole pressure system of sponsors, agencies, and networks. These are the only ones who can appreciably alter the conception of TV drama and widen its horizons of theme selectivity.

Television drama is probably at no crossroads at the moment. It can go on and on, improving or not improving, and still remain a pretty important fixture on the American scene. But what can happen is that it can, in a sense, commit itself to its own creative rut by not fighting for something a little bit better, and not looking for something that is new. Radio drama, after 20-odd years as king, left no lasting imprint of any importance. It left no legacy of particularly memorable moments in drama. It produced very few talents who could be remembered uniquely for their contributions to radio drama. From the point of view of the writer, there were no Chayefskys and no Roses and few anybody elses. Beyond Norman Corwin, Arch Oboler, and perhaps Wyllis Cooper, go look for a known name among radio writers. I don't think there are any, at least that the public knows about. Radio, in terms of its drama, dug its own grave. It had aimed downward, had become cheap and unbelievable, and had willingly settled for second best. It is quite conceivable that the television drama may well get stuck tighter and tighter into a mold of mediocrity. Creative people, particularly writers, can only be censored, sat on, and limited so much and for so long. After a time, fighting back seems relatively unimportant. The sponsor may continue to sell his soap just as the radio soap operas did for him, but by then the television drama will be a dull, sloppy old man who sits contemplating his widening paunch without interest, without energy, and with no horizon left at all. . . .

41. Remarks Before the National Association of Broadcasters

by E. William Henry

These comments were delivered before the National Association of Broadcasters in Chicago, Illinois on April 7, 1964. They are reprinted with Mr. Henry's permission.

I am doubly honored on this occasion. It is not only a privilege to address this distinguished association, but a high honor indeed to speak to the same group that heard Governor Collins yesterday and who will tomorrow listen to Dr. Billy Graham. Collins to Henry to Graham—if that combination doesn't give you religion, I suppose nothing will.

Several months ago I was chatting with a well-known syndicated columnist. We agreed that sometimes criticism—of a political candidate or any prominent figure—proves to be a blessing in disguise. This columnist confided to me that some years ago a Congressman, up for re-election, had come to him and pleaded to be criticized in one or two articles which would be carried in the Congressman's local newspaper. "He figured," said the columnist, "that if I criticized him, he would surely be re-elected!"

The other side of that coin we all know—praise from one source can result in condemnation from another. So, when LeRoy Collins faced your Board of Directors last January, I remained tactfully silent. Now, as I speak to you for the first time as Chairman I feel obligated to say to you—as my first order of business—that in Roy Collins you have one of the most able and conscientious men that it has been my privilege to know. A lawyer by profession, a great state's chief executive by popular demand, and a first-class human being by every sound standard, he is a wonderfully effective advocate for your cause. You are fortunate indeed to have him and I am proud to share this platform with him today and to call him my good friend.

One of the three luncheon gatherings at your annual convention is traditionally set aside for the Chairman of the FCC to speak his piece. Without changing that format, I'd like to put the spotlight briefly on my fellow Commissioners seated here at the head table.

These Commissioners are collectively bipartisan but individually independent. Their differing philosophies reflect their diverse backgrounds. Over-all, they represent an amalgam of experience and ability which this industry and the public have seldom if ever had at their service.

You all know Commissioner Lee Loevinger, our newest acquisition. He is the highly competent ex-member of the Minnesota Supreme Court and ex-Assistant Attorney General of the United States, whose abilities compensate for his retiring, shy, soft-spoken manner. He assures me that he is not really against multiple ownership, but his wife also confides that each of their children is allowed but one toy.

Ken Cox is the only man in the history of the Commission to be "dropped in" from Chief of the Broadcast Bureau to Commissioner, and you broadcasters are fortunate to have a man on the Commission who knows as much about your industry as he does. He is our expert on television allocations, and the only Commissioner who thinks as much, if not more, of local live programming than I do.

Fred Ford has served the Commission with distinction in many capacities, ranging from general attorney through Chief of the Broadcast Bureau's Hearing Division to Chairman. He is the only member, who, at the drop of a hat, can deliver a well-rounded speech on every major section of the Communications Act. He is a man of high principles—but whose golf score is not as low as he claims.

Next in line up the ladder comes Robert (Ultra-High Frequency) Lee, the only Irish Catholic Republican Commissioner from Chicago on any federal agency. As you know, his wit is as unique as his background. As a ten-year veteran of Commission battles and a former FBI agent, he is the FCC's candidate for broadcasting's Big Brother.

Bob Bartley was once one of you—before he decided to go straight. A former professional broadcaster, he remains an unprofessional Texan, a dedicated public servant, and our Defense Commissioner who needs no defending. All Commissions should have one Bob Bartley.

Rosel Hyde—who has been affectionately dubbed "the old grey fox"—completes the picture. He is Mormon with only one wife, who doesn't have to smoke or drink to have fun, a career public servant, and a high-class individual in every sense of the word. As Senior Member of the Commission, he votes just before I do—but not always the same way.

The members of this distinguished group occasionally fail to recognize the brilliance of their Chairman's proposals, but I am delighted to take this occasion to express publicly my appreciation, and I'm sure yours as well, for the manner in which they discharge the high public trust reposed in them.

Having talked about the Commission as individuals, let's discuss them for a moment as a voting body. What has our record been since we saw you in Chicago last year?

The regulation of advertising time standards—better known as overcommercialization—has been in the forefront of late. You all know what happened, but let me suggest to you again that there may be more in our notice terminating the proceeding than readily meets the eye.

I would have preferred a rule. A Commission majority, however, preferred to continue the case-by-case approach, and in a democratic society compromise is a way of life. The Commission still has not indicated the number of commercials or the percentage of time devoted to advertising that will cause some application, some time, to be honored with a designation for hearing. But it has made a start by letting you know that renewal applications from stations showing the highest levels of commercial activity will be given a thorough going over. Having looked closely at a few of these applications, we have found it necessary to write to some licensees about the nature of their commercial policies, how they mesh such policies with public need, and the reasons for their departures in practice from the policies they profess to follow.

In this way we are moving toward the development of sound regulatory standards in this field. The process may be painful to some—more painful, I think, than a rule would have been—but it is healthy and necessary.

I would also remind you, when and if the Commission decides to look reality in the face and put overcommercializers to the test of a hearing, you may want a specific rule then as much as you shun it now. You might even then recognize that legislation to prevent rule-making in this field is not a protective barricade, but a sandbag in disguise.

In this same vein, we have instituted an inquiry into loud commercials. In my mind there is no question but that action should be taken by the Commission, and that a solution can be found. I am also pleased to note that your Code Board, under the able direction of Howard Bell, has the problem under study.

Recorded commercials are louder than program content—99-44/100 per cent of the time. You may slice them lengthwise or crosswise, color a discussion of them with technical terms, and put meters in front of them that register different things at different times, but most radio and television commercials are simply too loud. You do not have to live in an apartment building and be aware of your neighbor's television set only during the commercials in order to recognize the problem. It is noticeable in your living room and mine.

A man from Florida recently wrote me an interesting letter on this subject. He said:

> Dear Chairman Henry:
>
> Television has had its share of criticism, but I want you to know that I get a great deal out of it. I was inspired by President Kennedy's appeal to this country for physical fitness, and I think the television stations are doing their bit.
>
> On the particular stations I listen to, the commercials are so loud that I must get up about 8 times every 15 minutes to go to my set—4 times to turn the commercials down, and 4 times to turn the program up. I am in

great physical shape because of it, and I want to compliment the television industry for its co-operation.

We know that television commercials are intentionally made loud for two reasons: first, to make an impression on the average viewer; and secondly, to hold the attention of Aunt Nellie who heads for the kitchen to stir the soup at every program interruption. We also know that loudness is primarily a matter of excess compression and that the elimination of such compression, though not the complete answer, will go a long way towards solving the problem. It remains for the Commission to require the elimination of excess compression in a manner consistent with sound regulatory and administrative practices. This is all that remains.

Here is a perfect opportunity for the broadcasters to be of help—to be positive rather than negative, to be progressive rather than reactionary. You won't lose a single customer and your sponsors can still make their pitch without the change of one single word. A few of you have been helpful; more should do likewise.

As William B. Lewis, Chairman of the Board of Kenyon and Eckhardt, has told you, "sooner or later the pitchman will pass in radio and TV as he passed on the midway. The faster you boot him out the easier your public relations will be, and the happier your image."

Also during the past year the Commission renewed the licenses of the Pacifica Foundation radio stations which had been on deferred status for over three years. Complaints originally came to the Commission from several sources and varied from disgust over allegedly obscene programming to charges of the possible affiliation of key personnel with the Communist party.

Now, when a regulatory agency is called upon to handle allegedly obscene Communists, it indeed has a hot potato on its hands. The Pacifica potato was admittedly handled gingerly for too many months, but I am proud that the Commission showed its calluses while I was its Chairman. It issued a forceful, broad-gauged opinion clearing Pacifica of the charges levied against it. In my judgment, this action will stand as a bulwark against the enemies of free broadcasting and free speech.

At every public meeting of broadcasters I have attended, a speaker has only to declare himself in favor of "freedom" to be rewarded by an automatic burst of applause. But oratory is easy; firm action is difficult. Surely, if ever there was a time when the freedom of broadcasting was at stake, this was it. Who took the action in this case?

Which state association sent delegations to Congress charging that the FCC had deferred the Pacifica licenses for an unwarranted period and was operating outside its jurisdiction? Which of you wrote me a letter urging the Commission to dismiss these charges and to re-affirm the Commission's time-honored adherence to the principles of free broadcasting? Where were your libertarian lawyers and their *amicus* briefs—your industry statesmen with their ringing speeches? Did the sound and fury reach no ears but ours?

If broadcasters or their advocates felt involved in this issue, there is no evidence in our records to indicate those feelings. Apparently, not one commercial broadcaster felt obliged to make his views known to the Federal Communications Commission. As it has done time [and] again the FCC itself staunchly upheld and protected broadcasting's right of free speech and expression for the people of this nation.

Your contrasting reactions to these two struggles—overcommercialization and Pacifica Foundation—cast a disturbing light on the basic motivations of an industry licensed to do business in the public interest. And you might similarly gain insight into the reasons why, for all your magnificent services to the public, your critics remain vocal. When you display more interest in defending your freedom to suffocate the public with commercials than in upholding your freedom to provide provocative variety—when you cry "Censorship!" and call for faith in the founding fathers' wisdom only to protect your balance sheet—when you remain silent in the face of a threat which could shake the First Amendment's proud oak to its very roots—you tarnish the ideals enshrined in the Constitution and invite an attitude of suspicion. You join the forces of crass complacency—in an industry and at a time in the history of this nation when complacency of any sort is both misplaced and dangerous.

There is no such complacency, I assure you, in the halls of the FCC.

For this is still an age in which we need all the excellence and all the greatness we can muster. Government cannot create excellence by fiat—businessmen cannot create it simply by spending money. But we can work together to produce the conditions under which excellence and greatness can grow. We can at least try to make our actions in the service of freedom match the high-flown words we use to praise her.

I could talk for quite a while on other subjects that have cropped up during the last year—our efforts to enforce the rules now on the books against those few who refuse to abide by them, multiple ownership, responsibilities under the fairness doctrine, our new program of collecting fees for the filing of license applications, the proposed new program forms for television and radio, increased activity in our inquiry into the control of networks over the sources of program supply—to name but a few. But I would rather devote my remaining time to a subject that concerns me perhaps as much as all the foregoing put together. It is a matter which lies shrouded in the mists of the future. Its exact shape is imperceptible, but its growth is of vital interest to everyone in this room. It approaches on cat feet but with the appetite of a ravenous tiger. It has been given a number of names, but they all mean the same thing. I refer, as you may have guessed, to pay television.

Pay television is an enigma. No one knows the answers to the questions it poses. But since "the beginning of wisdom is an admission of ignorance," let us examine the subject—not as experts—but as neophytes in search of the best method of realizing the vast potentialities of the television medium.

You all know that the Commission has authorized a program of experimentation with subscription television, that one experiment is being conducted in Hartford, Connecticut, and that another has been approved for Denver, Colorado. Several years ago there was a considerable hassle over the question of whether there should be any experiments at all with pay television and, if so, what the scope of the experiments should be. That controversy was settled when the experiments were limited to prevent competitive damage to our advertiser-supported television system as a whole. Indeed, some proponents of pay television claim that the restrictions are so stringent that they prevent any significant test of what pay-TV can accomplish. In any event, our experimentation with pay television via broadcasting stations is proceeding cautiously and under carefully controlled conditions.

But life in a free society is never that simple. As Thomas Jefferson said, "The boisterous sea of liberty is never without a wave." The technology of modern communications does not stand still, and neither do the imaginative entrepreneurs who look for ways to serve new and unmet public needs. The communications industry, of which broadcasting is a part, is the most dynamic in the nation. A broadcasting transmitter was once the marvel of the age. But now communications satellites soar overhead. Lasers and masers are workaday concepts. And the wire—thrust from the center of public attention by the glamour of radio communications—prepares for a comeback in a new form.

Transmission of communications by wire and microwave is becoming increasingly less expensive. Low-cost cable may soon be available. Experiments with the transmission of television signals over a pair of telephone wires are under way. It may soon be economically feasible to transmit television signals by microwave between metropolitan areas and to distribute a multiplicity of these signals to every home in every metropolitan area by wire. "Talk-back," or two-way communication between one's living room and the transmitting studio may soon be commonplace. Your wives may one day respond to the electronic huckster by pressing a button on the television set to order the home demonstration of a new refrigerator. An electronic scanner mechanism attached to your television set may perform a variety of supplemental jobs—such as reading your gas meter, providing continuous stock market quotations, and sounding a remote alarm in case of fire. And for all of that, you will watch now and pay later on easy credit terms.

Now, are these merely intriguing daydreams, far removed from present reality? Let's take a close look at present developments.

A 25 million dollar wired closed-circuit television system is underway in California. As I understand it, entrepreneurs in charge have as a financial base, a contract allowing the broadcast of local major league baseball games on a closed circuit. Los Angeles will broadcast the Los Angeles home games; San Francisco will broadcast the games played in Candlestick Park. Other programs—first-run movies, off-Broadway shows, instructional courses for credit, BBC productions, operas, etc.—are also planned. Due to the high density of homes in both areas, family dwellings may be wired for a relatively low price per unit. No licenses are now required from the Commission because no transmission of radio energy is yet involved.

However, let us assume that these separate systems in Los Angeles and San Francisco prosper. The inevitable next step is for the system's owners to apply to the Commission for a microwave license to connect the Los Angeles audience with the programs originated in San Francisco, and vice versa. So it is only a matter of time before we shall be called upon to render decisions vital to the California project.

Let me now shift your attention to the East Coast (and it takes longer for this shift than it does for a microwave signal to go from Los Angeles to New York). The heavyweight fight between Sonny Liston and Cassius Clay was viewed by an estimated 475,000 people on closed-circuit television. Gross receipts from television rights are estimated to be $2,375,000. That, without more, is pay television. It is different from the usual concept of pay television only in that it was not available on the same television receivers that carry free television.

But that's not the whole story. The Liston-Clay fight was available on an estimated additional 200,000 home television receivers. Unlikely, you say? Not at all. These television homes were subscribers to the 88 community antenna television systems which were themselves subscribers to the closed-circuit fight. Therefore, the inevitable conclusion is that, in addition to the Commission's authorized experiments in Hartford and Denver, one brand of pay television is already here—via the community antenna system.

At the present time approximately 1,200,000 homes are wired to receive television on a paid subscription basis through CATV systems. Community antenna homes are increasing at a rate of about fifteen thousand per month. Community antenna systems not only provide service to those communities which are without local broadcast stations, but may now bring additional signals from large metropolitan areas to communities with more than one local station. Signals from Los Angeles are being carried to places as far away as Yuma, Arizona. Signals from New York City now reach Laconia, New Hampshire and Wilkes Barre, Pennsylvania, and further expansions are planned.

Thus, as the Commission and the Congress labor in Washington on many pressing matters, the world is being wired for sight and sound. Technical know-how is being acquired, and vested interests of king-sized proportions are building up. And as importantly, the concept of paying for television has become an accepted fact in over a million homes.

Let me say to you as strongly as I can that it is not enough to meet these developments with slogans such as "Free enterprise must prevail!" or "Free TV must be preserved!" We will not help ourselves or our country by attempting to make a blind choice between over-simplified formulas. Nor is it enough to make speeches decrying the outworn dogmas of the past and advocating original thought. We need the commodity itself. And in the fast-paced modern world, where there is so little time for thought of any kind, original thought is a scarce commodity indeed.

So let us analyze the fundamental issues now.

The basic problem is the competitive effect of wired television upon our present advertiser-supported system. Does community antenna television limit or block the growth of television broadcast stations in outlying communities? Will it stunt or thwart the growth of UHF stations, upon which our hopes for greater diversity within the television industry so largely rest? Would pay television provide the additional program choices which its supporters claim and which we all desire; or would it gradually erode and eventually destroy the free television system?

To approach these questions intelligently, we must first examine the merits of our present system. What is it we are afraid of losing?

First of all, the programs of our present television system are free to the viewer. Of course, free television is supported by advertising, and advertising adds to the cost of the products we purchase. But free television's programs are not withheld from the viewer who lacks the ability to pay.

Today we hear much on the subject of the inability to pay—otherwise known as poverty. Twenty per cent of the families in the United States earn less than $3,000 per year. But it is a striking fact that only 8 per cent of the homes in the United States are without a television set, and some of these homes are in Palm Beach. So poverty itself is rarely, if ever, a barrier to the powerful beam of entertainment and intelligence called television. For millions of Americans with low incomes, the glow of the television set is one of life's few lights. That glow may often be dull, with programs to match, but who can say that its extinguishment is a matter of no concern?

Additionally, the signals of television transmitters in our present system travel to areas which are economically beyond the reach of cable. The rancher in the isolation of Wyoming's mountains—the Indians on Arizona's outlying reservations—the inhabitants of farm houses all over the country—all are entitled to share in the benefits of the national television system. We cannot abandon them merely to provide a variety of services to people in the core of our urbanized society.

Finally, our present system is built upon the principle of local station ownership and local program origination. Local news, local public affairs, and other local programs may be only a fraction of the total programming produced by our system. But it is that fraction which justifies the vast frequencies now allocated for TV services. It is a growing fraction which we cannot give up.

But, wired television also has its merits. Community antenna and pay television systems are not the inventions of money-mad moguls. They are primarily a response to expressed public needs.

Community antenna television started, and still has its heartland, in areas which lack adequate off-the-air television service. In modern America, people are not content to go without television, to rely upon inadequate signals or upon only one decent signal. If they cannot obtain the benefits of our present system any other way, they are more than willing to pay for them. And the men who bring those benefits to them perform a real public service.

I must say, in all frankness, that I think the growth of community antenna television reflects an inherent weakness in our advertiser-supported system—a failure to make its basic program services available to the entire country, including outlying communities too small to support more than one station, or any station. The CATV systems moved in to make good that failure.

The same thought applies to the newer trends in both community antenna and pay television. The men who seek to bring the programs of New York and Los Angeles stations to smaller communities via microwave relays and community antenna systems—the man behind the California pay television venture—are betting that the American viewer wants something more from his television set than the present system makes available. They think the viewer wants and will pay for more variety and more choice—and they see no reason why he should wait longer to get the variety and the choice we have thus far failed to provide.

On balance, what conclusions should we draw? First, the time to grapple with the problems raised by wired television is now. There is always a lag between the thinking of the scientists and engineers who create new technology, the businessmen who apply it, and the government officials who try to decide whether the public interest requires control. We must shorten that lag as much as possible. Legislation is clearly required on some points, and action by the Commission or the industry on others. Congress, the Commission, and the industry must make critical decisions about wired television before events make them for them.

Second, community antenna television provides a useful, though limited service which is not provided by free television. We should seek ways to integrate it fully into our national television system. We should not shrink from hard choices when community antenna television threatens to thwart basic purposes of the Communications Act. It must not be permitted, for example, to hamper the development of UHF. But we must make a strenuous effort to find measures which will allow both kinds of television to survive—to keep open for the public all of the choices which a free economy makes available.

Third, pay television—though closely related to community antenna television—raises problems of a different order. Here we are dealing, not with an established industry, but with one that is struggling to be born. Moreover, here we deal, not with a system aimed primarily at distributing more widely the programs produced by advertiser-supported television, but with a wholly new source of programs for the television audience.

Fourth, the problems of pay television over wire are the same as the problems of pay television over the air. It makes no sense to have tightly controlled pay television experiments using broadcast frequencies while giving carte blanche to the development of pay-TV over wires. It may well be that the California venture could provide an appropriate testing ground for the development of wire systems. But it should be a real test subject to government control and regulation. Wire pay-TV should not be permitted to grow like Topsy.

For pay television in any form must have a public interest justification. It cannot be so justified unless it brings to the public a greater variety of choice through specialized, high-quality programming. It must be a supplemental service, not a substitute service. And in my opinion, if pay-TV is to come into our homes, it should not be allowed to bring the sponsor's commercials with it. If viewers must pay for additional programming, they must not pay twice.

Finally, the time has come for the television industry to heed the advice of Macaulay to "Reform if you would conserve." You cannot get protection without providing a service worth protecting. You cannot beat something with nothing.

Your claim to protection against competition from community antenna systems and from pay-TV rests largely upon the fact that you provide local service. If that claim is to be persuasive, your service to local needs and interests must be real and substantial. It cannot be limited to one or two regularly scheduled news programs and a very occasional local "special." Riding the network and relying upon the projection of old movies may be as easy a life as riding the rails, but in the long run it may be just as dangerous.

Moreover, the search for maximum profits in the short run does not represent maximum wisdom in the long run. With every dollar that drops into your till from the sale of advertising, you are creating a greater public willingness to pay for programs without commercials. There is such a thing as pushing the long-suffering public too far, and those who do so may find the eggs that are laid far from golden.

But above all, wired television systems challenge you to bring more variety and more choice into the present system. In addition to emphasizing program diversity in order to give the public more choice on present outlets, the Commission has sought in every way to encourage the development of additional outlets on UHF channels. Your long-range interests and the interests of free television generally are dependent upon the success of UHF television.

If free enterprise cannot come to television via UHF channels, it will do so by means of wires and microwave relays. The result may not be all that you—or I—would hope. But in the long run the choice is inexorable.

For modern technology and the modern marketplace afford us no easy resting place. In today's communications industry we are propelled around the bend and into the future not to the steady chug-chug of a Mississippi riverboat, but to the piercing whine of an Atlas Agena rocket. How we will navigate depends both on the natural forces within the industry and our response to them. Can we create conditions under which the constructive forces of the market place will flourish and the destructive forces be controlled? That, I suggest, is the challenge that confronts us today.

42. The Solid Gold Egg

by Leonard H. Lavin

Mr. Lavin is President of the Alberto-Culver Company. This speech was delivered before the Television Bureau of Advertising on November 20, 1963. It is reprinted with the author's permission.

Of all things, my talk today has to do with the Alberto-Culver Company and television.

My speech is divided into three parts. In the first part, I confirm something you probably all suspect . . . that television has done miraculous things for the Alberto-Culver Company. This part of my talk is intended to disarm you and make you smile and feel glad all over.

It is also intended to convince you . . . that I am a fine and perceptive and appreciative fellow indeed.

In the second part of my speech, I put my arm around your shoulder and lead you to the head of a long marble staircase. Then . . . in a friendly fashion . . . I give you a shove. For while television has helped do marvelous things for my company . . . has helped make it rich and renowned . . . I suspect that this is just stage one of a dark and devious plot hatched by you owners and managers of television stations.

In the second and final stage of this dark and devious plot . . . *you* get all the money!

This part of my speech is calculated to convince you that I am not and never was a fine and perceptive fellow, and that glad-all-over-feeling you had was just induced by this marvelous Chicago weather and a brace of Chicago-style martinis.

In the third and last part of my speech, I repeat with even greater emphasis, all the wonderful things about television that I had said in part one of my talk.

I make the point that the services performed by television on behalf of Alberto-Culver and American industry are as nothing, I am convinced, to the contributions that television will make in the future.

This last part of my speech is intended to disarm you once more. I am sure this is precisely what I will have to do . . . disarm you. It is also intended to convince you that your first impression of me was only partly correct. That is, you will think I am an even finer, more perceptive, more appreciative fellow than you thought originally.

Because while you will realize that I am on to your dark and devious plot to end up with all my company's money, you will know that I, nonetheless, hold you in the highest regard. After all, if I could end up with all *your* money, I hardly think I would let the opportunity pass.

Now, to part one of my speech. . . .

It is the fashion today to take a stand on television. My stand is a definite one. My point of view is a clear one. My conviction about the medium is rock-like.

My stand, my point of view, my conviction, all are dictated by one simple fact.

Television . . . or more precisely . . . you television owners and managers . . . have had the common, ordinary, everyday decency to help make my company successful. I am grateful. And I can't tell you how much I like television.

You will find that fully seventy-five per cent of the advertisers that television has helped, approve of the medium. The remaining twenty-five per cent were too busy counting their money to voice an opinion. In the first days of television, everyone was mad about the medium. The strongest criticism I heard in those days was voiced by a little old lady who had just seen the Milton Berle show. She thought he really didn't belong in television because he was a show-off.

But today, the criticisms are quite bitter. I don't need to tell you what these criticisms are. You are aware of them. All criticisms have one thing in common . . . high moral indignation. This puzzles me.

We send our children to school at five or six years of age. We keep them in school until they are 21 or 22. Or preferably even older. We let them imbibe the culture of the ages.

When they graduate and come out into the world, they nonetheless seem to prefer the wisdom of Doctor Zorba to the wisdom of Aristotle. So, what is the conclusion?

The conclusion is that something is wrong with television and that nice old man, Doctor Zorba. I wonder if something is not wrong with the conclusion. But, no matter. This isn't what I want to talk about in part one of my speech.

My view of television is a businessman's view. My view is that of an advertiser. I look on television as a selling medium. I am a producer of mass packaged goods. I, therefore, use the medium to reach mass audiences. I do not create nor mold the tastes of the ordinary citizen. I am out to *reach* the ordinary citizen. To communicate with him . . . to tell him about my products . . . to brag about my products.

Now, if someone were to ask me if I thought the trend in television today was against . . . quote "good programs" unquote . . . I would answer yes. The cause is economic as I see it. The cost of television advertising keeps going up. The consequence of this is that the advertisers who can afford television are the more established and successful manufacturers of goods sold to the mass public.

Such manufacturers must, of necessity, conform to the tastes of the mass market in order to communicate with it.

Now, let me tell you about Alberto-Culver and television. . . .

This year, my company will spend more than 30 million dollars in advertising. The vast bulk of this money will go into television advertising. We have bought into 21 daytime and nighttime network shows. The remaining third of the television budget is set aside for spots.

We have been told by many people that this money is being spent most foolishly. We have listened carefully to what these people have had to say. We have been extremely interested in what they have said. However . . . and obviously . . . we did not follow their recommendations.

We buy television because we believe in television. We believe in television advertising because we have seen it work . . . work for us, that is. It has worked from the beginning.

We started Alberto-Culver in 1955. Our beginning was auspicious: we owed a great deal of money . . . a half a million dollars. This was the price we paid for our first product, VO5 Hairdressing and Conditioner.

We compounded our problems . . . if you want to call them that . . . by putting all our advertising money into television. Three spots a week in Philadelphia. I made the buys personally.

Our first year, we spent 75 thousand dollars in advertising. Our net sales that year were about 400 thousand dollars.

Our net sales in 1956 were a million and a half.

In 1957, 2.8 million.

In 1958, five and a quarter million.

In 1959, 10.3 million.

In 1960, 14.9 million.

In 1961, 25.3 million.

In 1962, 57.4 million.

Our earnings per share of stock went from six cents in 1956 to two dollars and 30 cents in 1962.

Our 1963 record looks quite good, too. At the end of the first six months, our net sales were 40.7 million. This is an increase of 47.5 per cent over sales the first six months of 1962.

I would enjoy telling you that Alberto-Culver owes its success entirely to television. Trouble is . . . it wouldn't be true.

But, I can assure you that we would not have been the success we have been without television.

I cannot give complete credit to television because advertising has been only one of the problems that we, as a new company constantly introducing new products, have had to face. Face and solve.

I refer to distribution and to the actual difficulties of creating new products. Television has created a demand for our products. Getting our products into the stores has been a whopper of a problem. This has involved creating a sales force of hundreds.

We have done it, but it has been difficult. It is a problem that we have always had with us. And that we hope we will always have. Because it is a problem, and a manifestation of growth. It is a problem created for us by television. If there was no demand for our goods, we wouldn't have the problem of distributing them. We wouldn't have any problems at all.

It has been said of us that we create products for television. This is a terrible over-simplification, but in a way, it is true.

The thirteen or so new products we have created have been meant for the mass market. If you watch television, you will know what these products are. . . . Hairspray. . . . Hand Lotion. . . . Shampoo. . . . our latest product is called New Dawn. It is a permanent hair color that you shampoo in. It has been successful in test market. We go national with it in December. It should turn the hair color business on its head. And heads we win.

I have already indicated what New Dawn and the other products have in common, is mass appeal. They are meant, as I've said, for the mass market. This is where television comes in. It allows us to reach the mass market economically. And it allows us to reach them the most effective way. That is, we can communicate with our potential customer more effectively when we use television. Television is simply a more effective means of communication.

What is it that we communicate? An advantage. A product advantage. This is the point of all our advertising.

Insofar as there is a secret Alberto-Culver way of doing business, this is the secret way.

Our aim is to both pioneer new types of mass-consumer products and to build better products in already established product categories.

Now, what television offers us is the opportunity to say and show and prove precisely what our product advantage is.

To sum up this part of my speech . . . we have invested heavily in television from the beginning because we are the producers of packaged-goods for the mass market and television has given us access to this mass market.

We have used television because we feel it is the best means of communicating with this mass market. Moreover, we have always felt that television is the most economical medium for the Alberto-Culver Company.

This concludes part one of my speech. Now to part two.

Let me put my arm around your shoulder . . . there's a long marble staircase I would like to show you. Here we are. See . . . isn't it a beautiful staircase?

While you're looking at it, let me read you something. It is one of Aesop's fables. You will find it on page 25 of the Grosset edition. I quote:

> A farmer went to the nest of his goose to see whether she had laid an egg. To his surprise, he found instead of an ordinary egg . . . an egg of solid gold. Seizing the egg, he rushed to the house in great excitement to show his wife.
>
> Every day thereafter, the goose laid an egg of pure gold. But, as the farmer grew rich, he grew greedy. Thinking that if he killed the goose he could have all her treasures at once, he cut her open . . . he found nothing at all.
>
> End of fable.

As a television advertiser, I must admit great sympathy for that goose. As a television advertiser, I know exactly how she felt.

That fable is my bridge into some straight talk about the cost of advertising on television. Some of the things I say will not make you happy. They are not intended to. However, they are not said to make you unhappy. They're said because they must be said.

I speak to you as a friend. As a friend who owes a great deal to you. As an advertiser. As a friend and advertiser who keeps tabs on every single television buy made by my agencies. As a friend who has personally been involved in every important negotiation with networks my company and my agencies have had. As a friend and advertiser who has talked with many other advertisers about the spiraling costs of television advertising.

Let me read you something. It is a memo prepared for me by one of my agencies. It has to do with the buying of spot announcements on television. The material in the memo refers specifically to purchases made or contemplated by the Alberto-Culver Company. I quote:

> Rates for spot announcements have been increasing steadily. On the other hand, rate protection period has dwindled from six months to three months in most cases, only two months in others, with a few stations granting as little as thirty days.
>
> Stations are also resorting to a number of devices to avoid granting any rate protection at all. For example, instead of issuing a new rate card, a station will assign a special participation rate to a given program. Then, if the program is replaced, a new rate is assigned to the replacement program. In such cases, an advertiser receives no protection against the new rate.
>
> Many stations grant rate protection on the existing schedule only. In other words, any schedule changes or additions must be made at the new higher rates, as soon as a new rate card has been announced.
>
> Rate increases seem to be based primarily on what the traffic will bear. An example of this is the inclusion of a pre-emptible rate on many cards. During the heavy buying seasons, when good announcements are scarce, stations can ask a premium rate, and when the availability situation softens up, the station can be competitive by offering these announcements at a lower rate and still be in a position to recapture the spot and sell it at the premium rate if an advertiser comes along who is willing to pay that higher premium rate.
>
> Most stations have tried to justify rate increases on the basis of higher operating costs and substantial investments in new product to improve programming and attract a greater audience share. However, market growth in terms of the number of TV homes within a station's coverage area is not in direct proportion to rate increases. As a result, an advertiser is paying a higher cost per thousand homes reached.

Examples of some rate increases which became effective this fall in the top 30 markets are as follows:

A station in Atlanta and one in Detroit have increased rates on one of their better time periods about 10%.

A Dallas station and a San Francisco station have issued new rate cards which reflect a 15% increase.

A New Orleans station and a Philadelphia station have increased rates on certain participating programs from 15 to 23%.

Rate increases of 20 to 25% have been announced by the following:

A Buffalo station, a Cleveland station, a Columbus station, a Detroit station, a Hartford station, an Indianapolis station, a Los Angeles station, a Miami station, a Milwaukee station, a New York station, a Pittsburgh station, a Sacramento station.

The following stations have announced rate increases of 28 to 40%:

A Cincinnati station, a Chicago station, a Portland station, a Detroit station, a San Diego station, a Washington station.

In two markets, increases have exceeded 50%. Rates on a Cleveland station from 11:00 to 11:20 P.M. went from $300 to $480 per announcement. Rates for announcements at 7:30 P.M. on a Sacramento station were increased from $100 to $185.

Unquote.

This memo does not touch network television costs. I think their pricing policy can be summed up in one sentence: As an advertiser, I expect an automatic 10 per cent increase in the cost of network advertising each year.

The effect of this increase in costs . . . both network and spot . . . is to slow down an advertiser. We advertisers have so much money to spend on advertising and no more. We obtain this money from sales. We obtain the sales from advertising. There is a relationship in my business between the amount of advertising we do and the volume of sales. More advertising, more sales. But if you force us to cut down on advertising, you affect our sales. Next time around, we have less money to spend on advertising. And next time around, we won't even get as much advertising out of our money as we did the time before.

I presume station owners and managers are searching for the golden mean. That is, how much money can you charge the advertiser before you drive him crazy. I suggest you're coming pretty close to that point with a lot of your customers right now.

To be frank with you, gentlemen, I think you're getting greedy. And I don't think this is smart. I don't think it's smart for you, and I don't think it's smart for the country.

I know that if this were 1955 and I had a product called VO5 Hairdressing that I wanted to sell to the public, I know that your rates would discourage me.

I know that if the rates you charge today were the rates in 1955, the business I have built up would not possibly have prospered the way it has.

Do you know what the big argument against television advertising is? It costs too much money. And it costs more each year.

I know they are right about it costing more each year. I don't think they are right about it costing too much right now. But if the pricing trend is unchecked, there is no doubt that there will come a time for Alberto-Culver when we will be forced to use more of other media.

That's the end of part two of my speech. The third part of my speech is the shortest.

I willingly admit that you gentlemen have been greatly responsible for the success of the Alberto-Culver Company. In my estimation, television is the strongest selling medium there is. Television is to selling what the transcontinental railroad was to the development of America. It is what the atom bomb was to warfare.

I just hope that television is adapted to peaceful uses. But I am sure it will be.

We advertisers know that we owe a great deal to you. And I think you, in turn, owe a great deal to the country. I'm not talking about program content. I am talking about your obligation to the economy.

Television is the greatest instrument for growth in this economy. Television has democratized selling. It's given everyone a chance to speak to the whole country at once about his product. If the product is good, the producer will make it. But, if you keep him away from the medium by charging too much, he may never make it.

I realize this won't affect you immediately. The big companies will keep advertising more and more. Your income will continue to get bigger and bigger. But, eventually, you will be hurt.

Gentlemen, learn a lesson from a salesman. Don't price your goods too high. You're just making business for someone else when you do.

Let me close by thanking you on behalf of Alberto-Culver for having helped make us successful. And, let me personally thank you for having asked me to speak here today.

43. Can Advertising Be Harmful?

by Fredric Wertham, M.D.

Following training in Germany, England, France, and Austria, Dr. Fredric Wertham served in various capacities as senior psychiatrist and consultant at Johns Hopkins Hospital in Baltimore, and at Bellevue and Queens General Hospitals in New York and Queens. In his various capacities, Dr. Wertham observed much mental and emotional illness and maladjustment, and began to seek causes for it in the environment, including the mass media. An abundance of articles and books resulted, among them Seduction of the Innocent, The Circle of Guilt, *and* A Sign for Cain: An Exploration of Human Violence.

The present essay was a paper read before the Association for the Advancement of Psychotherapy at the New York Academy of Medicine on March 31, 1962, and later published in the American Journal of Psychotherapy. *It is used here with Dr. Wertham's permission.*

The question of the effects of advertising on people leads directly into one of the major controversies within present-day psychopathology. According to the most widespread dogma, only individual intrapsychic and interpersonal factors have serious effects. Social and economic factors are regarded as secondary and negligible. On the other hand, the point of view which I represent has found by concrete analysis that environmental social factors are just as potent and often decisive. Psychodynamics is incomplete without the social dimension. We do not distinguish schematically between primary and secondary, or between causal and contributing, but consider contributory factors also as causal.

From this viewpoint advertising is a most important subject. There is no doubt that advertising can be informative and beneficial. What is not sufficiently recognized is that large segments of advertising are harmful to the individual, physically and mentally. Industry produces commodities; advertising produces needs. The underlying rationale is well expressed in a widely read book by a business man. He says that the successful advertiser "sends his thought through all the instincts and passions of men, he knows their desires and their regrets, he knows every human weakness and its sure decoy." This reference to deep study of human nature for promotional purposes is not from a recent book. It was published almost half a century ago. Since then mass advertising has made tremendous strides. It has become all-pervasive and inescapable. It is not only part of business but part of our life. It faces us in our living rooms on the television screen and in the landscape we view from our automobiles. Advertising has become a real public education. Our instincts, such as hunger and sex, are innate, but the shaping of our needs is learned.

What is at issue is not advertising, but *over*-advertising. The difference is like that between drinking and over-drinking. We may not know precisely when a man becomes an alcoholic—but we do know there are too many of them. Confronted with the ubiquitous pressure of advertising, the average individual feels—and rightly so—that he is exposed to a superior power. This contributes to his sense of isolation and alienation. Advertising does not appeal to our strength; it builds on our weaknesses. People are asked to believe that they choose when they are really being manipulated; to believe that they are buying when they are really being sold. If a lot of current advertising would cease, people would have more chance to be themselves. The pressure helps to create the supposed antagonism between the individual and society.

How does the individual react in this situation? There are two ways: resistance and submission. Resistance, the attempt to protect ourselves from being too gullible, has led—especially among the young—to a general attitude of cynicism: not only ads, but *everything* "is hooey." In the second, submissive, type of reaction individuals respond with complaisance. They have been conditioned to jump for the product if the bell rings often enough. They have learned from childhood on that singing commercials are a basic form of communication and cannot give their attention to any sober exposition that takes more than two minutes. They don't want to be told, they want to be sold. Many have been processed actually to accept absurdities: that perspiration in a steam cabinet can be prevented by an ointment; that a beer is "nonfattening"; that one application of a toothpaste kills bacteria for all day long.

A great deal of advertisement is really pervertisement. I shall briefly document this in three areas: alcohol, tobacco, advertising to children.

Alcoholism is currently one of the most important social problems of this nation. It is our No. 3 health problem, coming right after heart disease and cancer. There are counties where 6 per cent of all adults are alcoholics. In some counties 80 per cent of the 14-year-old pupils have been introduced to drinking—many of them with parental permission. Overdrinking would be a vast opportunity for preventive medicine. What is the greatest obstacle? Alcohol over-advertising. This advertising is designed not to inform but to seduce. We are supposed to believe that whatever we do, alcohol makes us do it better. This is expressly stated for all spheres of life: parties, homemaking, family life, Christmas, friendship, sports, patriotism, gracious living, romance, status, travel.

Recently there have been two major breakthroughs in this field which seem to have been completely overlooked.

1. More and more alcohol advertising is addressed directly to the female sex. You are shown a young woman glass in hand, or with a highball on the table or tray beside her. This is in part responsible for the current marked increase of drinking among women, especially harassed young housewives.

There used to be a tradition that women's magazines do not carry alcohol advertisements. Now that ice has been broken by one of the biggest women's magazines. It carried a Schenley advertisement advising homemakers to serve liqueur *"always, with any meal."*

2. The second breakthrough involves the tradition that broadcasting stations do not carry commercials for hard liquor. Recently more than 60 stations (including some TV stations) started the practice of using such ads.

These two instances present the dilemma in its sharpest form. The financial interest of a corporation is in direct conflict with the personal interest of the individual. The good sense of the people created a tradition; advertising breaks it.

What about tobacco? It is now scientifically well established that too much cigarette smoking is a contributing factor to lung cancer and to coronary heart disease. Here, as with alcoholism, is a large field for mass medical prevention. It is practically impossible, however, because of the tremendous amount of money spent on cigarette advertising. Much of this advertising is especially persuasive to young people. As a result, cigarette smoking is actually increasing—a fact for which advertising is in some degree certainly responsible. The belief in the much-advertised filters has no more validity than the medieval superstition that a substance from the horn of a unicorn has healing power. There were no unicorns, and there are no safe filters. Here again is an instance where the advertising interests run counter to the personal interests of the individual.

Advertising directly to children, including even the youngest ones, is common on television. They are also used as high-pressure salesmen to sell products to their parents. The child is considered not as a child but as a consumer or a salesman. And it has to be a very strong child who resists the commercials directed at him! The collective education by advertising is in grim competition with the personal education by parents. Just try to convince the little boy that he does not need the toy gun or the special dry cereal (actually less nutritious than bread, milk, or an egg), or the little girl that she does not need a home permanent or the cosmetic shown by the little Coty girl (which actually is bad for her skin). These things have been photographed on children's minds. If a parent tries to resist buying them, there is trouble in the family.

Skilfully devised advertising tells parents that children "2 to 4 years" old should have "pearl-handled guns" and "two-holster belts" to play with. It is supposed to be good for them, to give them, as one expert put it, "a sense of power and freedom"! Experts and advertising men (it becomes increasingly difficult to distinguish between them) do not discover needs, they create them—and then they call them "basic needs."

One particularly clear example of harmful advertising directed at children is furnished by switchblade knives. They were advertised and illustrated in tens of millions of comic books. We were told that they were the expression of youth's natural instinct of aggression. These knives have no constructive use whatsoever. They serve only as weapons for threats or sudden attacks. When we pointed out the harm done by these knives, however, the advertisers paid no attention. The advertisements did not stop until the switchblade knives themselves were forbidden by law (to adults as well as to children).

What is the remedy? There are two contrasting social philosophies. One advocates an ideal of general welfare with an enormous consumption without any inhibitions. Present-day advertising fits perfectly into that. The other aims at the satisfaction of human needs according to healthy normal and scientific principles. For this, present-day over-advertising is a great obstacle. A recent article in the *Harvard Business Review* says that the consequences of marketing as far as the individual is concerned are "none of the businessman's business." That may be true. But in my conception these consequences for the individual are part of the *doctor's* business, or as Dr. Carl Binger calls it, the doctor's job. For he is there not only to try to cure but to help to prevent.

44. Control by Advertisers

by Jack Gould

Mr. Gould is the New York Times *television critic. This article is reprinted from the* New York Times *for July 12, 1959, by permission of the author.*

Agency Officials Tell FCC Who Is Responsible for What On Home Screens—It All Adds Up to a Primer on TV

How advertising agencies operate in television—their strict supervision of shows and the business factors that influence or limit the choice of programs that the public sees—was explained by agency executives testifying at a hearing held in New York last week by the Federal Communications Commission.

The initial witnesses—C. Terrence Clyne, senior vice-president of McCann-Erickson; Robert L. Foreman, executive vice-president of Batten, Barton, Durstine & Osborn; Richard A. R. Pinkham, senior vice-president of Ted Bates and Co.; and Dan Seymour, vice-president of the J. Walter Thompson Company—spelled out the advertising side of TV with uncommon vividness and forthrightness.

Indeed, seldom has there been a more revealing summary of the conditions, apart from purely theatrical matters, that govern the day-to-day operations of the video medium and affect its substance.

The testimony of the executives was part of the continuing inquiry by the FCC into the practices and policies of the TV networks. Very probably it will be a matter of months if not years before the Commission decides whether any new regulations are necessary in the TV broadcasting field.

What was made abundantly evident was that advertising agencies, which never solicit billing on the screen in practice, may be virtually the actual producers. The Theatre Guild, David Susskind, and Desi Arnaz may take the public bows, to judge by the testimony, but they don't make an important move without an approving nod from the agency men.

In the case of shows in which they are active, for instance, the agencies said that they review all scripts in advance, scrutinize dialogue and story lines, and have their "program representatives" on hand to check each day's production work.

The Theatre Guild, one of the more independent institutions of Broadway, agrees in the case of television to let BBDO sit down and jointly review what dramatic property to do, the wisdom of the casting, and "each revision" of script, Mr. Foreman testified.

The agencies readily acknowledge that considerations of advertising dictate limitations on subject matter.

Mr. Dan Seymour (V.P., J. Walter Thompson) testified that "on dramatic show after dramatic show" the advertising agencies delete material deemed contrary to a sponsor's interest. "An advertiser cannot afford to lose any segment of society," he said. Any political mention is prohibited in drama supervised by his agency, he added.

Mr. C. Terrence Clyne (V.P., McCann-Erickson), speaking for clients of his agency, said as a matter of company policy a sponsor does not want to leave a viewer "sad and depressed" about the one-tenth of 1 per cent of the country that knows desolation and misery. The sponsor is not in "the business of displeasing" and wants to leave with the viewer "a pleasant and favorable impression," he said.

A program that displeases any substantial segment of the population, Mr. Foreman said, is "a misuse of the advertising dollar." Most advertisers do not want to spend their money to arouse controversy that might cause customers to think ill of a sponsor. "It's just bad business," he said.

Mr. Foreman added that even a relatively small volume of critical mail can make a sponsor "apprehensive." He recalled an experience of seeing the head of a very large corporation personally reading each letter received. Because of the nature of their business in dealing with the public, advertisers are "extremely sensitive," he said.

The agency executives agreed that the policies of sponsors did not usually lead to difficulties with most writers, producers, and directors. The creative folk are "hep," Mr. A. R. Pinkham (of Ted Bates Agency) observed; they know the headaches to avoid. One result, Mr. Seymour reported, is that script conflicts have become fewer and fewer.

Mr. Pinkham noted that a manufacturer of nonfilter cigarettes wanted the villains in the drama to be shown smoking a filter cigarette. Similarly, he said, a filter manufacturer wanted the villain to be depicted as preferring nonfilters. An aspirin company, he noted, would not stand for a drama that showed a suicide committed by swallowing too much aspirin.

Despite the high degree of agency participation, the executives stressed that in their opinion their companies often improved programs in terms of theatrical effectiveness and that their supervision did not dampen creative spirit.

POLICIES

In deciding what show to sponsor, the agency men also were in general agreement on the manifold factors that are regarded as pertinent.

Mass circulation was a dominant objective, particularly for low-cost package goods such as detergents and cigarettes. Another approach may be to implant an image of a product—an automobile, for example—against the day when the consumer is ready to buy.

Mr. Pinkham observed that a cigarette company, which sells mostly to men, should choose action shows and sports. A food company, selling mostly to women, should avoid newscasts, which have a large masculine audience, and pick situation comedy or drama, he said. Mr. Foreman said that *Lassie* had proved very satisfactory for Campbell's Soup because it reached a family audience.

Other advertising considerations may be to enhance the "image" or "profile" of a sponsor and not have a direct sales purpose. Also the development of a sales personality, such as Dinah Shore's pitches in behalf of Chevrolet, can be sufficiently important in itself to warrant settlement for less than a maximum audience. Stimulating dealers, both on the retail and wholesale level, to sell a sponsor's wares also is an influence in program choice.

Near the end of his testimony Mr. Foreman (Robert L., V.P., BBDO Agency) remarked that broadcasting was the only advertising medium where the advertiser had a voice in determining content. The only exception, it was noted, was the newscasts, which networks keep firmly under their own control. Documentaries also are free from agency participation in their preparation, though agencies and sponsors see them before deciding whether to put them on the air.

The sum of the first days of the FCC hearing is to establish a fresh factual record of the limitations of commercial TV inherent in advertising sponsorship.

In the lucid questioning by Ashwood P. Bryant, counsel to the Broadcast Bureau of the FCC, it also was made clear that, in one way or another, the function of selling, not the show, is paramount. Even when a sponsor may offer a public service attraction, according to Mr. Clyne, it is to fulfill an advertising purpose of some sort.

The agency men properly observed that a great deal may be accomplished within the existing TV relationship between commerce and culture and that often the best interests of advertiser and audience do coincide. Moreover, it was brought out, some sponsors, seeking an institutional image rather than a direct sale, may offer an increasing volume of programming intended for "the lace-curtain audience" rather than "the bread-and-butter audience."

But the testimony left no doubt that, if TV is to achieve its full potential as a medium giving vent to the limitless thoughts and ideas of mankind, the task cannot be left only to commercially sponsored TV. If the taboos, apprehensions, and anxieties of sponsors and agency men may have application in the marketplace, it does not follow that such a set of mores should govern one of the country's major platforms for human expression.

Interestingly enough, the agency men stressed that it was only the network that had both the ultimate and practical authority to determine the composition of today's national TV. Mr. Pinkham, for one, urged the networks to try experimentation and, if necessary, withdraw from sale enough evening time for the types of programming that most sponsors could not undertake for purely business reasons.

PROBLEM

The effect of the FCC hearing, therefore, has been to stress at a timely moment the crucial matter of balance in programming. Hysterical outbursts against agency men or sponsors will not offer any solution. In all entertainment forms there always is a preponderance of producers interested in pleasing most of the people most of the time and avoiding anything that might prove unattractive; the average sponsor's attitude is nothing unique in mass show business.

But the urgent lesson to be drawn from the agency testimony is that there must be companion forms of TV that are free from worries over sellers and consumers, marketing and distribution, corporate profiles and product images. In a healthy TV medium there must be room not only for happy entertainment but also for the play that does deal with misery and desolation or the documentary that sets the public agog over a controversial issue that requires the nation's immediate attention.

A mixture of commercial and sustaining programming always has been basic to the American concept of broadcasting. In recent years economic pressures and FCC indifference have thrown this approach out of kilter. The major service performed by the agency men was to illustrate that sponsors can render many excellent services in meeting public needs but they are clearly not in a position to do everything. It is a point that has needed to be made for a long time.

IX. International Television and Radio

WHAT ARE the omens of the satellite age? Is it to be used for propaganda or for understanding; for peaceful or warlike purposes? Is it to emphasize any one aspect over another of commerce, education, or culture?

Is it to serve the underdeveloped countries of the world, regardless of ability to pay? Or is it a medium which, because of economic demands, is destined to serve only those who can afford it? What is the image of the United States as seen on the television screens of the world?

What are the uses and alternatives considered by other nations in their uses of television and radio?

Considered here are a representative group of writers and spokesmen who consider U.S. television abroad as a lucrative form of business; the implications of satellite communication; pay television; and a challenging statement presented by CBS's Frank Stanton.

45. Big Brother's Television Set

by William Benton

Senator William Benton, Publisher and Chairman of the Board of Encyclopaedia Britannica, besides serving as Senator from Connecticut, has been an Assistant Secretary of State, and was a founder of the Benton and Bowles Advertising Agency, from which he retired in his prime to devote himself to writing and public service. He has been active in UNESCO and many other groups interested in education and world understanding. This article appeared in Esquire *for March, 1964. It is reprinted by permission of that publication, which holds copyright, and by permission of the author.*

> We recognize clearly the enormous potential of radio and television for education. These incomparable media must not be just a waste of time. They must be intellectually stimulating, vital, full of ideas. We shall utilize these media to educate our people, to raise their aesthetic tastes and to help make them more fully developed human beings.

If these statements had come from the presidents of the three big American networks they would be cause for national rejoicing. Ominously—they did not. They were made to me in Moscow by an intense, vigorous, youthful-looking cabinet minister who was describing the broadcasting plans for the people of the U.S.S.R.

The official is Mikhail Kharlamov, formerly Chairman Khrushchev's press officer. Kharlamov's name is largely unknown to Americans. Yet he occupies a position of enormous potential influence and power. Pierre Salinger had urged me to call upon him. As chairman of the State Committee on Radio and Television, Kharlamov is not far behind Gromyko in the Council of Ministers. And he is hurling at us a new challenge to which the Soviet Union gives the highest priority.

Nine years ago, on my first tour behind the Iron Curtain, I found the gap between Russia's commitment to education and our own alarmingly wide. Russia is devoting a much higher percentage of its gross national product to education than are we. It is true that except in certain areas—correspondence courses at university level, number of engineers in training—Russia still may be behind us. But the Soviets are determined to surpass us in every project. Following my fourth visit, I can now report that the fervor for teaching and learning within the Soviet Union has grown even more intense. And we Americans have been unaware of the extent to which the U.S.S.R. plans to employ a weapon that can prove to be the most potent in its entire educational armament—broadcasting.

Dr. Thomas Clark Pollock of New York University said not long ago: "Television offers the greatest opportunity for the advancement of education since the introduction of printing by movable type." The new Russian leadership understands this. They understand the potential impact of television just as they understand and respect the power of the nuclear bomb. That is why the astute Mr. Kharlamov and his able staff are bustling with plans for the future.

When I visited with him he was supervising the design of a great group of buildings to form a "Moscow Television Center." This is to have the latest and finest equipment. A 1700-foot TV tower is under construction. The nation's entire administrative structure for broadcasting, he tells me, is to be reorganized from top to bottom. Six channels are to be used. Plans are being made to insure good TV reception for the whole of the U.S.S.R., which embraces eleven time zones. Under study is the possibility of bouncing the signals from four Telstar-type Sputniks—but the more conventional cable and microwave hookups also are to be employed.

By the beginning of 1963, according to Kharlamov, there were 130 stations equipped with studios and capable of originating programs, plus 220 relay or booster stations, all serving areas with a total population of 90,000,000. There were nine million receivers in use, he said, with five thousand being added daily. Studio-equipped stations originate 850 "program hours" a day, compared with only 150 five years ago. This is still, of course, only a small fraction of U.S. totals—but the growth rate is impressive.

And by far the most significant aspect of the Russian TV system is to be its emphasis on education. For example, Kharlamov plans to set aside one full channel entirely for visual support of correspondence courses. Already English lessons and instruction in a variety of home, factory, and farm skills are being televised. A year or so ago 52,000 farmers in the region surrounding Moscow clustered around their TV receivers in the evening hours as part of a correspondence course in scientific agronomy. Students were divided into small, manageable groups. Attendance was taken by an ingenious monitoring system and instructors checked the required written homework. This program, a special enthusiasm of Chairman Khrushchev, was said to be such a success that plans are underway to expand it throughout the Soviet Union.

There are of course serious deficiencies in Soviet television. So far they have only a fraction of the receivers we in the U.S. have. They are years behind us in production techniques. Most of the programs now broadcast over the government-owned and -operated stations are like most other Soviet manufactured products—simple, serviceable, and often dull. Much time is devoted to Chairman Khrushchev's comings and goings, party meetings, political addresses, lectures, and major sports events. Entertainment is supplied by feature films, plays, operas, the great ballet performances, dance programs, and musical concerts.

But the directors of Soviet broadcasting are now eagerly studying and adopting the techniques—though not the content—of American TV. They are even introducing the capitalistic system of competition between networks in a major effort to improve performance. "Let the different networks fight for the people's attention," Mr. Kharlamov told me. Each of the five existing radio networks in the Soviet Union is to operate under this new competitive system. The same principle is to be applied eventually to the six television networks now under construction. Of course, centralized control will never be relinquished fully. "We cannot allow all the stations to put on talk programs at one time," Mr. Kharlamov points out. Nor (he did not mention this) can he allow stations to put on talks—or films or plays or instruction on anything—that do not fall into the framework of state policy.

Let me concede also and at once that Soviet planning and Soviet publicity often outrace Soviet achievement. Nevertheless, we must face a chilling reality. Even if the Soviets accomplish only half of what they have set out to achieve in television, the result may be remarkable. For the Russians, far poorer than we in almost every way, are richer in zeal for education. They have begun to grasp what the controlling interests of U.S. broadcasting do not accept as a primary goal—the superlative potential of television to broaden a man's knowledge, deepen his understanding, and enrich his life. Our programs are improving only somewhat, if at all, in intellectual quality. Newton Minow, before he resigned as Chairman of the Federal Communications Commission to join Encyclopaedia Britannica, told me: "There are now more patches of greenery visible here and there throughout the wasteland, but not enough to convince me to withdraw that designation completely."

Entertainment should of course have the major place in American network TV—no thoughtful person would dream of suggesting otherwise. But programs that stretch a man's mind and enlarge his horizons are far too few. The slick and the merely palatable still have a stranglehold on the commercial airwaves. And the commercial airwaves have a stranglehold on TV.

Commercial television may claim it is functioning in the "public convenience" and perhaps in the "public interest." But no one can argue successfully that it is indeed functioning in the public "necessity." These three words—the public's "interest, convenience, necessity"—are the key words in the Communications Act which authorizes the present radio and TV setup; and these three words establish the obligation all stations supposedly assume when they accept a license.

Professor Harold Lasswell of Yale, former president of the American Political Science Association, asks this question about television: "Suppose you were an enemy of the United States and were hired to demoralize the American nation, what TV strategy would you use?" Dr. Lasswell answers thus: "In all probability you would do what you could to keep the present situation as unchanged as possible."

Commercial television executives in effect deny the deep thirst of many Americans for education. These many Americans, in the present system, don't constitute a profitable audience. It is not conceded that sizable minorities with serious interests also have rights—the right, for example, to turn the dial past *The Beverly Hillbillies.* Today there is indeed nowhere for a viewer seeking mental stimulation to turn, little to choose at prime viewing time among variety show, 1946 movie, police thriller, and 1935 gangster film.

Thus American television for the most part steers safely along the easy and profitable road, concentrating on what it has learned will attract the largest percentage of set owners. It ignores the remarkable cultural revolution that is producing more inquiring minds than ever before in our history.

Yet we have some tremendous advantages in the TV competition. We have the transmitters and receivers. We have the networks, the resources, and the skills. We have something else—a "trained" audience that has seen more movies and more TV than any other population. What we lack is diversity in our programming—the diversity which will give millions of willing people a chance now denied them in the uniformity of the commercial stereotype.

To remedy this lack, the FCC in 1962, under Chairman Minow's leadership, successfully sponsored an Act of Congress which can affect profoundly the future use of television. After April 30 of this year, all TV receiving sets manufactured in the United States and shipped in interstate commerce must be equipped to receive eighty-two channels, not merely the twelve channels for which most sets are now equipped. Each year, starting in May, between six and seven million new eighty-two-channel receivers will flow into American homes. It is believed that most homes will have such new sets before 1972. This should stimulate greatly the use of the seventy so-called Ultra-High-Frequency (UHF) channels, now largely neglected because of lack of reception.

Mr. Minow has predicted a far greater diversity of programming in consequence—including serious programs. Further, he hopes for the creation of a fourth commercial network "appealing to higher rather than lower common audience denominators."

My own hope is that the projected multiplication of stations will make possible a chain of "subscription" stations catering to minorities with serious interests—for a fee. The subscription technique, called "Pay-TV" for short, involves a home installation which "unscrambles" advertising-free programs the set owner is willing to pay for; it carries a coin box or makes a record for billing purposes. The station can thus afford to produce programs for groups much more limited in size than the audience demanded by advertisers.

With commercial television now devoting itself to entertainment, one would logically expect that educational TV—known as ETV—would be carrying the torch for enlightenment. Is it?

Almost eleven years have passed since the first ETV station, KUHT, went on the air in Houston, Texas, in May of 1953. Now eighty-three such stations speckle the land. Most of these beam instructional programs to classrooms in the daylight hours and present cultural and civic programs in the evening.

Despite the fine things that must be said about it, and the brave announcements of things to come (one forecast is that there will be two hundred ETV stations within a decade), a particularly painful fact about ETV remains unrefuted: the overwhelming majority of ETV stations are floundering in a financial morass, struggling along from month to month against steadily rising costs of operation and maintenance. As a result, they are unable to prepare or procure the adult programs which desperately need to be prepared.

ETV stations are understaffed and underequipped. Normally they must employ inadequately trained people and, as one study reported, "Too few staff members must wear far too many hats; they do not have time to mount a program or rehearse talent and crew adequately." While some programs are excellent, local ETV stations frequently offer, in *Time* magazine's words, "yawning forums and tediously detailed state histories."

ETV's major financial support in its earliest years has been the Fund for Adult Education, established by the Ford Foundation. Help—though not much—has come in recent years from other foundations, from business and industry, and, on a quid pro quo basis, from tax funds of local school systems. Senator Warren G. Magnuson, chairman of the Senate Commerce Committee, after a long effort secured passage of a bill authorizing federal money for construction of ETV stations.

When an ETV station is authorized by the FCC, private commercial ownership, commercial sponsors, and profits are prohibited. Operating money must thus be raised through gifts, raised coin by coin and dollar by dollar by patient, dedicated men and women who sense that ETV can become a great force for good in their communities. The typical ETV station today, according to National Educational Television, gets along on an annual budget of about $400,000, plus a few gifts of services, equipment, and materials. This is perhaps a dollar per year per evening viewer. The eighty-three educational television stations spend less on programming in an entire year than is spent via NBC, CBS, and ABC in a week.

Mr. Minow told the tenth-anniversary convocation of the Fund for the Republic in New York in 1963 that the "lighting up" of the new UHF channels "will make possible a truly nationwide educational television system through a network of stations devoted to classroom instruction during the day and to broad cultural adult programming in the evening." The key word here is "possible." But is such a development *LIKELY?* Where will the money come from? Will advertisers pay for the higher quality fourth commercial network? Mr. Minow doesn't tell us. What we know for sure is that ETV's crucial need is a sound economic base.

Out of some thirty-five years of experience with commercial and educational broadcasting, and with the Voice of America, I have arrived at two principal conclusions: On the one hand, we Americans can try to stimulate commercial television, under its present setup, to program for the high common denominator as well as the low. On the other hand, we can undertake to give educational television an infusion of new strength. I envisage two major steps that might take us a long way toward both objectives.

First, let us now and at once, by Congressional action, create a National Citizens Advisory Board for Radio and Television. This commission would be composed of leaders in the civic, educational, cultural, and religious life of the nation, and of men experienced in communications. Its members would be charged with responsibility for making findings on trends, problems, and opportunities in broadcasting, and making recommendations about broadcasting accordingly. The Board would function somewhat as a U.S. equivalent of the Royal Commissions employed so effectively in Great Britain. It would have no power other than that given in its title—the power to advise. It would have no share in the authority of the Federal Communications Commission to grant, withhold, renew, or revoke broadcast licenses, no judicial or legislative function. It would make an annual public report.

The influence of the Board could be great. It could help provide leadership to public opinion about broadcasting. It could suggest alternatives. It could examine the problem of financing educational television, and recommend solutions. What network, what station, could wholly ignore the reports of such a Board? They would be front-page news—where news of television belongs.

When I was in the Senate I introduced a resolution to create such a Board; it was shelved. Later Mr. Minow, while he was still FCC Chairman, lent his considerable prestige to the plan. "The Board was never created," he said in an address. "I think it should have been. It is not too late." Now a new group of Senators is planning to revive the project. If this Board had been created in 1951 the pattern of TV today, in my judgment, would be different.

Second, let us act now to put ETV on a self-supporting basis. My strongest recommendation is that the ETV stations currently and in the future authorized by FCC, and the new high-quality commercial UHF stations envisaged by Mr. Minow, be encouraged to adopt the "subscription technique" I have described above. Originally, the proponents of ETV hoped the stations could finance themselves by gifts, as does the Red Cross. It should now be clear that ETV will be unable to perform its massive and vitally important tasks—including improvement of the programs—if it must rely for support on local fund-raising. It must collect from the customer.

Is there, after all, any real doubt that millions of Americans would willingly pay small sums for new cultural and educational opportunities? Consider what has happened to the book-publishing business in the U.S.—it has rather suddenly become a billion-and-a-half-dollar-a-year industry, with reference works leading the rise. Consider the sale of recordings of serious music. Or the new art-appreciation courses. Don't these show the willingness of people to pay?

ETV itself has produced encouraging symptoms of this willingness. I do not believe ETV can produce a flow of revenue consistent enough, or adequate to its needs, by selling course materials or examination services. But I do believe the following instances suggest that a substantial number of viewers might become paying subscribers to complete ETV programs:

1. In Chicago a "TV College" is now in its eighth year of operation over WTTW. Audience surveys report that regular viewers range between 5,000 and 100,000. Thousands buy study guides.

2. In Denver and Chicago, many thousands paid fifty cents and a dollar for foreign-language guides to follow lessons over ETV.

3. In Cleveland, many hundreds paid $3 each for a syllabus with which they could audit a course in elementary psychology given by Western Reserve University.

4. In New York a hundred forty-two persons ranging in age from seventeen to seventy-three showed up at New York University to take a stiff two-hour final examination for college credit in a course in Comparative Literature which they attended for fifteen weeks via TV. Each paid $75 tuition for the course. For five days a week they had risen early to go to "class" at 6:30 A.M. For homework, they read sixteen books. About 120,000 others had watched the sessions, WCBS-TV officials estimated.

5. In New York, seven thousand people bought the textbook for a college-level course, "Russian for Beginners"; in the first two months the course was carried by Channel 13.

6. Throughout the country, an estimated one million education-hungry viewers arose at dawn to sit before their television sets and absorb a course in Atomic Age Physics. This was presented over NBC's *Continental Classroom,* which was originally financed in large part by the Fund for the Advancement of Education. Housewives, businessmen, working people—Americans from every group in our society—were avid students. Each year many hundreds made arrangements with universities in their communities to obtain college credits for the course. In the very first week the course went on the air 13,000 textbooks were sold. Reports the Ford Foundation: "Parents marveled at the sudden alertness of formerly late-waking teenagers—Catholic institutions rearranged Mass schedules to permit viewing by students and clerical teachers. . . ." In all, an average of 400,000 persons daily watched the course the first year it was telecast.

7. Last year hundreds of thousands in all parts of America watched a course called The American Economy presented by the Columbia Broadcasting System's *College of the Air.* In 1962 other thousands tuned in on a course in The New Biology. Some three hundred participating colleges offered credit for these courses, when special arrangements were made by students. Most interestingly, some 33,000 copies of a student guide offered for sale with the American Economy course were bought by viewers at $2.95 each.

Finally, a study by the National Opinion Research Center in Chicago claims that 25,000,000 adults in the U.S. are "following some plan for adult education." They are meeting and studying in every possible setting—in public schools, universities, libraries, business establishments, religious centers, union halls. By the hundreds of thousands they are taking courses in the liberal arts, the sciences, the professions, and all the crafts and hobbies. The Book-of-the-Month Club is said to have paid the Metropolitan Museum of Art over $860,000 in royalties on its Seminars in Art.

The potential audience for subscription ETV can be limitless as Americans are persuaded to realize that education does not stop at age fourteen or eighteen or twenty-one, that it continues for a lifetime.

Though the use of the subscription technique seems to me to be the single most promising way to finance ETV (and perhaps also Mr. Minow's "higher level" commercial network), I have three additional ideas for discussion. These may seem unorthodox to many—to educators as well as others:

1. Today all ETV stations are not-for-profit operations. But this need be no bar to their acceptance of commercial "patrons" to help finance expensive programs. During 1962 the not-for-profit National Educational Television, which then provided ten hours of programs a week for ETV stations, received "underwriting" of more than $500,000 from business sources for specific programs. In most instances this money came from the public-relations budgets of the Humble Oil & Refining Company; International Business Machines Corporation; Mead, Johnson & Company; Merrill Lynch, Pierce, Fenner and Smith; the National Association of Manufacturers, and other business sources. These "underwriters" were credited, at the opening and close of each program, with having made the program possible. There was no direct selling, no "middle commercial," and of course no program control by the "underwriters." Although its ETV license prohibits the use of regular advertising commercials, the FCC has approved these "credits" or form of commercial support.

I have no fear that the boards who control ETV stations—take, as an example, the board of station WGBH in Boston, which is headed by the distinguished Mr. Ralph Lowell—are going to be corrupted by the temptation to commercialize their stations or debase their program standards. They would not and should not permit a sponsor to determine program content. Thus, I would be willing to consider giving the patrons more than a mere "credit line" on the air. And surely, the competition for advertising dollars ETV stations would give commercial TV would be no more worrisome than the competition *The Atlantic Monthly* and *Harper's* provide for *Life* and *Look*. If we trust the boards of directors of NBC, CBS, and ABC to deal with sponsors in the public interest, surely we can trust the boards of our ETV stations. Let the latter use their own judgment on what they permit their patrons to say on the air.

2. Because ability in communications can often command high financial rewards, I would ask whether ETV can find formulas which would attract outstanding creative and management talents. One way to achieve this might be for the nonprofit ETV stations to enter into contracts with private managers and producers to take over part of their programming. Because considerable capital is required to install a subscription system in any community, the "Pay-TV" part of an ETV station's schedule might be contracted out, with the contractors sharing the earnings, if any, with the station.

3. There is of course one other way to finance educational and cultural television and that is through the taxing power; for example, the British

technique of financing the BBC through an annual levy on home receivers. I confess I do not share the horror such an idea seems to evoke in the U.S.—so long as independent and nonsubsidized systems remain in competition.

I do not foresee the development in the discernible future, as suggested by Walter Lippmann, of a U.S. government-financed network: there is no audible movement in that direction—and a decade of campaigning would probably be required to produce action in Congress. I do believe there is one way by which federal financial support might be developed for ETV in the next five years, given an organized effort. The Congress has now established a precedent by authorizing matching grants to the states for *construction* of ETV stations. There is now in the statutes a federal excise tax of ten per cent on TV receivers. Should not the receipts from this tax be earmarked for grants, via the Department of Health, Education and Welfare, to the states for support of ETV *programming?* I prefer taxing the customers to levying a tax, as has been suggested, against the commercial stations.

I began this article by reporting what Mr. Kharlamov told me Russia proposes to do with television. He told me how the Soviets plan to expand present instructional programs for farmers, workers, and technicians; how they plan to devote one entire network to support of correspondence courses for professional people; how they plan to use TV to train engineers and advanced students; and how they mean to use the entire system to make the Soviet people "more fully developed human beings." Above all—and this is consistent with their record as well as their pronouncements—I reaffirmed how intense is their devotion to education itself. The Soviets know what they want. And if television is a weapon in the Cold War, they are taking aim—zeroing in on a target. We in the United States have never thought of TV as germane to our national strength. We have been using television as a kind of fowling piece, scattering shot wildly.

I believe the competition between the Soviet Union and the United States is likely to turn on which society makes the best use of its brainpower. For most adults, this means the best use of communications media.

We have neither the wish nor the need to imitate the Soviets. We can meet the Soviet challenge in our own way. But if we are to live up to our own great pioneering tradition of universal education, we should employ television for education on a scale even more vast than the U.S.S.R. We should do this even if the U.S.S.R. were to sink suddenly into the sea. We should do this because it is indeed not only in the tradition of the American Dream—it is potentially the very essence of the Dream.

46. U.S. Television Abroad: Big New Business

by John Tebbel

John Tebbel, former Chairman of the Department, and currently Professor of Journalism at New York University, is one of Journalism education's most prolific writers. He has been affiliated in executive or writing capacities with American Mercury, Newsweek, *and the* New York Times, *and other periodicals, and is still a regular contributor to* Saturday Review *and other magazines. He is also author of a number of books, including a biography of David Sarnoff.*

This article was originally published in the July 14, 1962 Saturday Review. *Although the figures would now need to be updated, the timeless nature of the problems and trends sketched justify inclusion here. The article is reprinted with Mr. Tebbel's permission.*

While the debate over television's role in American culture is being argued on the playing fields of Washington, the entrepreneurs of the prerecorded small screen are quietly building themselves an international industry. With tape and film they have introduced the American wasteland, if indeed it is, into the programming of every country that has any television at all, and they are anticipating the needs of emerging nations that do not yet enjoy this blessing of civilization.

In little more than a half-dozen years, international television sales abroad have reached a total of more than $25 million. The center of the new industry is London, where the three major American networks and several producing organizations have sales offices busily selling United States television in Europe and on other continents.

London is the center for European, African, and Near Eastern distribution, but the networks and other distributors also have salesmen in Toronto, Mexico City (covering Central and South America), and Sydney, Australia, which is presently the distributing point for Asia.

There are sound business reasons for this expansion, which has only begun. One reason is the startling fact, little known outside the trade, that during this year the total of television sets owned outside the United States will surpass the American total. By the end of 1962, we will have fifty million sets; the remainder of the world will have fifty-three million, and this is one gap not likely to be closed. It will, in fact, tend to widen steadily as time goes on. As one might expect, the United Kingdom has more receivers, twelve million, than any other foreign country, but West Germany can boast four million and the Scandinavian countries nearly two million.

There is a tough competitive scramble to reach this juicy market. Of the American networks, NBC was first to realize and exploit the situation, and consequently has an impressive head start on the others as a supplier of television film. CBS has proceeded more cautiously, but it easily ranks second in sales, while ABC, a relative newcomer to the struggle, is third. Then there are the several large non-network suppliers: Revue (MCA), Screen Gems (Columbia), Ziv (United Artists), ITC (Independent Television Corporation of England, formerly an American-British company but now wholly British-owned), Desilu Productions, Warner Brothers, and Four Star. The BBC, too, is active all over the world; many of its shows are seen on American networks, who do not, however, distribute them through their international divisions. These strong competitors—producers and distributors—have one thing in common: they are big and getting bigger.

There is now some American television on every service in Europe, and it will be seen on the newest group of small TV services, just going on the air this summer in Sierra Leone, Kenya, and Gibraltar. For TV has become a status symbol. If a new nation wants status these days among the developing countries, it must first have an airline and then a television service, which will undoubtedly run at a substantial loss. In most of these countries, advertising revenue will be sought immediately to help offset the expense. The new African nations carry commercials on their services, as do some in the Middle East, but old established Western European countries like the Netherlands, Belgium, and France, along with some in Scandinavia, permit no commercialism.

The chief usefulness of television in every country is propaganda: it gives the government a means for communicating instantly with the people. While such usage does little for the quality of programming, it justifies a cost which few of these nations would otherwise consider worth meeting.

In spite of government propaganda, however, television remains an entertainment medium. This is particularly true in Britain, where the decision in 1955 to introduce commercial TV opened the way for American companies to establish their new industry. It was, in some ways, a painful introduction, accomplished through the Independent Television Act of 1954, which created the Independent Television Authority and ended the BBC's 27-year monopoly. This act was pushed through Parliament by a determined group of Conservative back-bench MPs, who later were accused of high-powered pressure tactics on behalf of the "commercial interests," meaning industry and advertising agencies.

Having been created, the new Authority found itself confronted with the formidable task of filling the air in competition with the BBC's established programs. Fortunately for it, at about this time American television began to be available on film, especially the high-rating Western shows and crime thrillers in the usual thirteen-week- and twenty-six-week series. This was exactly the kind of fare the Authority needed to create and reach the mass audience it believed was waiting in Britain. There were complicated questions of clearing rights to be solved, but the American networks were only too willing to help solve them, and on September 22, 1955, the new Authority put its first telecast on the air from a London station. By the end of last year, 95 per cent of Britain's population was within reach of ITA's operating stations.

Skeptics on both sides of the Atlantic who predicted that Britons would not like or understand American television films proved to be profoundly wrong. To experienced observers this was hardly surprising because the viewers were only looking at a condensation of something already thoroughly familiar to them, the American motion picture. Hollywood, in fact, has conditioned the whole world to easy acceptance of the rival medium. When the BBC, countering ITA's success, imported the Perry Como show on film, British intellectuals protested in the press and even in Parliament, but BBC audiences plainly enjoyed not only Como but the familiar motion picture and stage personalities who were his guests.

As America's TV film salesmen moved from Britain to the Continent, they had to deal at once with the language barrier. Again the movies had broken trail, because American motion pictures had long been dubbed in French and other languages, besides employing subtitles. Soon our TV films were being brought directly to France and dubbed there, although not without difficulties. "Lip synching," as dubbing is called, is a long and tedious process, and sometimes the French simply don't bother.

Not surprisingly, General de Gaulle has taken an austere view of dubbing. He believes French television ought to be French-made, and there is a quota on American film coming into France. Quotas, in fact, exist nearly everywhere. In Britain, the independent companies have agreed on a quota of 14 per cent of total time for foreign film. The BBC, making its own rules, has a quota of 9 per cent. In Gibraltar, Kenya, and Sierra Leone, the quotas will be 50 per cent for the British, because that is the audience; 25 per cent from other countries, and 25 per cent live.

The West Germans have been exceptionally hospitable to the American film salesmen, and are already the second largest consumer. They can easily afford to buy the foreign product because of their substantial advertising revenue. After some early frostiness, the Bonn government has become much more receptive, and audiences have long been getting the benefits of expert German dubbing.

Germans have always loved the old-fashioned Western melodrama in books and motion pictures, and they love it no less on television film. This passion they share with the people of every country. Television sales executives say the most popular show in the world today is *Bonanza,* which is telecast by every nation that has a TV service. The obvious reason for the Western's enormous popularity is the perfect escape it offers to an anxious world. Some television experts say the particular success of *Bonanza* rises from the fact there is no mother in the cast, thus offering a comprehensive psychological outlet to both men and women.

The pattern of popularity is almost exactly the same abroad as it is in America. Escape is the universal need. Dramatic suspense shows follow Westerns, succeeded by crime and mystery stories. Exact ratings, however, are virtually impossible to determine anywhere except in the United Kingdom, the only place which has ratings as we know them. Other countries make spot checks at the discretion of the program board or the buyer, but the resulting judgments are unavoidably subjective. Most boards and buyers display some sense of responsibility and try to maintain relatively high standards. American ratings mean nothing to the foreign buyer, however. He is always certain his audience is different—although he is likely to be uninformed about working-class preferences, because he so often judges his programming by talking to viewers who share his own interests.

In any case, the foreign buyers know they must depend on America for Westerns and crime stories. "We can do good dramatic shows of our own," says one, "but we cannot duplicate your gangsters and your cowboys and Indians."

Thus far, the tremendous supply of American film has guaranteed that the flow of business goes from West to East, but already there are indications that television film selling is not going to remain a one-way trans-Atlantic street. There is increasing activity, most of it in Britain at the moment, directed toward developing shows for international export. Contrary to popular opinion, this summer's miracle of world-wide satellite transmission is not expected to alter present patterns. Most experts believe satellite TV will be used for spot news—what Europeans call "actualities." There is no foreseeable reason to use satellite transmission in place of far cheaper tape and film, particularly when time differences make live programming so difficult.

The reverse flow of international business is presently two-pronged. One comprises the work of television producers who are making international shows, like the Inspector Maigret series based on Georges Simenon's novels, which is a British venture, and one-shot shows like the Jo Stafford special by Britain's ITC. The other prong is the formation of international producing companies like Intertel, conceived by Associated Rediffusion in Britain and involving Canada, Australia, and the United States as partners. Intertel's plan is to produce shows in all these countries and then circulate the product among themselves—a kind of television Common Market. This enterprise has been in business about a year and has already turned out "Decline of an Empire," made in Britain by an American producer; and a Canadian documentary, "Living With a Giant," about Canadian-American relations. TV's cooperative move to market an international product parallels the joint movie-making enterprises of Italian, French, British, and American producers.

Musical shows still offer a stumbling block to international television producers, because of the difficulties involved in clearing music rights and the legal tangles presented by personalities who are under contract to appear in participating countries. In the news field, by contrast, all countries draw on each other's resources, and television news is daily becoming more international.

International business is also branching out to include American participation in the ownership and management of foreign stations. Thus NBC has become a partner of the Nigerian government in setting up a new television chain. The network has sent key executives and technicians to assist in starting the service and to help maintain it for the next five years, during which time they will train Nigerians to take over the entire operation. Along with these developments, inevitably, an international advertising business is growing, based on such world-wide accounts as Coca-Cola and Lever Bros. Performers, too, are beginning to think in terms of international audiences.

American television, having saturated time and priced itself to the top in this country, is on the way to conquering the Western Atlantic nations and Africa. Tomorrow, no doubt, the world.

47. What Will the Satellites Communicate?

by Robert Lindsay

Dr. Lindsay is an Associate Professor of Journalism and Mass Communication at the University of Minnesota. His background includes affiliations with the U.S. State Department, UNESCO, the Radio-Television News Directors Associations, and numerous other organizations. He teaches courses in broadcast journalism, international communications and comparative foreign journalism, and public relations. His speech was delivered before the American Council for Better Broadcasts in Columbus, Ohio, June 12, 1963, and subsequently published in the NAEB Journal, August, 1964. It is reprinted with the author's permission.

We say of the communication satellites that used wisely they should prove of great value in achieving world understanding. I am not entirely certain what is meant by "world understanding"; I suspect it means something different to each of us. I do know that for several decades there have existed channels of communication which, if they but carried the appropriate messages, could have achieved this desirable if loosely construed state. The problem seems to be that we have yet to determine what communication—what messages—should go into the channels. And while we grope for the solution, the engineers and scientists invent more and better means of communicating. I should think that the first order of business is to close the immense and widening gap between technological advance and cultural lag. Unhappily, too few of us in education, in government, and elsewhere appear interested in this challenge.

The first experimental communication satellites have been aloft and functioning but a short time. Unquestionably, they represent a singularly important milestone in the developmental history of physical communications facilities. Their impact on man's continuing effort to improve his message-delivery capability is typically described in the bromidic language of wish-fulfillment. Dr. Allen B. DuMont, in his recent address to the Third International Television Symposium, spoke of the "vital importance" of such devices as the communication satellites "for the widest possible exchange of programs, information and culture" among all countries. "True understanding," he said, "usually leads to goodwill and peaceful existence."

I hope Dr. DuMont is right. I hope with all my heart that one day—soon, before it is too late—we may be able to assert this not as a matter of faith or trust or conviction but as a matter of fact. For the moment, however, I do not see that we are justified in describing our present thinking about the role of the communication satellites as much more than hopeful optimism. The mere capacity to deliver a message does not mean a like capacity, or willingness, to receive it on the other end. The capability to empathize does not necessarily result in this "true" and "world understanding" of which we speak so glibly in these forums. It is not without significance that the "hot line" between the White House and the Kremlin is perceived as an emergency forestaller of disaster when the Cold War heats up to fever pitch. Would that it could be an open line over which might flow less urgent but at least as important segments of an open-ended dialog. Even so, perhaps even this thin line will become the first strand among many to come in the service of a less tension-laden purpose.

Meanwhile, we look toward the heavens where, if we look carefully, we may be able to distinguish the communication satellites from the spacecraft, the "spy-in-the-sky" monitors, and assorted man-made debris. The assumption is that the communication satellites constitute a notable improvement over present global communications systems. However, it may be some time before they will surpass in utility and economy the present and planned submarine cables and other facilities. Yet no one can doubt that the satellite eventually will be the principal reflector or repeater of global communications. (And after that, we shall live to see the LASER relegate even the satellite to a secondary, if not obsolete, role in the inventory of communications technology.) But once the potential of communication satellites has been postulated, the area of agreement ends and we find ourselves in *terra incognita*. The fact of the electronic marvel stands in stark contradistinction to the vast no-man's-land which describes the debate of just how the marvel should be used in the service of civilization. Not yet in the United Nations, and certainly not in Washington, has there been delineated a clear-cut policy for utilization of the satellite communications system soon to be operational.

This pusillanimity in approaching the policy issues posed by the imminent era of satellite communications is understandable. The rush of technological breakthroughs and rapidly changing political mixes in the years since World War II have placed immense stress on the human decision-making process. And after all it is no more than fifty years since Tesla offered the earliest serious advocacy of the feasibility of communication satellites. British author-journalist Arthur C. Clarke published his famous article little more than fifteen years ago. Not until the late 1950's was it suddenly realized that our engineers and scientists—most of them in the employ of the communications industry—had in the span of a very few years moved to within launching-pad distance of perfecting that remarkable by-product of rocketry and missile research, the communication satellite.

Whether it be the low-orbit, passive reflector or the high-orbit, active repeater, or the synchronous or "Roman candle" type, the communication satellite is a matter of national concern. Quite literally in the closing hours of his second administration, President Eisenhower urged development of a satellite communications system to be supported by the government but operated as a corporate enterprise by the common carriers. The National Aeronautics and Space Administration, the Federal Communications Commission, and various congressional committees began serious consideration of space communications in 1961. Proposals and applications from commercial interests were received, hearings were held, and in general the nation—or some of its citizens—became aware that satellite communications would soon be a reality. Unfortunately, it also became apparent—again, to only a few persons—that the scientific achievement contained the seeds of exceedingly complex and totally unprecedented issue. An index of these issues would include, as a minimum, these headings: Social, Political, Economic, Cultural, Legal, Psychological, Language, Sovereignty, Ideological, Propaganda, Sociological, and many others. Many of these issues were raised and courageously and cogently debated in the summer of 1962 by a handful of concerned Senators and Representatives. At a time when the Congress was being criticized as never before as an inefficacious legislative machinery, fewer than 25 of its members stood tall in their inspired attempt to safeguard the public interest.

But as we know, the enabling legislation proposed and urged by the Kennedy administration was finally rammed through. On the last day of August, 1962, the 87th Congress enacted Public Law 87–624—the Communications Satellite Act of 1962. I have tried without success to determine why it should have been—of all Presidents—John F. Kennedy, who could find in his heart rationalization for this badly conceived device for operating our satellite communications system. It is interesting to note that at least a few Congressmen who supported the Administration's proposal at first have now had second thoughts about the corporation established by the Act. Nevertheless, we are stuck with this awkward instrumentality which elevates private, commercial interest above an already ill-served public interest.

Obviously, I believe that the Communications Satellite Act should be subjected to thorough examination and discussion. Among a multitude of things I wish for is a national survey to determine how many Americans know how much about this Act and the incredible corporation it creates. Perhaps one day ComSatCorp will be as familiar an acronym in the American lexicon as SNAFU—but I doubt it. Comsat is a venture of transcendent import. I submit that as a legislative solution to a new and uncharted dimension of the Space Age it is of highly questionable validity.

Above all, this unique corporation collides with the traditional American concept of the public weal. Today, as never before, it is imperative that the nation use every means at its disposal to ensure that we can serve the global community in the way it must be served in the grim contest for achieving peace through understanding. It may be that those who will be given the use of the satellite communications system will assume this task. But first, they must be willing to do so, and motivated to do so. And they must be committed to the national interest as the paramount consideration at all times.

In signing the Act on that fateful day in 1962, the late President Kennedy declared that, "in this way, the vigor of our competitive free enterprise system will be effectively used in a challenging new activity on the frontier of space." I wish I were not so uneasy on hearing what seems to be a faithful echo of the late President's words in this passage from a public relations blurb of the Bell System: "Above all else, Project Telstar is a tribute to the American free enterprise system. Through its own initiative, spending millions of dollars of its own money, the Bell System is exploring new voiceways in space to help bring better communications to the nation and the world." This of course disregards the fact that no

Telstars could ever be launched were it not for the billions of tax dollars expended on research and development in achieving the satellite-launching capability.

But this challenge of AT&T's self-congratulatory back-patting is another aspect of the issue. In terms of the use to which the communication satellites are to be put, the question must be asked—and asked over and over again: *Whose* "new voiceways" are being explored? In our Federal Communications Act we pay lip service, at least, to the concept that the airwaves are a natural resource and therefore belong to the people. And what, precisely, is meant by "better communications" for the United States and the rest of the world? Is the "better communications" for the common carriers benefited by the corporation necessarily what is best for the national interest? U.S. program producers are crowing about the expanding overseas markets for such syndicated television series as *The Untouchables* and *I'm Dickens—He's Fenster*. If this is the game a better mousetrap will attract, then I think we should pay very close attention to the specifics of how satellite communications can aid in creating world understanding.

Mr. Kennedy described the Act's purpose as being "to establish a commercial communications system utilizing space satellites which will serve our needs and those of other countries and contribute to world peace and understanding." But what *are* these needs in terms of the utilization of satellites for a truly global broadcasting? The President of NBC International said in March, 1963:

> There are cold war implications in many areas of our activity. We know that the uncommitted nations, whether for prestige reasons or true development, will quickly establish radio and TV services. We feel that if the U.S. and other Western countries permit a vacuum to exist in assisting these nations in broadcasting, the Communist bloc will surely take advantage of this powerful medium to influence minds.

Now I have no quarrel with the spirit expressed in this statement. But I wish we did not have to rely upon network executives to discuss the need to fill the broadcasting vacuum in uncommitted nations. It has long been my understanding that articulation, if not implementation, of foreign policy is a responsibility of the President of the United States, not of a broadcasting network (which isn't so much as licensed to operate by the government). I simply do not comprehend why the government should hand over to profit-seeking program producers the job of "influencing minds" and, to use that unfortunate expression, "projecting the American image." Of course, the networks and syndicators are already filling the air in countries around the world with their programs on film and tape. (The U.S. Information Agency does this, too, but to a deplorably lesser extent.)

And when the communication satellite system is in being, what then?

Suppose that by some miracle we could formulate an acceptable policy statement defining the kind of programming to be fed into it. Who should exercise responsibility for this programming? Who would have policy control over it? Should the National Association of Broadcasters extend its so-called Codes into space? Shall the networks and syndicators have free rein to show Indians and Africans and Peruvians whatever programs they choose to sandwich between commercials for Preparation H and Geritol? Or should perhaps our educational broadcasters and the U.S. Information Agency be permitted to have some access to the system? On the domestic scene just now, some members of the FCC are finally talking about steps to establish a truly national educational television network. Extrapolate this kind of thinking—late in emerging as it is—to the international level and I think one can appreciate that we are far, far away from the kind of thinking necessary to meet the challenges posed by communication satellites.

David Sarnoff was reported a while back as saying that with the laying by 1966 of AT&T's transistorized undersea cable, capable of carrying a greatly increased traffic, it would appear that the "economic need" for a communication satellite system could perhaps be placed far in the future. In April, 1963, an official of the National Association of Educational Broadcasters said:

> By 1980, the number of messages transmitted across the Atlantic from the United States is expected to increase at least 25 times. That means that most of these messages will have to be sent via satellite. Satellite communications will be jammed. There just won't be room for relaying TV programs, let alone educational TV programs.

I suspect that Vernon Bronson, NAEB's research and development director, was making an accurate prediction. And I say *No* to Senator Fulbright's recent question: "Is it such an intolerable hardship for Europeans to have to wait 24 hours while video tapes of American quiz shows and horse operas are flown across the Atlantic in jet planes?" Excepting only "hard" news coverage—which the networks do so well—and weather and health advisories and the like, there will for a long time be little need to transmit live programs to foreign audiences. And of course there will always be the problem of the time differential.

Given these conditions, I think I see a possibility for making intelligent use of our satellite communications system as soon as it becomes operational. Could we not require the Communications Satellite Corporation to reserve at least some of the system's transmitting time and facilities to the U.S. Information Agency and our educational broadcasters? The corporation will have custody of the hardware—although presumably at the sufferance of the citizenry. The corporation should be required to release the channels on a stipulated and regularized basis to nonprofit programs serving the national interest, rather than that of corporate stockholders and advertisers.

I should think that the American people would want at least some of the program content delivered globally by satellite communications to be of such nature as to picture them, and their institutions and culture, honestly and faithfully. If it must be, then let *The Beverly Hillbillies* ride around the world on microwaves awash with detergent commercials. But let us also have the best of our educational programming, and the best of our USIA programming, bouncing up to the satellites and into the villages and cities of all countries. I state this as a minimum desideratum. But if it develops that our commercial broadcasting interests have no great need to use the satellites—if they can just as well use the jet airplane and the cables for conveyance of their products—then let us require them to do so. As I read the Communications Satellite Act, the government has the authority to do this; if it doesn't then Congress ought to pass its first amendment to the Act.

That staunch defender of the American way of broadcasting, *Broadcasting* magazine, of course anticipated such proposals as I have been offering. In an editorial back in October, 1960, *Broadcasting* said: "The natural inclination of government to assume authority over the content of international transmissions must be resisted. With government in command, television would inevitably be perverted into an instrument of propaganda."

To this I reply by stating what *Broadcasting* must well know: that the United States necessarily has been engaged in propaganda activities of all kinds, using every conceivable means of communication, for a long time. Sad to say, it is likely that we'll have to continue our propaganda effort into the indefinite future. But there is nothing untoward about this. Quite the contrary, the propaganda function is in many ways the most important and potent weapon of deterrence at our disposal. There is nothing "perverted" about our national effort, a function of the U.S. Information Agency, to compete with the Communists for the hearts and minds of men. Unless, of course, *Broadcasting* really believes that *The Untouchables* projects a better "image" of the American society than does the Voice of America. Furthermore, it is, I suggest, far better that our propaganda effort be openly supported and conducted, rather than covertly as is the case with the Central Intelligence Agency's role in Radio Free Europe.

The *Broadcasting* editorial went on:

> The burden of international programming must fall where the burden of domestic television now reposes. It will be up to the television broadcasters of America to give meaning to the instruments whirling through space.

This assignment is the most complicated American broadcasters have faced. It is infinitely more complicated than the business of fashioning a domestic network schedule.

I agree with the point made in the last sentence of the editorial. To attempt to give meaning—intelligent, public-serving, peace-seeking meaning—to the programming conveyed by the communication satellites does indeed present problems never faced by our commercial broadcasters in shaping up their seasonal phantasmagorias of Westerns, whodunits, and weirdies. The fact is that the devising of a viable satellite programming concept, one which will serve the national as well as commercial interests, is infinitely too important a responsibility to be placed exclusively and without some measure of policy guidance in that hands of profit-motivated entrepreneurs.

I do not begrudge Desilu Studios the some $5 million it says it grossed last year in its foreign and domestic sales. But I care very much that 700 million people on this planet can neither read nor write—and that most of them are Asians, Africans, and Latin Americans. This is a "tragedy," as the recent UNESCO report on world illiteracy described it. Yet it is one which the people of the United States can do specific things to help alleviate and eventually erase, with the communication satellites an obvious and marvelous instrument of opportunity. Unfortunately, we seem reluctant to take action. For example, in all of Africa today there are but a few thousand television receiving sets. I deem it truly tragic that our Agency for International Development has quietly buried its plan to supply 1,000 transistorized television receivers that were to have been used for educational purposes in developing nations. It is reported that this project would have cost little more than one and a half million dollars, had Congress approved it. This occurs at a time when an American free enterprise entrepreneur thinks nothing of spending nine times that amount to introduce an old toothpaste in a new tube.

Far too little attention is being paid to the problems and portents to which I have referred far too sketchily. It is to be hoped that the Federal Communications Commission, in particular, will address itself to the responsibilities stipulated for it, although in broad terms, in the Communications Satellite Act. The Act commands the FCC to "make rules and regulations to carry out" the various purposes of the Act. That familiar phrase, "the public interest, convenience, and necessity," appears in the Act. Nondiscriminatory use of and access to the system, effective competition, rate-making appropriate to the services—these are critical areas for concern and it will be interesting to observe how assiduously the Commission concerns itself with them.

I have referred elsewhere to satellite communications as providing "a spectacular avenue to a truly empathic world community." This is indeed the "great promise and the high hope" of the Telstars, Relays, Echoes, Syncoms, and others to come.

And it is the declaration of Congress, in the Communications Satellite Act of 1962, that it is "the policy of the United States" that its system, "responsive to public needs and national objectives," will "contribute to world peace and understanding."

Let us hope that this policy will be guided by men of wisdom and vision.

48. Less Declaration and More Revelation

by Frank Stanton

Dr. Stanton is President of the Columbia Broadcasting System. He speaks frequently and articulately on the broadcasting industry, its place and its responsibilities in a contemporary society. His remarks are reprinted here through special permission of the network. His speech, the major address at the annual award banquet of the National Civil Service League in 1965, presents some provocative, thoughtful challenges.

It is a great pleasure for me to emerge under Mr. Minow's auspices from his vast wasteland to your vast wonderland.

The idea of the National Civil Service League was the offspring of that zest for reform that is a refreshing part of our national character. I have reason to recall that Mr. Minow has had some strong reforming impulses of his own. It is most reassuring—I guess—to know that he recalls me at all. On the surface, it speaks well for his genial temperament that he has seen fit to do so before this chaste and unimpeachable company. For my own part, I find it infinitely more delightful to respond affirmatively to his invitations than to his demands—although even the command performances were stimulating. In any event, I find it enjoyable indeed to be alongside Mr. Minow—reasonably sure of no searching interruptions—rather than opposite him.

While this is not the first time that the former Chairman of the Federal

Communications Commission has provided me with a forum, it is the first time that he has introduced me to such a distinguished forum—in excess of seven—and such a pleasant occasion. And if he thinks that either his eloquence or his magnanimity will stay me from trying to make the most of it, he has indeed been strangely affected by the strains and stresses of private life.

I congratulate the public servants whom we are honoring tonight, and I salute all of you here for the wide reach of your labors and for their effectiveness—not only here but all around the globe.

The National Civil Service League, as it moves closer to the century mark, can look back on changes in the scope and nature of governmental services that are no less striking than those that we have seen take place in communications. In 1881, when the League was founded, communications for the most part consisted of the mails, newspapers, seldom exceeding eight pages, and some crudely illustrated magazines. Western Union was 25 years old, but the telegraph was still used sparingly, and as practical things the automobile, the Linotype, motion pictures, phonographs, radio, and all but local telephone lines were still to come. The corps of men and women who made up the federal civil service in 1881 offers a dramatic contrast with today's. There were, for example, only seven executive departments compared with ten now; no independent agencies compared to 65 now; and serving all these and the rest 100,000 civilian employees compared to over 2½ million now.

But all this quantitative growth, dramatic as it is, seems to me far less significant than the striking and revealing changes in the nature, the objectives, the range, and in fact, the whole spirit and tone of federal government activity today. One major component of this change, in one way or another, can be summed up in the single word "international." Our increasing involvement as a nation with the rest of the world became technically plausible with that burst of inventiveness that characterized the 1880's, and it became politically inevitable with the First World War.

Now, we have in the federal government, not including participating units in the United Nations, over 60 departments and agencies dealing primarily with international aspects of our life as a nation and as a people. All this activity is no longer a matter solely or even largely—as it was in the nineteenth and early twentieth centuries—of diplomatic maneuvering. It involves economics, science, health and human welfare, the exchange of ideas and experiments, the whole fabric of life. Internationalism has come out of the staid enclosure of political positioning and into the crowded arena of an infinite variety of human needs, and hopes, and capacities.

BROAD CONCERNS OF HUMANITY

These broad concerns of humanity around the world that have given new dimensions and new depth to the professional lives of many of you, have not only created such relatively novel agencies of governmental action as the Office of the President's Special Representative for Trade Negotiations and the U.S. Information Agency, but have also revitalized and enlarged the responsibilities of offices in executive departments as old as the Republic—for example, those of the Assistant Secretary of State for Educational and Cultural Affairs and the Fiscal Assistant Secretary of the Treasury, who know that their decisions are going to have profound repercussions no longer only in American newspapers and financial centers, but in London and Rome, in Tokyo and Calcutta.

This past month, when the first commercial satellite in synchronous orbit began operations, television communications in America also moved, for the first time fully, into vibrant and instant international life—because, for the first time, the voices and the presence of peoples around the earth could be simultaneously seen and heard without barriers of time and distance at any time of day or night. When Early Bird was first successfully launched into orbit and then maneuvered into permanent anchorage 22,000 miles over the equator, a new potential began for all mankind. The mountains have been leveled and the oceans dried up by an 85-pound piece of scientific jewelry transmitting a six-watt signal.

POWER OF EARLY BIRD

The power of Early Bird's signal is not more than a tenth of that of any light bulb in this room. But it has great potential in terms of its capacity for generating a world community of understanding and dialogue among people and among statesmen. It brings to fruition the promise of its predecessor satellites for a new kind of internationalism in communications.

I do not think that, before this audience, I need to dwell upon the potential of all this for the future of civilization. But you may want to recall with me some words of Woodrow Wilson that pointed up inadequacies that were already putting a heavy strain on democracy in his time and could destroy it in ours.

Commenting on the size and complexity of modern societies, he said,

> It makes the leaders of our politics, many of them, mere names to our consciousness instead of real persons whom we have seen and heard, and whom we know. We have to accept rumors concerning them; we have to know them through the variously colored accounts of others; we can seldom test our impressions of their sincerity by standing with them face to face. Here certainly the ancient pocket republics had much the advantage of us; in them citizens and leaders were always neighbors; they stood constantly in each other's presence. Every Athenian knew Themistocles's manner and gait and address, and felt directly the just influence of Aristides. No Athenian of a later period needed to be told of the vanities and fopperies of Alcibiades, any more than the older generation

needed to have described to them the personality of Pericles. Our separation from our leaders is the greater peril because democratic government more than any other needs organization in order to escape disintegration; and it can have organization only by full knowledge of its leaders and full confidence in them.

THE DEMOCRATIC DILEMMA

Wilson was speaking of the people of a single nation, but his statement of the democratic dilemma—as he would be quick to recognize—has even more forceful application to a world in which, for better or worse, there are no islands any more and no longer any impassable borders or impenetrable barriers.

Many of the world's turmoils, today as throughout history, can be laid at the door of distrust—one people distrustful of another's actions and intents, distrustful of the other's leadership and institutions. The great opportunity of the new age we are entering—the age of full and immediate international communications made possible by the satellites—is to diminish and in time, we can have reason to hope, to demolish that distrust.

THE SHORT-TERM GAINS

We cannot, of course, move towards that goal if the use of Early Bird, and its successors when they come along, is misdirected or impeded. The temptation is always great, when new and effective communications media appear, to give priority to propaganda—to seek to impose, by conscious advocacy, one group's or one nation's ideas and institutions on the peoples of another. In some of these instances there may be short-term gains or, more commonly, the appearance of such gains. But in the long run, the indiscriminate and repeated use of propaganda not only falls on fallow ground but boomerangs badly, for eventually it becomes recognizable—and the more immediate and direct the medium the quicker and more certain that recognition becomes. The quicker and more certain, too, will be suspicion about everything else disseminated by the medium regarded primarily as a vehicle of propaganda. The stakes that we and all humanity have in this ultimate weapon of truth and mutual understanding are far too great for us to allow it to be debased at the very outset of an age that can be the most promising the world has ever known for the overcoming of ignorance, distrust, and distortions.

PROPAGANDA . . . NATIONALISM

Some other governments have shown in the past a tendency to regard satellite television communications as nothing but an extension of familiar tools of government policy, of propaganda—an instrument of flagrant nationalism. A memorable and disturbing example of this attitude on the part of the French government occurred at the time of the very first *Town Meeting of the World* broadcast by CBS News via Telstar II in 1963. On that occasion, four great and respected statesmen of both hemispheres participated in a discussion of world affairs. They were former President Eisenhower speaking in Denver, former Prime Minister Anthony Eden in London, former West German Foreign Minister Heinrich von Brentano in Bonn, and the father of the European Common Market, Jean Monnet, who was to speak from Paris.

To make this four-cornered broadcast, it was essential for technical reasons to obtain the cooperation of the British and French governments, which have jurisdiction over the satellite ground stations in England and France.

After intricate preparations, and just eight days before the scheduled date of the first live transatlantic television discussion in history, the French government pulled the rug out from under us. We were refused use of the French ground station at Pleumeur-Bodou, the only one on the Continent, thus preventing the two-way exchange of pictures and sound. We were refused even the use of a studio and telephone lines in France, making it necessary for M. Monnet to speak from Brussels. And French officials refused to transmit the discussion to the people of France, who had to read it later in their newspapers. The reason: according to the French Ministry of Information, it was apt to be "too political and controversial." What had become scientifically possible—bringing leading statesmen together in open discussion—became politically impossible.

We went ahead with the first *Town Meeting of the World* anyhow, even though only the American audience could see all the participants. Not only the Continental countries were barred from witnessing the event, the participants overseas could not even see their American counterpart or each other. But thanks to privately owned AT&T lines, they could at least *hear* each other through the good, old-fashioned, underseas cable.

PRESS REACTION

The reaction of newspapers, both here and abroad, including French papers, was well exemplified by the lead editorial in the *Washington Post*, which said in part:

> The *Town Meeting* lacked the power of actual decision which invests real New England town meetings with a vital spark, but it possessed other virtues that recommend its frequent repetition in a world that needs to hear dispassionate and friendly discussion of the problems of greater Western unity. It is too bad that the voices of these citizens of the Free World were not heard in France because of that country's decision to foreclose the program's reception.

Such episodes as France's banning the use of Telstar purely for political expediency make it all the more imperative that we, the world's most powerful advocate of freedom of communications, reject by our enterprise and our example all cynical uses or arbitrary restrictions on the use of Early Bird. There is an urgent need, beyond any doubt, for the peoples of the world to have the opportunity of seeing, and hearing, the leaderships of the world's nations stating their cases directly and honestly—and showing, thus, what manner of men these leaders are and what manner of ideas and institutions they represent. The Wilsonian longing for combining the simple directness of the Athenian democracy with the strength of the huge, self-governing societies of today can only be achieved by such direct revelation. But this must be done in an atmosphere of freedom, with openness and in candid discussion. Early Bird should not be construed by any government as just another door to be opened when there is a self-serving point to be made and a door to be slammed when that point is in danger of being questioned.

And that, precisely, is the focus of the concern which all of us must feel. Early Bird must not be transformed from the unprecedented opportunity into the most universal and pervasive censorship—both affirmative and negative—ever known.

NATURE OF AMERICAN FREEDOM

In this connection, sending overseas last Saturday's inter-university "Teach-In" with its stimulating discussions of our policy in Vietnam would have been far more revealing of the nature of American freedom, of the thoughtful criticism of American foreign policy here at home and, on the other hand, of its consistency and strength of purpose, than all the one-sided declamatory rationales imaginable. It would have done more than give us, as a government, a natural and provocative occasion to reassert and enlarge upon a policy of deep interest and genuine concern overseas. It would have given us a chance to show—not merely to state—that we are willing to subject our official policies, however grave, to unofficial scrutiny and free dissent.

Nor should the use of Early Bird be subject to arbitrary restrictions that unnecessarily limit its availability. Last week, for example, there occurred in England one of the most profoundly symbolic events in Anglo-American history. An acre of Britain's most historic land—where the Magna Carta, the source of our common democratic heritage, was promulgated—was given the United States as a perpetual memorial to President Kennedy.

A HAZARDOUS PRECEDENT

CBS News requested use of Early Bird to broadcast the ceremonies live so that all America as well as Britain and Europe might witness this event as it was taking place. But the decision was made by COMSAT that Early Bird could not be made available to bring these dramatic events at Runnymede to the American people unless all three television networks requested it. As a result, you and I had to watch film and tape reports on our television screens 14 or more hours after the event, whereas we could have been there at the very time of the ceremonies if Early Bird had been made available to us. This seems to be both senseless in a practical way and a hazardous precedent to set. No useful purpose can be served by insisting on saturation as a condition of sight and sound reporting of an event from one country to another. There is, to be sure, only one two-way television channel available in Early Bird, but it seems to me poor logic to conclude that all television networks must simultaneously want to use it for a single purpose and that otherwise none can use it at all. What we need is a variety of interchanges between the world's peoples—not all of which will interest all the people at the same time.

HIGHEST USE OF EARLY BIRD

The world is, I suspect, sick and tired of proclamations, manifestos, ultimata, and communiques. Some are necessary, of course, but there has been an unending flood of them in our tempestuous times. What the peoples of the world yearn for—and what peace for the world needs—is less declaration and more revelation. This is the highest use of Early Bird. It should reveal us to one another on many occasions and on many levels—science, the arts, the conflict of ideas, the ways and customs and diversions of a people—and not just on occasions of state and on a political level. Political purposes and objectives are not to be ignored, but they have little meaning in themselves. They have reality to the great body of mankind only insofar as they have roots in the matrix of those elements that make up the daily lives and the constant hopes of the men and women with whom we wish to establish more effective contact.

In speaking of all those areas of the world with which we seek deeper and stronger associations, one of the most respected and effective of our public servants, Eugene Black, said, in *The Diplomacy of Economic Development*, "Nor is it enough to talk about an integration of political aims and ideals between the West and these parts of the world; there will be no such integration unless it grows out of a long period of constructive contact in tasks of common interest."

To us in broadcast communications the advancing of political aims, however generous, and of political ideals, however lofty, requires—if it is to be effective—the background of a free, constant, and, from time to time, spontaneous interchange relating to fundamentally human problems. It is in this sphere of human problems, I am sure, that the majority of you here in this room are spending a major part of your professional activity—many of you at the international level. It is in this sphere, too, that I think you in government and we in communications must make common cause. And it is certainly ultimately in this sphere that if we all bring imagination and vision and courage to our tasks, we will achieve a common progress that can indeed move the world forward—an inch or two.

X. New Problems and Alternatives: The Future

WHAT STANDARDS should be used in deciding which world interests as well as national interests should be served by broadcasting? What changes in present structures should be considered if our broadcast system is to be responsive to the needs of contemporary society?

The future of television and its role in the latter quarter of the twentieth century is and will be faced with some critical problems, many of which are upon us now: i.e., lasers, satellites, color television, new nations developing their own broadcast systems, commercial pirate stations broadcasting to dozens of nations with noncommercial systems, and community antenna systems.

These are only a few of the more immediate areas which must be critically and objectively analyzed if the great promises and potential of television as an instrument of mass communication are to be fully realized.

49. Broadcasting in the Interest of the Free Society*

by Roscoe Barrow

Mr. Barrow is Professor of Law at the University of Cincinnati and was director of the 1955–58 FCC network study, well known in the industry as "The Barrow Report." As noted below, the views expressed here are Mr. Barrow's own, and do not necessarily reflect past or present Commission policy. These remarks were presented before the National Association of Broadcasters in Chicago, Illinois on April 1, 1963, as part of a panel discussion on the place of broadcasting in a free society. It is reprinted here with Mr. Barrow's permission.

There is substantial disagreement as to whether broadcasting adequately serves the public interest and whether stricter regulation of the industry is necessary to assure that the programming needs of the free society are fulfilled.[1] Some say that broadcasting is fully serving the free society and that competition and self-regulation can be relied upon to effectuate such change as may become desirable. Others say that the free society is ill served by broadcasting and that adequate service can be achieved only through firmer governmental regulation. This paper expresses the view that broadcasting is contributing much of value to the free society but that industry initiative must be supplemented by additional regulation if it is to serve adequately the interest of the free society in this time of trial.

Our free society, as envisioned by the founding fathers, is a society in which the people participate in deciding the issues of their day. The procedure by which the people are represented by elected officials does not controvert the fact that progress in a free society depends upon the social, political, and cultural maturity of the people and their ability to contribute to the making of decisions, including the election of qualified officials. Our society is accustomed to time for problem-solving. The problems are growing more complex and the time for decision-making is shortening, increasing the margin for error. The Cold War with Communism, the emerging nations which are making a choice between the free society and totalitarianism, advancing technology, the population explosion, and many other factors combine to make this a "time of trial" for the free society. In an increasing number of decisional contexts, the capacity of the people to contribute to the decision is being questioned. The adequacy of our decision-making machinery, the quality of the decisions, and the confidence of the public in them are being tested. Yet, the extent to which the people are given an opportunity to play a part in decision-making is a significant gauge of the vitality of our democracy. Will the responsible participation of the people in decision-making be discouraged and the decisions be made in a few centers of control—a trend toward the authoritarian way? At hand is one of the greatest tools for furthering the American way that genius has provided—broadcasting. Here is a Forum of Democracy in which all Americans may deliberate simultaneously on the issues of the day. However, comments by the Commission on National Goals, appointed by President Eisenhower, regarding broadcasting service are troublesome. The Commission states:

> . . . Entertainment has almost swallowed up information and education in the operation of the mass media. Television, for example, is fast becoming the Circus Maximus rather than the Forum of Democracy, and we are missing out consistently on a rare opportunity to hold popular discussions of a range and influence unmatched in all history. . . . [*F*]or the most part television has not even come within hailing distance of the press as a device by which Americans can communicate sensibly with one another. . . . Sooner or later we are going to have to face up to the harsh fact that the democratic dialogue is in real danger of being smothered. . . ." [2]

The orbital flight of Friendship Seven, the presidential conventions, and the Kennedy-Nixon debates are a reminder of the heights which television can reach in informing the people on national issues and significant news events. For the contribution which broadcasting has made to the democratic forum of the air there is deep appreciation from all who are concerned with the progress of the free society. In recent years, there has been substantial improvement in public affairs programming, as the result of more vigorous administration of the Communications Act. However, the amount of such programming regularly scheduled at times when there is substantial audience potential continues to be inadequate and the network line-up forecast for next season indicates little improvement.[3]

The primary concern of the free society is with the life, liberty, and happiness of the individual citizen. The characteristic of a free society which places it above all other forms of government is that the free society encourages an environment within which the individual citizen can achieve by free choice the maximum personal development. Thus, the role of broadcasting in the free society must be examined from the viewpoint of the individual citizen in the intimate setting of the home. The essence of broadcasting is the programming which he sees and hears. This programming is many things: education, news and public affairs, sports, theatre, cinema, concert, religion, entertainment—altogether it should represent a substantial contribution to our cultural life. It comes to us over airwaves which are owned by the people. It is provided by licensed trustees who have been selected on the basis of qualifications to serve the needs, tastes, and desires of the people. The free society is properly interested in having broadcasting provide programming which will give the individual citizen an environment within which he can seek cultural and social development. Television—combining sight, sound, motion, and simultaneous broadcast to all the people—probably has the greatest potential of all communications media for stimulating the individual's social and cultural development. A single broadcast of "Oedipus Rex" is viewed by more persons than have seen the play in theatres since the days of Sophocles. Through broadcasting, the best that the genius of the free society creates can be made available to all the people, contributing to their social and cultural growth. However, the use of television for mass marketing has limited television's contribution to culture in the free society. Regarding the effect of the mass market on broadcasting, the President's Commission on National Goals reports:

> Thus far, television has failed to use its facilities adequately for educational and cultural purposes and reform in its performance is urgent.[4]

It is further observed in the Report:

> In the field of television we see the problem in its most acute and disturbing form. Here, more than anywhere, is cause for concern that the level of popular culture in America is being lowered. Third and fourth-rate material seems increasingly to replace the better shows as the merchandiser reaches out for a wide market. The managers of the broadcasting companies seek the same large audience in order to sell their broadcasting time.[5]

Industry spokesmen usually seek to justify the existing imbalance of programming having maximum audience appeal on the grounds that (a) the industry provides the programming for which the majority of people, by turning the dial, have voted, and (b) it is impractical to broadcast programming substantially above the cultural level of the maximum potential viewing audience. For example, in the hearings on responsibility for broadcast matter, a representative of the industry testified that television must be geared to the level of formal education and assured that as the level of education and program taste rises the quality of television programs will rise.[6] The thesis that only formal classroom-type programs educate is a strong indictment of television. It assumes that the talent and capital expended on television programming leaves the viewer exactly as it found him. If this were the case, the advertiser would abandon television. The program, no less than the advertisement, affects the viewer. Television is an educational force regardless of whether the program bears the label "educational." The many action-adventure series educate, although the education disciplines the mind in the wrong direction. Regarding this effect of programming, Dr. William E. Hocking testified during the hearings on broadcast matter: "Without intention, the radio-TV pablum becomes weighted in favor of the animal end of the emotional scale; and the incidental education moves not from the primitive to the advanced, but from the advanced to the primitive. . . ."[7]

It may be noted that the network lineup forecast for next fall indicates that the *DuPont Show of the Week* will be the only remaining live, original drama.[8] The imbalance of mass appeal entertainment in broadcasting is eroding our culture. This is occurring in a time of trial for the free society when broadcasting should be doing its share to stimulate excellence.

Children are the young leaders of tomorrow and it is important that the program fare to which they are exposed stimulate their imagination and contribute to their social development and maturity. Children's programs consist largely of action-adventure vehicles, cartoons, and old comedy films. With rare exception they are time-wasters and make no contribution to the child's knowledge or understanding of the world about him. From the commercial standpoint, the lack of suitable programming for children is understandable. The products which can be sold through advertising exposure to children are limited. Hence, inexpensive programming is used and it is scheduled in time slots when there is no potential adult audience. By contrast, in other free societies—for example, the United Kingdom, Canada, and Australia—particular attention has been given to the programming needs of children. Surveys indicate that children spend as much time viewing television as they spend in school, taking into account viewing time on weekends and vacations. The opportunity here for stimulating the imagination of the young, giving them an awareness of the world around them, and contributing to their social growth is great. This opportunity has been substantially wasted.[9]

The television spectrum is a publicly owned resource which, the Congress has declared, shall be used in the public interest. Television should contribute to the resolution of the fundamental issues of the day through educational and public affairs programs, should provide programming which represents broadcasting's fair share of support of an environment in which the individual citizen can seek cultural, social, and political development, and should do its part in stimulating the imagination and social growth of children. The use of television as a marketing mechanism should not be permitted to prevent television from serving its communications function in the free society.

The importance of broadcasting to the free society was recognized by Herbert Hoover when he pointed out the potential contribution of radio to education and the public welfare and recommended that broadcasting be regulated in the public interest. It was recognized by the Congress when the broadcast spectrum was declared to be public property and the Commission was empowered to license private enterprise to serve as broadcasters and to regulate broadcasting.[10] The regulatory authority is broad, extending to prescription of "the nature of the service to be rendered by each class of licensed stations and by each station within each class." [11] Thus, qualified entrepreneurs in private industry were commissioned as trustees of the publicly owned spectrum and were placed under a nondelegable duty to ascertain and fulfill the tastes, needs, and desires of the people. Anticipating that the growing chains might impinge upon the ability of the licensee to exercise his duty, the Commission was granted authority to make special regulations applicable to stations engaged in chain broadcasting. In considering the appropriate role of broadcasting in the free society, it is to be remembered that broadcasting is a licensed industry. Whenever the Commission grants a license to use a channel, the Commission confers a monopoly, denying the use of that channel to other entrepreneurs. It is appropriate that the holder of this privilege be deemed a "trustee" and be under a duty to serve the public interest.

The nub of the problem with broadcasting today is this: If broadcasting is to serve the interest of the free society, it must be primarily a communications medium. Programming decisions should be made on the basis of their qualifications to fulfill the needs, tastes, and desires of the community served. The fact that broadcasting is an advertising medium, supported by revenue from advertising, should be given only secondary consideration. If the advertising factor is controlling, the potential of broadcasting as a communications medium cannot develop.

The Commission's hearings on responsibility for broadcast matter established that the principal influence on the character of broadcasting is the utilization of broadcasting as a marketing instrument.[12] As the advertising dollar turns the wheels of the industry, the industry structure has been built to serve the role of advertising medium. The merchandiser seeks a national market and the networks seek a national audience for the advertiser. A program is sought which will attract a maximum number of viewers. The audience ratings indicate a few types of light entertainment attract the largest audiences. This oracle foretells the future of television programming. Television programming is costly and advertiser, agency, and network may hedge on the risk of a new program type. Programs with the highest ratings become stereotypes for imitation. The producer and script writer select a subject and format which has a proven record of performance.[13] Thus, television programming is limited in variety. Programs which seek to cope with the exigencies of our time are deemed too gloomy to attract viewers or leave them with a happy image of the sponsor. Hence, the serious side of life is seldom treated. Gresham's Law operates in television to drive out programming of interest to substantial minority audiences, such as children's programs, serious drama, and public affairs. The licensed broadcaster places a practical reliance on networks to select programming for his station. The broadcaster has been substantially insulated from the program selection process insofar as network programming is concerned. In fact, there is no practical way to give the broadcaster a significant voice in choosing programming provided by the network. In this way, commercial motives have dominated the program selection process. The communications function—imaginative and creative development of programming to fulfill needs, tastes, and desires of the community—has taken a back seat.

In the hearings on responsibility for broadcast matter, there was considerable testimony that self-regulation may be depended upon to assure broadcasting in the public interest and that additional regulation is unnecessary.

Industry self-regulation is the most desirable form of regulation. However, it is highly unlikely that the industry, without supplemental governmental regulation, can overcome the commercial motives which have resulted in imbalance of mass response programming. The code of good practice exercises substantial influence on advertising and program taste. It proscribes profanity, obscenity, derisive references to race, attacks on religion, the use of horror for its own sake, and has held the line on advertising of hard liquor. This is a form of private censorship but it has contributed to good taste in broadcasting. Not all broadcasters are members of NAB and withdrawal of the seal of good practice from an offending member is not a strong sanction. Moreover, networks may deny the Code Board opportunity to preview a controversial program. Thus, a request by NAB to see the controversial episode in "Bus Stop" was denied by ABC.[14] Also, the network may exhibit programming which uses horror or sex for audience appeal rather than to delineate character or plot.[15] In such instances, it is difficult to hold the individual broadcaster responsible. While the NAB Code states that television is accountable to the public for "the special needs of children, for community responsibility, (and) for the advancement of education and culture," the Code Board has done nothing about the imbalance of mass appeal programming.[16] Statements by the President of NAB regarding stricter enforcement of the Code, restrictions on advertising, and supporting authority of the FCC to request information from licensees regarding programming have evoked considerable criticism by the broadcasting industry,[17] and it seems unlikely that the industry will be inclined to seek balanced programming through self-regulation.

Even were self-regulation of program balance attempted, it is unlikely that the impact of advertising support of the industry on the character of programming could be overcome. Competition in advertising products and in network service limits the extent to which a public-spirited advertiser or network can provide programming suited to the needs and tastes of substantial minority audiences. Competition requires that the sponsor of a network program reach as wide an audience as his competitor reaches. Similarly, each network competes with rival networks for audience and considerations of audience flow encourage an unbroken schedule of programming having maximum audience appeal. The Network Affiliates Committees have not attempted to exercise control over the programming provided by networks.[18] While the Committees are previewing more shows now, it is unlikely that this will have a significant effect on the character of programming. Enlightened broadcasters should encourage all components of broadcasting to provide programming which meets the tastes, needs, and desires of all significant publics within the community served, as provided by law. Also, networks should meet their asserted voluntary obligation to program in the interest of significant publics which compose the national audience. To the extent that self-regulation serves the interest of the free society, the need for governmental regulation is reduced.

From the beginning of broadcasting, governmental influence on programming has been minimal and indirect. To the extent that self-regulation fails to achieve broadcasting in the public interest, regulation in the public interest should prohibit industry practices which limit the ability of the broadcaster to exercise his duty to ascertain and fulfill the programming needs and tastes of his community. In *Network Broadcasting*, the report of the FCC Network Study Staff which was released in 1957, several recommendations were made which, if adopted, would aid the broadcaster better to serve the public interest.

In the areas of affiliation, rates, and compensation it was found that the network has a superior bargaining position because the station is dependent on the network programming and attendant national advertising revenue, the network sets the station's rate to network advertisers—a lever for favorable clearances, the network has the experience of bargaining separately with approximately 200 stations, and there are differences in the share of compensation received by affiliates of the same network. It was recommended that in the areas of affiliation, rates, and compensation industry self-regulation should be encouraged through publicity. Affiliation contracts, affiliation criteria and decisions, and rate-making procedures and changes in rates were to be filed with the Commission and opened to public inspection. In recent years, the Commission has adopted a further recommendation that compensation plans having a high clearance incentive factor be prohibited because they impinge on the nondelegable duty of the licensed broadcaster to select programming and are, thus, contrary to the public interest.

With regard to option time, the Network Study Staff found that the practice is not necessary to the operation of healthy networks; that the mutuality of interest of the network in integrating a network of stations and of the station in obtaining access to attractive program service and attendant advertising revenue is the foundation of the partnership; that pre-emption by networks of the desirable viewing hours handicaps the development of non-network program sources; that the practice violates the Antitrust law; and that the practice impinges on the exercise by the broadcaster of his duty to select programming. This recommendation is still before the Commission. Initially, the Commission found that option time is reasonably necessary to the successful conduct of network operations but reduced the period subject to option from three hours to two and one-half hours in each segment of the broadcast day.[19] The trade press predicted that this reduction would have no effect on network clearances.[20] Comparison of programming by networks during a typical week in 1960–1961 and 1961–1962 shows that all networks increased their programming in prime time in the season following the reduction in option time. This is further evidence that networks can operate a healthy service without the leverage provided by option time. In the period since *Network Broadcasting* was submitted, production of programs for first-run syndication has virtually collapsed.[21] Most important, it is inconsistent to place the licensed broadcaster under a duty to select programming and then permit the network to pre-empt time on the stations. If the network is to have the power to pre-empt time of the station, the network, like the station, should be under a duty to provide a balanced program service in the public interest.

In the area of multiple ownership of stations, it was found in *Network Broadcasting* that multiple ownership, particularly in the large markets, is increasing; that the growth of multiple ownership decreases broadcasting viewpoints and opinion sources; that emphasis on a past record of performance contributes to the growth of multiple ownership; and that, under the limitation imposed by statute against considering the qualifications of third persons in transfer cases, the Commission's licensee selection function, to a substantial extent, has passed into the hands of the industry. To restore to the Commission the licensee selection function and to stem the tide of multiple ownership through the transfer route, the Network Study Staff recommended (a) that the Commission scrutinize more rigorously broadcast performance in considering renewals; (b) that the statute be amended to permit consideration of applicants other than a prospective transferee selected by the seller; and (c) that, if such amendment should occur, the Commission should hold a comparative hearing among those meeting the terms of the proposed sale and, in the hearing, apply presumptions favoring local ownership and diversity. The adoption of such procedures, in time, would winnow the trafficker in stations from broadcasters who have a bona fide communications interest and would contribute to the concept of broadcasting stations as local institutions.

The answer to the existing problems in television is not to be found in Telstar and the now potential development of UHF. The satellite system of transmission will not make a significant contribution for several years and even then it will not change the problem of dominance of individual broadcasters by concentration of power in program services. In fact, the satellite system of transmission can be another focal point for concentration of such power unless care is taken. Commercial development of UHF is several years away. The value of the all-channel receiver will be first felt in educational television. Most educational television assignments were made in the UHF and the failure of educational television to develop has been due, in large part, to the lack of receivers equipped to receive educational programs. It will be several years before there are sufficient all-channel receivers in the hands of viewers to make another commercial network possible. Moreover, the facts [that] the VHF has been long established, experienced networks and broadcasters are operating in the VHF, and advertisers and agencies have long regarded UHF as inferior to VHF, render uncertain the time when a viable UHF may be expected to take up a competitive position vis-à-vis VHF. A viable UHF will bring its own problems and may well entail further regulation.

In 1960, the Commission adopted a statement on program policy in which it recognized that regulation of stations alone will not ensure broadcasting in the public interest; and recommended legislation granting the Commission authority to classify networks, to require networks to file information regarding network policies and practices, and to regulate network practices which restrict the ability of broadcast licensees to operate their stations in the public interest.[22] In the statement, the Commission cautioned that control of programming should remain the primary responsibility of the licensed broadcasters and that any responsibility which might be placed on networks should be complementary to the responsibility of stations.

Representatives of the industry have expressed the view that regulation by the Commission of programming would contravene the guarantees of free speech and press contained in the First Amendment.[23] The constitutionality of the Commission's authority to concern itself with programming was excellently stated by Chairman Minow at Northwestern University two years ago.[24] The report by Attorney-General Rogers to President Eisenhower on "Deceptive Practices in Broadcasting Media" found that such valid authority exists. In every instance in which a question as to the Commission's authority with respect to programming has arisen the Commission has been sustained. The essence of broadcasting is programming. Whenever the Commission applies the public interest standard or the balanced program standard in the comparative hearing and renewal contexts, it concerns itself with programming.[25] The equal time, rigged quiz show, and obscene language legislation is legislation regarding programming. The Commission's policy on editorializing, providing for discussion of public issues and giving the public an opportunity to express opposing viewpoints is a programming matter.[26] In classifying systems of communication as military, educational, governmental, and commercial, the Commission says much regarding programming. When the Commission grants a channel to one applicant, it excludes others, and every excluded applicant is in a sense denied the right of free speech by radio. However, such action does not violate the Constitution and it does not constitute censorship within the meaning of the Communications Act.[27] The authority of the Commission to concern itself with programming was recognized by NAB in 1934 when the Chairman of the Legislative Committee testified:

> It is the manifest duty of the licensing authority, in passing upon applications for licenses or the renewal thereof, to determine whether or not the applicant is rendering or can render an adequate public service. Such service necessarily includes the broadcasting of a considerable proportion of programs devoted to education, religion, labor, agriculture, and similar activities concerned with human betterment.[28]

In recent years, the Commission has acted in significant respects to improve program service. It has denied a construction permit for a radio station on the ground that the program proposal was not based on a determination of the needs of the community to be served;[29] in a number of cases has limited renewal of a license on the basis of substantial deviation of performance from promised performance;[30] has denied renewal on the ground of lack of qualifications to provide program service and failure of programming to conform to program proposals;[31] has denied renewal in the case of a station using substantial off-color broadcasting, on the ground that the public interest was not served;[32] has announced a rule-making proceeding which, if implemented, would require statement to the Commission of the measures taken by the applicant to determine the needs and desires of the community served;[33] has conducted hearings in

Chicago and Omaha to obtain information as to the extent to which broadcasters are serving local needs and tastes; and has required that, in the case of application for original grant, renewal, or transfer, the applicant give notice thereof in his local community.

Denial of application for renewal and renewal for less than the maximum term should have a salutary effect im improving the service of marginal broadcasters. Holding of hearings in communities to ascertain whether the needs have been served and requiring the broadcaster to state the measures taken to ascertain and fulfill those needs should increase the effort at the "grass-roots" level to provide service in the public interest. The people are not organized to represent themselves before the Commission regarding their programming needs, tastes, and desires. More contact between the people and their representatives on the Commission should aid the Commission in its important work.

The Second Interim Report of the Office of Network Study recommends that the Commission adopt rules, in effect, prohibiting networks from engaging in syndication program service. When the Commission prohibited networks from representing their affiliated stations in National Spot, it recognized the importance of having non-network program sources compete with the network service. Independent producers have found the networks to be the assured market for their programs, the networks providing financing and obtaining network exhibition rights and profit participations and other subsidiary rights.[34] The networks operate divisions or subsidiaries which sell, to their affiliates, programs for non-network exhibition in station time.[35] Most of the programs now offered in syndication are film series originally exhibited over the networks. Thus, the concentration in the networks of available programming and the importance of the affiliation relationship discourages the development of non-network program sources and competition by non-network systems of advertising. The recommendation that networks be excluded from syndication has great merit.

The most important viewing period, in terms of service to the free society, is, of course, prime time. Prime time is substantially blanketed by programming supplied by the networks. The present system of regulation, limiting the direct application of rules to stations and not to networks, is of limited effectiveness. Impingements on the public interest by undue concentration of control in networks can be better handled through the application of rules directly to networks. Moreover, administration of a rule applicable to a few network entities is more efficient than a rule directed to hundreds of stations. Greater flexibility in regulation would be provided by adding to the present authority to regulate stations the authority to apply rules directly to networks. The Commission has informed the Congress that it should be authorized to apply directly to networks rules regarding network practices which impinge upon the broadcaster's exercise of his duty to serve the public interest. This authority should have been granted from the beginning. While the FCC has not requested the authority for such purpose, in the opinion of the writer, if such authority is granted, the Commission should apply to networks the balanced program standard which has long been applied to the stations.

An indirect approach to improving broadcasting's service to the free society is the establishment of an advisory committee of national stature to make an annual evaluation of broadcasting's service to the free society and to report to the people. In the opinion of the writer, such a committee should be established. The people are not organized for the purpose of presenting their views to the Commission. The letters received from the public are unlikely to be representative. An advisory committee composed of eminent Americans in broadcasting, business, labor, the arts, professions, education, religion, and government would be able to appreciate the free society's tastes, needs, and aspirations and the degree to which these are being fulfilled by broadcasting. The interest of the free society is not a static concept but is an evolving one, sensitive to changing national and community tastes, needs, and desires. Such a committee would be more effective if appointed by the President. Its report should be to the public rather than to the Commission and broadcasting industry.

Broadcasting is contributing substantially to the interest of the free society. Actions taken by the Commission in recent years have achieved significant improvement in broadcasting service. However, the dominance of the commercial motive in broadcasting and the undue concentration of control of the industry by networks, prevent the industry from adequately serving the public interest. Additional regulation, along the lines of the above recommendations, is necessary in order that broadcasting may fully serve the interest of the free society in a time of trial and growing complexity.

NOTES

* The writer is a sometime consultant to the FCC and, accordingly, makes the customary disclaimer. The views expressed herein are the views of the writer and do not purport to represent the view of any member of the Commission or the Staff.

[1] The conflicting viewpoints are recorded in the hearings on responsibility for broadcast matter, FCC Docket No. 12782, as well as in the general literature on broadcasting.

[2] *Goals for Americans* (Prentice-Hall, 1960), pp. 72–73.

[3] *Broadcasting*, February 11, 1963, reports the network line-up for next fall at pp. 24–26.

[4] *Supra*, fn. 2, p. 9.

[5] *Ibid.*, p. 132.

[6] *Supra*, fn. 1, Transcript, pp. 8000–1.

[7] *Supra*, fn. 1, FCC Mimeo 82087, p. 4.

[8] *Supra*, fn. 3, p. 26.

[9] The Senate Committee on Juvenile Delinquency has pointed out the scarcity of programs designed for children, the early age at which children watch adult fare, and the excessive violence in programming viewed by children. Opinions differ as to the effect of violence in television programming on the young and impressionable. However, there can be no question that we are wasting the opportunity provided by broadcasting to contribute to the social growth of children.

[10] Communications Act of 1934, s. 303.

[11] *Ibid.*, s. 303(b). The only exercise of this authority has been the restriction against use of educational stations for commercial purposes.

[12] Interim Report of the Office of Network Study, FCC Docket No. 12782, *passim.*

[13] The hearings on responsibility for broadcast matter disclosed considerable "censorship" by advertisers and advertising agencies of programs included in the network service. However, a more important impact of advertising support of television is that it predetermines the character of the programming to be a type having a proved track record.

[14] *Supra*, fn. 1, Transcript, p. 9483.

[15] *Ibid.*, pp. 5527, 5368, 5398, 5445, and 5491.

[16] *Supra*, fn. 12, pp. 201–206.

[17] For example, see *Broadcasting*, August 7, 1961, p. 64, August 13, 1962, p. 30, and March 4, 1963, p. 42.

[18] *Supra*, fn. 12, p. 102.

[19] FCC 58–37, issued January 9, 1958, Docket No. 12285. Thereafter, further notice of proposed rule making in the option time matter was issued. FCC 61–582; Docket No. 12859.

[20] For example, see "Is the FCC's Option Time Cut A Mirage," *Sponsor*, 21 November 1960, pp. 33–35, 54.

[21] *Broadcasting*, September 18, 1961, pp. 19–21. See also *Broadcasting*, August 7, 1962, p. 27 and January 7, 1963, p. 31.

[22] Comments of the FCC on H.R. 11340, 86th Congress, Document No. 88411, adopted May 4, 1960. Such a regulatory approach is incorporated in S.2400, 87th Congress, 1st Session.

[23] Statements of Whitney North Seymour and W. Theodore Pierson, submitted in FCC Docket No. 12782.

[24] *Freedom and Responsibility in Broadcasting* (Northwestern University Press, 1961), pp. 15–33.

[25] For a review of the Commission's decisions in which evaluation of program service was involved see *Network Broadcasting*, pp. 54–64, 127–141, and 145–156.

[26] 13 FCC 1246, 1257–8.

[27] Among the many cases involving the Commission's concern with programming are the following: *National Broadcasting Company* v. *United States*, 319 US 190, 226–227 (Free speech not impinged by denial of license to person who violated regulation intended to ensure broadcaster's freedom to select programming); *Regents of the University of Georgia* v. *Carroll*, 338 US 586, 598 (Commission could weigh the character of the licensee's broadcasts in determining whether license should be granted); *Simmons* v. *FCC*, 169 F2d 670, cert. den. 335 US 846 (Commission could deny a license to an applicant who proposed to broadcast all programs offered by a network); *Bay State Beacon, Inc.* v. *FCC*, 171 F2d 826, 827 (Commission's consideration of amount of sustaining time reserved by applicant does not contravene free speech or press); *Johnston Broadcasting Co.* v. *FCC*, 175 F2d 351, 359 (Commission could consider the nature of the programs in determining whether the program service was in the public interest). For a complete review of the cases, see FCC Mimeo No. 8674, prepared by Mr. Joel Rosenbloom, Legal Assistant to the Chairman, FCC.

[28] *Interim Report*, Docket No. 12782, pp. 26–27.

[29] Suburban Broadcasters, Inc., FCC Docket No. 1332, released July 5, 1961.

[30] KORD, Inc., 21 Pike & Fischer, R.R. 781, FCC Docket No. 14003, released July 12, 1961; for a review of other cases, see *Broadcasting*, August 6, 1962, p. 60.

[31] Eleven Ten Broadcasting Corp., 22 Pike & Fischer, R.R. 699; Leo Joseph Theriot, *ibid.*, 237.

[32] Palmetto Broadcasting Co., 23 Pike & Fischer, R.R. 483.

[33] FCC Docket No. 13961, document FCC 61–863, adopted July 6, 1961.

[34] *Second Interim Report*, Docket No. 12782, pp. 154–155.

[35] *Ibid.*, p. 161.

50. Tides of Change

by Sterling C. Quinlan

At the time of this address to the Headline Club in Chicago, April 25, 1964, Sterling Quinlan was President of Field Enterprises, which he still serves as a consultant. As a former writer, actor, and executive in broadcasting, connected with various stations in the Chicago, Gary, and Cleveland areas and with ABC and NBC, TV executive and author Quinlan knows the problems of broadcasting intimately.

This article appears with his permission.

Sometimes your best friends won't tell you that CHANGE is going on. Often that's because most of your friends don't know. And the few who *do* know choose to be silent out of a sense of courtesy . . . caution . . . or even self-interest.

I submit without further preliminary that, in the communications field, which certainly touches the lives of everyone in this room, a great, rolling, subtle adjustment is going on.

I further submit that the face of communications may change drastically in the next five years.

Now that I am no longer employed in the broadcast field, I can for the first time freely express my thoughts as an individual rather than as a representative of the industry. These thoughts I have not communicated before publicly. I have not even submitted a preliminary plan to my new associates at Field Enterprises, Inc. So accept these thoughts, if you will, as those of a creative man who is intensely interested in the arts. A man who cares about the future of television and views the tides of change with great expectation.

And historically, a few years from now, perhaps we may look back and say that the kickoff . . . the official opening whistle in this new game, blew at the last convention of the National Association of Broadcasters. This convention convened in April of 1964; already it has become clear to most thinkers that the issues, diverse and complex as they are, were squarely joined at the Hilton Hotel only three weeks ago.

Trial balloons were launched all over the place. The National Association of Broadcasters, that excellent lobby of the broadcasting industry, has called for legislation against pay television, and to a lesser extent, CATV. The owners of these community antenna television systems gave up trying to keep a secret of the fact that there are 1,295 of these systems in the United States and that, with these systems, we already *have* the rudimentary structure of a pay television network.

Right now these 1,295 systems have a potential of one million homes. The recent Liston-Clay fight saw 88 of these systems interconnected for the purpose of serving 200,000 viewers who paid to see this fight in their homes.

Conservative estimates disclose that within three years, 20 per cent of the country may be wired to receive *some* form of pay television. Within five years—50 per cent—of population, that is; not simply geography. In other words, by 1969, it is estimated that 125 million people will have another television choice—that of paying for certain entertainment not available on free TV.

Right now, some 250 CATV systems ranging the West Coast, lured partly by the Subscription Television Enterprise in Los Angeles and San Francisco, are able to be interconnected to receive the baseball games of the Los Angeles Dodgers vs. San Francisco Giants. This potential audience comprises 25 million people, or 10 per cent of the population.

We are talking, ladies and gentlemen, of *a fourth network.*

The Federal Communications Commission, which has been accused in the past of trying to foster too much regulation, continued its policy of laissez-faire—that is, of letting all of these forces have free play—and there is fine irony in the sudden zig-zag of the National Association of Broadcasters, which never wants regulation, but now is calling for various kinds of rule-making that will protect the great American institution of free broadcasting. But the idea of regulation has become frightfully complicated because many of the free broadcasters have invested in pay *community systems* and *these* systems must be kept free, while pay TV must be regulated, don't you see. But the NAB is a master in the art of ambivalence and they will work it out.

For once our seven very bright, very busy, very dedicated FCC commissioners must be chuckling up their sleeves.

So now thanks to the recent convention, all the issues are on the table. Billy Graham came in with a nice, final change-of-pace on the last day with a rousing speech about morals, the conscience of broadcasting, what broadcasters can do to help our country, etc. . . . all of which is very fine and very true. It gave the industry something else to think about, at least until they got home. Then, a few days later, the trade magazines confronted them; and these trade reporters are experts. They capsuled the convention succinctly in 18 or 20 paragraphs.

I'm sure, to many broadcasters, the issues raised by Billy Graham seem easier to solve by comparison.

I will not be surprised if the NAB asks Billy Graham back next year and makes him their keynote speaker.

One thing is sure; they won't ask Pat Weaver!

But let us get down to particulars. And let us begin by posing some questions about the possible shape of a Fourth Network.

Questions like: Will community antenna television systems, which I will refer to from now on as CATV—will these systems become the tail that wags the Pay-TV dog? And will they comprise a link-up of this country, modest in numbers of people, yet portentous enough to convince sports tycoons that they should help this budding upstart by giving this network *first* preference in selling off the rights to the great sports attractions of America?

And will a Fourth Network emerge in a shape totally unlike networks as we know them today? A literal hodge-podge of pay-TV systems, CATV's, and UHF stations?

And can such a Fourth Network exist?

My answer, ladies and gentlemen, is I believe it can if pay-TV is a substantial part of its base.

Without pay-TV, FORGET a Fourth Network.

The great question that will be debated throughout this new period of change, will be: how will the public fare in all this?

Well, I submit, that, as is usually the case with vast technological change, the public will fare *both* better and worse.

A case can be made *for*, and *against*; just as, today, a case can be made *for* added leisure time and *against* more leisure time. *For* automation and *against* automation. . . .

In this case the networks surely will make a dramatic case *against* a fourth, pay-TV-based network; and they will be helped, oddly enough, by theater-owners who are beginning to like their own brand of pay-TV in closed-circuit theatres. Theatre-owners won't mind lifting major attractions from free TV pipelines as long as it suits their purpose; but for the present they are likely to side with the networks in an uneasy truce against other forms of pay-TV.

Now, to be sure, as a broadcaster who has been on the free TV side for many years, and who is only attempting to look at the full picture of the tides of change that confront us . . . I must admit that, if pay-TV comes, many presently free attractions WILL be lifted from the free pipelines. Here, there is no question but that the public may suffer somewhat in the pocketbook.

But at the same time, if a Fourth Network emerges, and its market place is established, there is equally no question in my mind that many new and additional program choices would be given the public at a price.

Yes, at a price.

What kind of price? I don't know. Size of potential audience will of course govern that. But I do know that the word "mediocrity," which cynics too freely use to describe free-TV, will be supplemented by the same cynics who will capsule the whole gargantuan business by saying something like: "TV consists of *free* mediocrity, highlighted by offerings of great uniqueness and excellence—*at a price!*"

And, as to free television being mediocre, I couldn't agree with that ten years hence anymore than I could agree with it today. Free television has a very great deal about it that is distinguished and dynamic and fantastic.

But . . . let's look at the present free-TV structure. The networks today are inextricably locked into a numbers game. A ratings game that has as its common denominator a mass audience. This is *one* system of TV. It has its strengths and weaknesses. Its problem is that as costs go up . . . entertainment patterns and values unconsciously, but inexorably, tend more toward a stereotype. Sameness becomes the norm.

Why?

Because originality, innovation becomes too CHANCY in this multimillion dollar numbers sweepstakes.

A Fourth Network will have a numbers game different from this. Networks today, while they plead for "more creativity," are not really able to do much more than point with pride at the sameness of their output, view with alarm the dissidents who don't fit into the numbers pattern, or cry havoc in their private board rooms. They are locked so irretrievably into this numbers game that the point of no return may be reached. The networks will find it increasingly difficult to mobilize the full creative forces and resources of this country.

As a result, these creative forces tend to languish. Some go to the West Coast or to the East Coast. They make good money quickly, and as quickly, become disillusioned. Some of them stubbornly stick to their dreams and turn those dreams into realities. Some of them make a picture like *David and Lisa,* which Frank Perry made for approximately $300,000 and, at latest report, had grossed about seven million dollars.

Would Frank Perry be happy in television today? You know the answer to that one.

Movies like *David and Lisa* are going to continue to be made. In Peoria. In Pittsburgh. In Chicago. Distribution costs are fierce for these brave "loners." A Fourth Network would be a natural pipeline for these "loners," and for all the other slaves in the vineyards who have that VSP—Very Special Project—in his or her pocket.

Consider the statistics of Chicago alone. A million pay-TV homes linked by wire, or to a UHF channel, would mean a gross for Chicago alone of a million dollars at a charge of $1.00 per set for an hour-and-a-half smashing documentary on our city; one that might cost $150,000 and take three months to make. In the numbers game of free television today, all this kind of thinking is out of phase. Doesn't fit. It's kooky to the *nth* degree.

What fun could be had on a Fourth Network, experimenting in truly indigenous types of programming.

Let me cite just one example:

When we had that problem a few years ago over that Denis Mitchell-produced documentary on Chicago—done from the British viewpoint, and financed jointly by the BBC and WBKB, I had the long-range idea of sending one of our crews to London and we would do London from our viewpoint. The BBC would run our output just as we were *supposed* to run their program about Chicago. A good idea, it seemed to me. Still seems like a good idea. To see ourselves through the eyes of others is always interesting, sometimes infuriating, and certainly enlightening. Then, I had hopes of doing Paris, and Paris would do Chicago. Then Rome. Moscow. Tokyo. Etc. One such big documentary a year. What fun doing such a thing. And what fun to watch the results.

But such a project was again "out of phase" for free television. Too expensive. It didn't fit the free-TV pipeline as we know it today. Nothing much worthy on a local level does.

In a way, therefore, you might say this idea was *too creative.*

Which is what I mean to say that the present system is too *locked* into its numbers game to mobilize and maximize the full creative forces that are alive and throbbing and fermenting in this great country of ours.

Proceeding with my analogy a step further, what fun it would be to carry out just this one, simple idea domestically. A home and home arrangement with New York. Chicago looks at New York, and New York looks at Chicago. Then Los Angeles. San Francisco. Philadelphia. The fur would fly, wouldn't it? But alive? Vital? Dynamic? Yes!

These would not be token efforts brought in at $7000. These shows would cost $100–200,000 and take months to produce.

And, forgetting the vital scope of a pay-TV-based Fourth Network for the moment, and asking ourselves the question: would a million people pay to see such locally produced indigenous fare?

Probably not a million. Probably only 900,000! And this is just *one* idea of dozens I have that I would personally put my chips on when this new pipeline is opened. No *one* person has a monopoly on ideas. Ideas will come fast and furious at the right time.

By now, I believe it is time to answer the question you might be asking of me—What am I for? Or—as the Irish would put it—who am I neutral against?

Well, I am against all forces that are static; forces that are undynamic and tend to surround this great miracle of communications with a stereotype pattern. I am for everyone making *all* the profits they can conscionably digest—networks, stations, agencies, clients—that's the American way and it's good enough for me.

But what I am really for is the creative process.

I am for all developments that will unleash new and stronger creative forces in this country. I don't care how offbeat they are, how untried or undeveloped. If a Fourth Network, with pay-as-you-see base, opens up a new world of creative talents, then I am for *that* in capital letters!

And, if the American public *pays* for this and the profits go to *artists, writers, actors, directors, producers*, etc., in a higher ratio than ever before, that is all right with me too.

Again, I am not even pleading for an upgrading of the financial base of the creative world. Artists frequently do their best work when they're hungry. At least they are supposed to. The main thing, it seems to me, is that they will be given a *chance* to be seen and heard. A chance to fit somewhere into this new pipeline.

The main thing, it seems to me, ladies and gentlemen, is that America be able to share, not just a fragment of the creative process in a locked-in numbers game, but that America be able to share as much more of the creative process as it is willing to pay for!

Indeed, this *is* a time to dream, ladies and gentlemen. I certainly claim none of the definitive answers. But our world of communications is in a state of flux. A state of grand and glorious flux. For that, I say, Thank God! Let's all dream and take the long view. Soon enough we'll get nailed down into *another* pattern, and penned in with other specifics. And our new specifics may be disappointing in some aspects. As I said earlier, with great technological change, things become both better and worse. But if we are motivated correctly, we tend then to dream of the pluses; the *better* side of things.

Now, how does all of this affect you? Well, as you know—and may have discussed here today—technology has some breathtaking changes in store for informational communications—meaning newspapers and magazines. I only know fleetingly of some of these possibilities—the use of electronics, for instance, in getting newspapers into homes faster and more economically.

I'm not at all sure that a Fourth Network, with pay-TV as its base, may not be able to provide, as one of its services, newspaper via a TV channel, with a reading time longer than you normally spend with your favorite paper.

Will this do away with the printing of your favorite paper? Of course not. A newspaper is like an old friend; you enjoy his company *more* when he is physically in your presence. You'll always want the real newspaper, whether it be by facsimile, offset, or letter-press. But a TV-supplemented reading service might increase circulation . . . might be a bonus to get you to subscribe. . . .

In any event, these electronic advances can only make your services as writers and journalists more important and worth greater rewards.

In conclusion, ladies and gentlemen, I say that a quiet revolution is going on in the world of communications.

Or rather, it was quiet as of yesterday. Today it is not so quiet.

Disquieting is more the word to some. To others it is downright alarming. To others—the government and the FCC, it is probably both perplexing and bemusing. And to a *few* people, as of this moment, this new revolution is comforting.

I number myself in this group.

Because change is a wonderful thing. Status quo, in my opinion, is a negative word, like "vegetate." And there is only a short distance from the phrase "status quo" to the word "ennui."

For those of you who have restless spirits and wide-open minds the next ten years will be a happy and busy time to be around. Others, unfortunately, will be hoist high on the petard of their own apathy; and that's all right too.

What parts Field Enterprises will play in this great renaissance, this dramatic change of pace—I cannot say, nor can anyone at this point. My associates at Field Enterprises and I have a lengthy period of study and investigation ahead of us before we can make any decision of that magnitude.

This much, though, I know: Field Enterprises is always committed to the daring and the venturesome. Ever since its founding it has taken part in innovation—the kinds of projects others shied away from through fear or caution—or plain lack of vision. With its resources, its interests in communication, education, and entertainment, Field Enterprises is in an ideal position to make a thorough, unhurried study of these new elements—and to take a major position in the industry if *that* should finally be indicated.

I shall not attempt to anticipate our own decision. Neither I nor my associates at Field Enterprises know today precisely what our course will be in the field of electronics. But, as I have said, historically Field Enterprises has been the kind of an outfit that responds and rises to challenge in its areas of interests—and I am convinced it will do so in this case.

Although the foregoing opinions have been strictly those of one Red Quinlan, even my brief association with the executives and the Board of Directors at Field Enterprises convinces me that there is a tremendous interest in entry by the Company into this field.

As for Marshall Field, Jr., himself—he thrives on the new, exciting, and the challenge that others muff or miss. I have never in my life known a man of greater personal courage and integrity. And although both these qualities are certain to be tested to the upmost in any electronic venture we may undertake, I am fully persuaded that there is in Mr. Field, and among his associates, enough of these qualities—and to spare.

So . . . damn the torpedoes, devil take the hindmost, full steam ahead, and all that sort of thing. On this note of sturdy old clichés, let me end with a final plea: Don't sell *short* in the market of ideas. Maintain a *long* position. And have fun dreaming the big dreams that lie directly ahead in the years to come.

51. Why Suppress Pay-TV? The Fight in California

by Sylvester L. Weaver, Jr.

Mr. Weaver has had a long and successful professional career in advertising, radio, and television. He has served in various executive capacities as vice-president, president, and chairman of the board for the National Broadcasting Company. Mr. Weaver describes in this article the attempt to introduce subscription television into the State of California.

The referendum referred to by Mr. Weaver in the article was held in November, 1964. The campaign preceding it was marked by misrepresentation, confusion of issues, and vicious tactics. Pay-TV was defeated, and the project described here was dismantled. The California Supreme Court has since declared that the referendum was invalid. However, Pay-TV remains as dead as if it had been legal and valid.

This article first appeared in The Atlantic Monthly *for October, 1964. With the deletion of the proposed program schedule, now of no interest, it is reprinted here in full, with the permission of Mr. Weaver and* The Atlantic Monthly Company, *Boston, Massachusetts, which holds coypright.*

We who are professionals in television have known for some time that two major developments are beckoning. One is the use of cable or wire to interconnect homes in order to improve the technical quality of sound and picture, and even more important, to provide virtually unlimited choice of programs to those who are dissatisfied with the commercially supported programs. The second is the development of the cartridge system, which will permit home movies to be handled like phonograph records: the picture and sound on a spool of film or tape will be inserted into the magazine of the television equipment, and with a push of a button the program will begin.

The two developments, to judge from my scrutiny, are interdependent for economic reasons: the cost of tape is so great that it must be reused, and the development of the cartridge and the technical necessities of transmission make it likely that we will deliver new tapes to homes over lines at what we call rewind speed and at night.

To the layman this is a long-awaited break-through in programming. Commercial radio and television have been predicated on the sale of advertising, and consequently built their business against a bulwark of criteria dictated by the needs of the sellers, not the needs of the buyers. For, the buyers were the public, and no easy reports on how the habit-viewing mechanism of radio and television became the most important selling and social force in the nation can overlook what *we did not do.* We did not offer to people with special interests any regular, convenient, in-depth, uninterrupted, full-length presentations. The hope that educational television can complement the commercial networks is unrealistic. Educational TV has tried, but it will never have the money to produce one-thousandth of the material we ought to have.

Radio and television were built by those of us who really knew advertising needs and who understood programming appeals and techniques, but I doubt if we expected that television would set up such an apathetic relationship between viewer and program. It was a sound, good kind of service, but it was extremely limited. I believe it was Alistair Cooke who called this service "audible wallpaper," a steady diet of Westerns, old movies, situation comedies, crime shows, and panels. It is there, turned on, hour after hour, observed or not observed, but present. It has values, powerful and good. And it has limitations, severe and built-in. Some have tried, as I did, to build a schedule with heavy accent on coverage of the real world, showing in prime time at night programs rising from our cultural heritage, informational programs, and great theater; but, basically, these commercially supported programs are usually superficial and constantly interrupted. Aesthetically, advertising is difficult to handle in combination with the more rewarding arts, and interruptions by commercials are at best upsetting and at worst sickening.

Subscription television will offer the variety we must have, the new usefulness we all want. In cable service, there will be no interference with ordinary television, either in the operation of the set (we merely connect our selector to the antenna leads outside the set) or in the commercial programs. The major influence we will have will be to encourage people to buy color sets and to upgrade their equipment. They will insist on decent hi-fidelity sound, which television does not give us with present equipment, but which we can deliver into hi-fi systems over telephone wires as part of our service.

HOW STV STARTED

Subscription Television, Inc., was formed in January, 1963, by the Reuben H. Donnelley Corporation, Lear Siegler, Inc., and William R. Staats & Co. Donnelley is a wholly owned subsidiary of Dun & Bradstreet, and its principal activities include the compilation and publication of classified directories for certain members of the Bell Telephone System, various marketing services, and the publication of business and professional magazines. Lear Siegler is a large, diversified manufacturer of military and commercial electronic equipment. Staats is a leading West Coast investment-banking firm. Each of the founding organizations had been exploring various facets of the subscription-television industry for a period of five to ten years. They became convinced that if the proven abilities of the Donnelley and Lear Siegler organizations could be applied to a satisfactory subscription-television system, an important new industry could be developed. Careful investigation into the various proposed subscription-television systems and techniques seemed to show that the system held by Tolvision of America, Inc., under license from Skiatron Electronics & Television Corporation, held the greatest promise. Acquisition of the rights to the Skiatron system was the first major milestone in the development of STV.

The basic concept of STV (a term used to refer collectively to Subscription Television, Inc., and its subsidiaries) was that a successful commercial venture in the subscription-television field could only be accomplished if it were done on a scale large enough to make adequate financing and programming possible.

The necessity for entering the subscription-television field on a large scale from the inception of the company is derived from the nature of the industry itself. It is obvious that viewers will not pay to see programs unless they are of a type or presented in a manner not already available on commercial television. On the other hand, owners of programming will not be disposed to offer their product on subscription television unless the potential audience is of a sufficient size to provide an adequate return. This interdependence of programming and consumer acceptance leads inevitably to the conclusion that subscription television can be successful only as a full-scale, profit-oriented venture.

Subscription television has been the subject of much controversy and some experiment since the early 1950's. It appears to be the next logical step forward in the evolution of low-cost mass entertainment and is uniquely equipped to offer programs which appeal to small groups with special interests. It is a supplementary service which presents programs for which the viewer would normally pay an admission charge at the scene of the event, or program material not now available on present television service. Subscription television in no way interferes with the ability of a customer's television set to receive all standard broadcast programs. Programs are presented without commercial interruptions and without the clock discipline of commercial television. They end when they are finished, as the writer or creator intended. In the convenience and comfort of the home, any number of viewers may enjoy an attraction at a cost of no more than one adult admission at the scene of the event.

HOW IT WORKS

Subscription television may be transmitted over the air or through a closed-circuit cable system. Over-the-air systems utilize broadcast methods to transmit programs but must scramble these programs so that only those sets equipped with a complex decoding unit can receive the programs as transmitted. Closed-circuit systems utilize a coaxial-cable distribution grid to transmit programs, providing signal security without scrambling and rendering better picture and sound quality. An over-the-air system is limited to one program at a time transmitted over a channel assigned by the FCC. A closed-circuit system may offer as many as three or even more programs simultaneously, thus multiplying program service for the viewers' selection and increasing the potential revenue for the system.

The STV system is a closed-circuit cable system in which three television channels are available for simultaneous programming, in addition to an information channel over which information is offered. The distribution grid to each subscriber's home is installed by local telephone companies, ending the drop-off with a jack near the subscriber's set. STV connects the jack to a program selector, which in turn is attached to the antenna leads of the subscriber's set. The program selector is actuated by a simple off-on switch and enables subscribers to select among STV's three simultaneous video channels and its information channel without interference with the subscriber's reception of standard commercial television programs.

An electronic interrogator periodically transmits coded signals back to the sender through the distribution grid. A program selector tuned to any one of the three STV channels automatically replies with a unique coded signal which is received and recorded by the interrogator. The interrogator forms the heart of an automatic data-processing system, which produces monthly customer invoices, computes royalties due the owners of programs, and records customers' program selections. It is also possible for STV to encourage viewing certain kinds of material by permitting sampling without charge.

The three channels are programmed simultaneously, and the subscriber receives a program guide telling him what is on, how long it lasts, how much it costs, and how long he can look for nothing. There is an installation charge of $10.00, plus $1.00 per month for service. Individual programs cost from fifty cents for a bridge lesson to $1.50 for a Broadway play. Each month the customer gets a bill for what he saw. We know, of course, that if it is too high for his rewards, he will take the service out. If he does not pay his bill, we will take the service out.

THE PROGRAMS

The range of programs is wide and includes much of the world's finest cultural fare: the very latest plays from Broadway; also off-Broadway shows, summer stock, repertory, civic theater, little theater, theater-in-the-round; musicals and revues; classic and modern operas—from Mozart to Bartók, from Puccini to Wagner to Verdi; the world-famous D'Oyly Carte Opera Company; the Bolshoi Ballet, the Royal Ballet, the Leningrad Kirov Ballet, and specialized national dance groups—Antonio and Les Ballets de Madrid, Moiseyev Dance Company, Ballet Folklórico of Mexico.

STV will present pianists of the stature of Artur Rubinstein and Van Cliburn; violinists, Isaac Stern and David Oistrakh; singers, Marian Anderson, Jan Peerce, Mary Costa, and Victoria de Los Angeles, along with famous chamber music groups. STV also plans to select from the great festivals of the world: Bayreuth for Wagner, Salzburg and Glyndebourne for Mozart; the great film festivals of Cannes and Venice; Shakespeare at Stratford-on-Avon; jazz at Monterey; the thrills and spills of the big rodeos; circuses, fairs, galleries, museums, and amphitheaters. Fiestas and spectacles, too: Mardi Gras in Rio or New Orleans, Fastnacht in old Basel, ice shows, aquacades.

In sports STV will cover, season-long, Dodger and Giant baseball. Football fans will find intense interest in special "Monday Quarterback" programs. In addition to basketball, ice hockey, golf, and tennis, STV will bring to the screen major games and matches and informative interviews with players and coaches, and will show many sports less well known in America, such as soccer, rugby, water polo, jai alai.

THE OPPOSITION

From its inception, the concept of subscription television has met opposition from motion-picture exhibitors and television broadcasters, together with a portion of the public which was led to believe that the advent of subscription television would eventually mean that the public would be required to pay for programs now carried on commercial television.

STV, as the first large-scale, profit-oriented venture in subscription television, has felt the full brunt of this opposition. The motion-picture exhibitors are the principal backers of the Crusade for Free TV, which is attempting, through an initiative referendum, to have subscription television outlawed in the state of California. In other words, the voters are being asked to decide whether this undertaking, which is in the best tradition of free enterprise, should be abolished. It is significant that the exhibitors, who have been the longtime foes of commercial television, should now be so vociferous in trying to "protect free TV." They are obviously aware of the public's desire to have a choice and variety of television programs and of its willingness to pay for that privilege.

The basic charge of the theater owners is that STV will make people pay for what they now see for nothing, that free TV, as they call commercial television, can be saved only by outlawing pay-TV. But this argument does not hold water. No viewer in his right mind would ever pay for series programs, which are being shown for nothing on more and more channels. And our opponents must realize that they can never present what we will use as our main popular fare—the new motion picture which is beyond an advertising budget now or in the future; the Broadway play, which, like other box-office attractions, can use pay television as an extension of the box office; the blacked-out sports events, which will never be seen on commercial television because the box office must be protected. Out of nine baseball teams, five of which have moved their franchises since television coverage began, all but the New York Mets have instituted various forms of blackout on television to save them from bankruptcy.

Although RCA, NBC, CBS, and virtually all respectable elements in the communications industry have come out opposing the use of the ballot in a matter of free enterprise, the California broadcasters have not repudiated the widely published statements that support the vicious distortions of the theater owners on the main point. And because it is a glib and easy slogan, the idea that we will actually end free television is believed by many people. Even the National Association of Broadcasters, to its everlasting discredit, has taken a self-serving position in opposition to a new competitor and is actively helping the theater owners in California to win the referendum.

Meanwhile, the heat is on at every organization in the state. The *Chronicle* in San Francisco, which has a substantial interest in commercial television and owns the NBC affiliate, has been reluctant to take our advertising, although all other newspapers in the state have run it without question. The *Chronicle* continues to oppose us editorially.

THE IMPACT ON CALIFORNIA

If permitted to function within the framework of the free enterprise system, Subscription Television, Inc., has the capability of exercising a highly beneficial influence on the California economy. Subscription Television, Inc., is expected to have a million subscribers by the early 1970's. At this level of subscription, the STV system has the capability of directly generating new employment opportunities for 38,000 workers, an annual payroll in excess of $315,000,000, and a cumulative capital investment approximating $170,000,000.

The most dramatic economic benefits are expected to accrue to the movie-production industry. The catalyst provided by the STV programming requirement to serve a million subscribers will, conservatively, generate a doubling of California feature-film production over the 1963 base. Employment to accommodate increased levels of production is forecast to increase by more than 32,000 jobs, and payrolls will increase by an estimated $272 million. In addition, increased production of other filmed television shows will produce an estimated $24 million in new payrolls.

The above estimates of economic impact, while impressive in themselves, do not indicate the full measure of the potential benefits to the California economy. These direct investments in men and capital will start a process of multiplied income and employment change. The multiplier effect is projected to generate a $1.9 billion impact on the California economy, which is representative of support for 237,800 new jobs. Based on the current ratio of population to employment in California, these new employment opportunities can support a population gain of 650,000.

While all the Hollywood unions and many others are for us because we obviously mean more jobs, more prosperity, and a better service to the people, the constant presentation of the old positions taken by those opposed to our service for whatever reasons and the new resolutions passed by unions and groups create a fog of uncertainty in which the easy acceptance of keeping television unchanged makes the fear of the unknown the chief ally of the theater owners.

Most astonishing is the opposition of the California Federation of Women's Clubs, whose spokeswoman is Mrs. Fred Teasley. She continually represents the women's clubs as opposed to our service, although no such opposition has ever been voted by the members and its constitution prevents the organization from speaking for its membership. The national federation has never gone further than to say that they favor keeping television free and healthy, which is as controversial as being for the flag and against the shark. Mrs. Teasley's contributions to the debate reach a high in irrationality that cannot be believed unless experienced.

The most disheartening, of course, is the opposition of intelligent broadcasters. Not only their opposition to government regulations in the field of ideas but also their own self-interest in a more vigorous communications system should make them welcome the challenge of a great new competitor. But, alas, too many eyes are fixed too rigidly on yesterday to understand that our service is opportunity as well as competition.

We expect to persuade the voters and to win in November, because we do have a great plan in the interest not only of the public generally, but especially of those who are less privileged economically and educationally and of older people. The greatest value our system has is that families can take the money saved from their entertainment budgets because of STV's single ticket price for any number of viewers, family and friends, and permit those same dollars, no matter how meager, to buy coverage of the wide world, the magic ticket of admission in first-class surroundings to many experiences that are denied to them in today's society. I am thinking of the opera house, the theater of Broadway and London, the concert halls, the festivals, the universities of the world. No longer will privilege be a matter of birth or money or influence, or even of having to be in the right place at the right time. STV lets each person follow his heart and mind to those attractions, services, and events that he wants.

52. The Inequalities of Television Franchises

by C. H. Sandage

Dr. Sandage is the former head of the Department of Advertising, College of Journalism and Communications, University of Illinois. A well-known consultant to industry, government, and university organizations, Dr. Sandage is one of the country's outstanding educators and statesmen. He has written several books, among them the leading advertising textbook used in colleges and universities throughout the country, and his articles appear in many journals and periodicals. This article was written for the Journal of Marketing *for July, 1959, and is reprinted with the author's kind permission.*

Is there a way to save U.S. television from the present quasi-monopoly position that stems from the scarcity of VHF channels? This brief article calls attention to some serious inequities that have developed from the manner in which this vital public property has been exploited, and suggests methods for reducing such inequities.

U.S. television is a combination of public property and private operation. It is a scarce "commodity" in which the number of stations that can operate is limited by space available in a crowded spectrum. This gives it many of the attributes of a monopoly, and so competition cannot function as an effective public protective device.

Competition is largely present in the industry only in the fight for the limited number of operating licenses available. In many markets a successful applicant has been made a millionaire almost overnight by the stroke of the pen of the licensing agency. In television, unlike other areas of the private-enterprise system, there is no freedom of entry. This encourages the use of pressure and unethical procedures in the fight for licenses. And such practices endanger the entire television industry.

Since television is so vital to marketing men in providing access to customers, marketing leaders should be concerned with the long-range economic and moral health of the industry. Such health cannot long exist under present conditions of "monopoly" profits to a favored few.

BROADCASTING A JOINT VENTURE

American commercial broadcasting is neither a private nor a public venture, but a combination of both. After all, the airwaves are public.

Yet private enterprise is "invited" to apply for a license to use a portion of the public property to provide broadcast services to the people. And it is understood that licensees shall not own any portion of the public property. Licensees, however, have been permitted to exploit their use of such public property without limitations as to rates charged for services or as to profits sought.

Formal regulations of these stewards of public property have been restricted primarily to matters concerned with providing broadcast program services to the citizenry, locally and nationally.

On matters other than program services, dependence has been placed on the efficacy of competition, in order to protect advertisers and stations against unnecessarily high rates and to keep profits at a reasonable level.

VARIATIONS IN FRANCHISE VALUES

Consider, therefore, some of the problems inherent in the broadcast industry and which influence competition as a regulatory device.

First, the cost of entry into broadcasting on the part of a private entrepreneur has no direct or apparent relationship to the quality of the public property assigned to entrants. Obviously a television channel serving a rich, populous market is potentially much more valuable than a channel serving a poor, sparsely populated area. Because of the relative scarcity of channels and the absence of complete freedom of entry into the field, normal competitive forces cannot operate here as they would in the case of such enterprises as food retailing and magazine publishing. Although there is wide variation in the basic value of channel locations, the "franchise price" is the same for all channels, namely, zero.

A second variation in the basic value of TV channels is related to their position in the spectrum. A VHF channel is considered to be much more valuable (other factors being equal) than a UHF channel. Yet there is no difference in the cost of these two to applicants.

The concept of competition as a regulator of business enterprise has undergone substantial changes in the history of American capitalistic enterprise. Basically the trend has been away from *laissez faire* to that of *fair* competition and the equalization of competitive opportunity.

Should the Licensee Pay for the Franchise? After all, if the concept of equalizing competitive opportunity in the television broadcasting field is to be enhanced, some system should be devised that would either erase or effectively compensate for the differences (often vast) in values of television channels and locations.

One approach to equalization might be to place a price on each franchise. This price could be determined by capitalizing profit potentials for each channel. In effect, this would "handicap" each television entrant in an attempt to provide an even race. Actually this is done in all cases of transfer or sale by the original licensee to a new operator. The market values of such franchises as measured by the sales price have been in the millions of dollars.

However, there might be substantial and logical objections to placing a price on each new TV franchise issued. The primary purpose of issuing franchises has been to encourage private enterprise to exploit a natural resource to the benefit of all the people. Under this concept it has been assumed that the placing of a price on new franchises issued by the government would reduce the quality as well as retard the growth of broadcasting services to the public.

Free licensing has hastened the growth of television facilities, and has provided revenues of sufficient opulence to permit the development of very expensive programs for the viewing public. An additional result has been to permit licensees of good channels in populous markets to reap profits on capital investment substantially higher than those in less favored localities and under less risk. At the same time licensees of channels of a less favorable quality must often compete with those operators blessed with a continuous flow of high profits available to throw into the competitive struggle. Other channels "go begging" for an entrepreneur and without success, because the prospects for eventual profits from operation seem to be nil.

Stated baldly, the policy of granting television channel licenses on a free basis results in a subsidy to those sufficiently fortunate to secure a "good" channel. Such subsidies may have been good in the early development of television; but in the light of current operations it may be logical to re-examine their place in the future of television.

Maximizing Benefits from Free Licensing. The wisdom of providing free access to public property may be questioned when such property is extremely scarce, and when no provisions have been made for limiting, recapturing, or directing the use of profits accruing to the licensee. Various possibilities might be explored for maximizing the benefits flowing from the issuance of channels on a no-charge basis.

One plan would involve the consideration of all television channels as a total unit, rather than as separate entities. If subsidies are to be provided, let them be on an industry-wide basis rather than on an individual-station basis.

The presumed purpose of subsidies is to encourage the development of television service to the public. Obviously, it is no longer necessary to subsidize the use of VHF channels in populous markets. But it may be in the total public interest to subsidize the development of television in sparsely populated and rural areas and on UHF television in specific concentrated population areas. To subsidize television development in these situations will require something more than offering channels on a free basis.

Consideration might also be given to a plan to recapture some of the excessive profits now accruing to operators of certain channels, and to use such recaptured profits to provide program service or other stimulants to users of "poor" channels or "poor" areas.

There should be no legal obstacle to a recapture of profits. Actually, each station licensee is using public property. Some royalty arrangement similar to arrangements in the case of minerals, oil, or timber could be tied into the granting of a license. Royalty contracts could presumably be written to correlate with profits so that no financial disadvantage would accrue to operators whose profits did not exceed the base from which royalties started to apply.

Another approach would be to follow the principle of an excess-profits tax, to be applied to all stations. Such a tax or recapture of profits could be graduated in a manner not to discourage enterprise, but also not to provide given licensees undue competitive advantages over others.

Funds obtained through the recapture process could be earmarked for retention in the television industry, and used to provide assistance of various kinds to enterprisers willing to exploit new television channels and lean market areas.

Other benefits to the public and to the industry might flow from the stimulation such a plan could give to seasoned operators. It is possible, for example, that holders of licenses that now return excess profits would be encouraged to invest more of their profits in program development, to experiment with satellite stations in poor markets, to operate subsidiary facilities to broaden the scope of their services, and to focus more attention on public-service operations.

UNRESTRICTED COMPETITION NOT THE SOLUTION

Competitive opportunities cannot be equalized by merely providing an abundance of channels for every market, when channels have unequal potential. They have unequal potential both in location and in position in the spectrum.

One of the most vital, natural resources of the American people is involved here. Protection of this natural resource for all of the people should not depend on competition alone for adequate protection, when scarcity of channels and vast variations in profit potentials among markets exist.

A CHALLENGE TO ADVERTISING MEN

Almost the total financial support of the television industry comes from advertisers. So, advertisers and advertising agencies have a real responsibility to exercise their influence to keep the industry healthy and the public well served.

If nothing is done to stop the withering on the vine of UHF television, and if new TV entrants into markets continue to be barred by lack of channels or pre-emption by a favored few, the problem becomes increasingly serious. The American public should not continue to tolerate the exploitation of their vital public property, when the financial rewards from such exploitation are so unevenly distributed and hold no necessary relation to the benefits rendered the public.

53. Global Television: A Proposal

by Robert W. Sarnoff

Mr. Sarnoff was Chairman of the Board of the National Broadcasting Company when his remarks were delivered before the European Broadcasting Union in New York City on October 22, 1962. They are reprinted here through special permission granted by the network.

I am greatly honored, in behalf of the National Broadcasting Company, to greet the European Broadcasting Union on the opening day of your unprecedented visit to the United States. In a sense you really made your first appearance here three months ago on the historic day when broadcasters on our two continents joined to extend man's sight farther than ever before.

The live program you sent us, ranging from the Arctic Circle to the warm shore of Sicily, will live in the memory of the millions upon millions who saw it. Yet, there is still something to be said for live appearances of the old-fashioned kind, and I know I speak for all my American colleagues when I say it is delightful to have you among us in the flesh, secure in the knowledge that you will not disappear over the horizon in twenty minutes.

The visit that begins today cannot compare in visibility, spectacle, or sheer excitement with the one we exchanged by satellite three months ago, but I hope that in the long run it may prove almost as significant a milestone in the development of intercontinental television. Toward that end, I want to take this occasion to offer a proposal for your consideration—a proposal intended to prepare the world's broadcasters to move together most effectively into the era of global television.

Any such proposal must be considered against the background of the unique nature of this revolutionary medium of communication and the role it is assuming on every continent. The universal scope of television is implicit in its very nature. For it brings into tangible reality some of the myths and fancies mankind has cherished through the ages—the flying carpet, the crystal ball, the occult feat of being in two places at once. This is the essential power of television as it exists in a single country or even a single community—the power that inspired the widespread phrase, "a window on the world." Small wonder that the whole world is crowding up to this window.

Television has taken root almost everywhere—in widely different cultures and under every kind of government; in newly emerging nations as well as sophisticated industrial countries. It can be found in every phase of growth, from the planning period to enterprising beginnings, from dramatic boom to the steady expansion of highly developed broadcasting services. It flourishes in different organizational forms—private, government-controlled, state-chartered, advertising-based, viewer-supported, and varying combinations of these.

Whatever these differences, television's broad aims are everywhere the same—to provide entertainment, information, and education to a vast public comprising the largest audiences in history. Today television is transmitted in some 80 countries by 3,000 stations to more than 120 million sets. And everywhere it has endowed our generation and those who follow us with the most vivid and powerful means man has ever devised for the extension of his senses.

The far-flung spread of television as a world-wide phenomenon has coincided in 1962 with exciting proof that we are swiftly mastering the means of achieving instantaneous communication between every television system on earth. Successful program exchange via Telstar, such as the recent orbital flight or the opening of the Ecumenical Council, has dazzled the world with the sense of a new era. It takes nothing from this achievement to remind ourselves that it is largely symbolic—that several years will still be required to develop and place into orbit a satellite system that will provide instantaneous transmission to every populated area of the earth at any hour of the day.

Even then it is difficult to predict to what extent satellite transmissions will figure in regular television fare. But whether or not satellites will be used only sparingly and on a specialized basis for program transmission, the rise of television in so many countries around the world, coinciding with the perfection of tape and film, means that the age of global television is upon us.

PRECIOUS INTERVAL

What lies immediately ahead is not only a period for continued exchanges through existing facilities but a precious interval of planning and preparation for the manifold realities of world-wide television as it matures. Man's hopes for this era are high; its potential is great. But so are its problems and challenges. Indeed, they are so complex and far-reaching in terms of practical procedures and international coordination that all of us with a stake in this glowing future should lose no time in laying the groundwork for a systematic approach to these problems. A number of broadcasters have perceived this need in recent years and months; it was touched upon in a speech in New York only a few weeks ago by Newton Minow, the Chairman of our Federal Communications Commission.

Clearly, we are faced with a difficult task of organized international cooperation. Part of this task—an indispensable though limited part—can be assigned to an organization already in existence. Within the framework of the United Nations, the International Telecommunications Union serves as a clearing house for agreements on frequency allocations and technical standards. This organization, with its specialized agencies, has a vital basic role to play. It has scheduled a special conference in Geneva next year to allocate frequencies in space communications.

But the ITU is an organization of governments, not broadcasters, and the functions it performs—vital as they are—do not deal with the myriad of new operational problems that will confront the world's growing number of broadcasters as technology draws them closer together. The experience of the EBU within Western Europe has demonstrated the need and effectiveness of joint international efforts in programming, legal, and technical matters. The need for similar cooperative arrangements among broadcasters of different regions is already being felt; it will grow year by year, and global television will not come of age until it can be met on an organized world-wide basis.

NEEDED: A NERVE CENTER

What are some of the needs that are likely to arise? One certainly is a need for authentic and readily available information about programs—a nerve center from which any broadcaster could learn what programs are available to him from all foreign sources. Such a center could maintain an up-to-date catalogue of such programs in every category—entertainment, information, and education—and of every type—live, film, or tape. At the same time it could compile a continuing record of what kinds of programs are needed or desired by various broadcasters, particularly in countries of limited production resources where television is in early stages. Prompt awareness of these requirements should in turn stimulate attempts to meet them, not only by the few countries now actively engaged as program suppliers to the world market but by many others which have barely or not yet entered it.

The developing exchange of live programs by satellite will pose a need for more specialized information as well, since these transmissions must cope with such factors as time differentials, scheduling clearances, and simultaneous translation. Ultimately, as we approach a world-wide television network—or even networks—comparable to the Eurovision network, international programming will involve major long-range planning as well as day-to-day coordination.

One of the most vital needs of the global television era will be to promote the closest international technical liaison. Where technical standards are not uniform, we must seek constantly better means of overcoming the differences. Another need can be served through personnel exchanges and training courses in all aspects of television to make the skills and experience of advanced broadcasters available to their newer counterparts abroad. Still another field for collaboration is in the development of uniform measures affecting artistic, business, and labor interests and rights in connection with internationally televised material.

A WORLD BROADCASTING UNION

To meet all these and other needs will require an international organization of broadcasters, global in scope, which does not yet exist. I propose that we set about creating a World Broadcasting Union. How can we best make a start in this direction? The question could not be put before any forum as qualified or appropriate as the European Broadcasting Union.

To most Americans, the phrase "international television" evokes the future, but for the EBU it also represents a distinguished past and present. Over a period of more than eight years, Eurovision has televised some 3,200 live programs throughout Western Europe, and it does so today with the resources of 21 television services in 17 countries. In making these exchanges possible, the EBU has tackled and mastered many of the very problems that will confront any organization devoted to program exchange on a global scale. Uniquely, the EBU's Eurovision has also exchanged live programs both with Intervision, its junior counterpart in the Eastern European Communist bloc, and with the broadcasters of the United States and Canada.

In the light of these imposing credentials, I respectfully suggest that the European Broadcasting Union form a study group to explore the prospect of a global organization of television broadcasters. I would hope that the study group would find it desirable and feasible to propose concrete steps for setting up such an organization. This is not a suggestion to enlarge the EBU nor to alter its identity or its present course in any way; it is a suggestion that envisages the creation of another organization, world-wide in character, in which the EBU—as a body or through its individual members—would play a significant role.

Through your clear record of leadership in international television, the EBU has earned the right, and indeed perhaps incurred the duty, to lead the way to the formation of a World Broadcasting Union. And you would be acting fully in harmony with the goals set forth in Article 2 of the EBU statutes; especially these two:

> (b) to promote and coordinate the study of all questions relating to broadcasting, and to ensure the exchange of information on all matters of general interest to broadcasting services;
>
> (c) to promote all measures designed to assist the development of broadcasting in all its forms.

The procedure I have suggested—the formation of a study group—is, of course, in accordance with Article 14 of the statutes, providing for the creation and functioning of such groups.

Having presumed to set this idea before you, let me presume briefly to offer some thoughts on the character of world broadcasting organization and the task of an EBU study group exploring how it might work.

The kind of organization I have in mind would consist of broadcasters rather than governments per se. In some instances, of course, the broadcasters would be direct agencies of national governments, while in others, they would be private companies and in still others, independent government-chartered corporations. It would be an organization open to broadcasters of every nationality on earth, for its usefulness would rest in large part on its universality.

Among the questions to be resolved is just how eligibility for membership would be based. Would it be by countries, by individual broadcasting companies, by national groups of broadcasters—or a combination of these? What would the relationship be with existing regional organizations, such as the Inter-American Association of Broadcasters and the EBU itself? What would the new organization do about financing, permanent staffing, a headquarters, methods of disseminating information? These are only some of the questions that suggest themselves.

Since television broadcasters in all parts of the world are vitally concerned, I would hope the EBU would see fit to invite appropriate participation in its study group by broadcast organizations not currently associated with the EBU as well as non-European broadcasters who are associate members. I can pledge to you the full cooperation of the National Broadcasting Company in such a project.

The recommendations of a study group in this field would be binding upon no one. But given patient and conscientious deliberations by a genuinely representative body of broadcasters, surely the findings upon which they could agree would commend themselves to television broadcasters everywhere.

OPEN CHANNELS

The organization that would eventually emerge cannot and should not be political or partisan in any sense. But I believe the nature of the broadcasting profession itself as well as the international character of the undertaking should commit it to the spirit of free expression and open channels between nations. In the last analysis, it seems to me, whatever television may be able to achieve on a global scale, that is the very least the world should rightfully expect from the custodians of this great medium.

In its short life so far, television has demonstrated a power more compelling than the written or spoken word alone to hold men fascinated, to influence their minds and shape their conduct. Its impact upon society has been enormous—on culture, politics, and the economy, on the way in which scores of millions of individuals order their daily lives.

In modern industrial society, television has become as ubiquitous as the weather. Sometimes it seems almost as much a subject of complaint. It would be unrealistic and unpersuasive to see television as an unmixed blessing. Yet, on balance, with all its shortcomings and fallibility, and in the remarkably brief span in which it has spread so far, television has vastly enriched the lives of growing millions of people by opening their horizons to a range of experience and learning opportunities unknown to any earlier generation. We may be far from fulfilling our potential, but again and again, we have demonstrated how great that potential can be.

And now we face the challenge of bringing this great potential to a whole new dimension of service. For all its technical virtuosity, the feat of transmitting television signals around the world will be an empty irony without the good will and wisdom with which men decide what to transmit. To the extent that those of us in this room will determine what to transmit—and it is a considerable extent—we must dedicate ourselves to be equal to this vast responsibility. Our first task, however, is to develop a practical means of using the wondrous new communications tools that await our grasp.

We are on the point of winning a gift that history has seldom yielded and never on such a scale. For centuries, men have dreamed of a universal language to bridge the linguistic gap between nations. In some measure, it existed in the fourth century before Christ when Alexander the Great spread one tongue from Greece to the borders of India. The expansion of Rome made Latin the universal language of Western Europe. It was the universality of Greek and Latin in their day that enabled Christianity to spread over so much of the world in a single generation. In the modern era in which Europeans alone speak 40 different languages, man's efforts to create a new universal language go back more than 300 years. They have resulted in such linguistic inventions as Esperanto, Interlingua, and Romanal.

Man will find his true universal language in television, which combines the incomparable eloquence of the moving image, instantly transmitted, with the flexibility of ready adaptation to all tongues. It speaks to all nations and, in a world where millions are still illiterate and semi-literate, it speaks as clearly to all people. Through this eloquent and pervasive universal language, let us strive to see, in the words inscribed over the portals of the BBC, that "Nation Shall Speak Peace Unto Nation."

54. The Public Be Served: Television for All Tastes

from the Carnegie Quarterly

James Reston, reporting early in 1967 in the New York Times, *said: "The Carnegie Commission Report on the future of ETV (Public Television) is one of those quiet events that, in the perspective of a generation or more, may be recognized as one of the transforming occasions in American life." Without a doubt, the Carnegie Report is the most important single document on broadcasting yet published in this country. The Carnegie Commission, composed of 18 leading U.S. scholars and citizens, after a year of comprehensive study of the values of ETV and its problems, concluded that massive support is needed for public television with strong and continuing financial support from both private and federal sources. A brief report on that study is included here. It appeared in* Carnegie Quarterly, *Winter, 1967, and it is reprinted with permission of the Carnegie Corporation of New York.*

We may call it "the boob tube" or "the idiot box," and Newton Minow did call it "a vast wasteland." The trouble, of course, is not with the medium itself, and the problem is not so much what has been done with it so far as what has *not* been done—yet.

For the salient fact about the American television system is not that it is of low quality but that it is incomplete, underdeveloped. What commercial television cannot do because of its need to reach mass audiences, non-commercial television cannot do because it lacks the money, facilities, and personnel. Hence in the technologically most advanced society in the history of man, the greatest technological device for informing, delighting, inspiring, amusing, provoking, and entertaining remains pitiably unexploited, and the American public is the loser.

In an eagerly awaited (at least by all segments of the television industry) report issued in January, the Carnegie Commission on Educational Television declared that "the American people have a great instrument within their grasp which they can turn to great purposes," and called for increased expenditure of public funds, particularly federal, in support of educational television. The Commission laid particular emphasis on the potentialities of what it calls Public Television, which "includes all that is of human interest and human importance which is not at the moment appropriate or available for support by advertising, and which is not arranged for formal instruction."

All television, the Commission points out, including commercial television, "provides news, entertainment, and instruction; all television teaches about places, people, animals, politics, crime, science. Yet the differences are clear."

Commercial television, financed to the tune of over two billion dollars a year by advertising, aims to entertain, and occasionally inform, mass audiences, the bigger the better. At the other end of the spectrum lies instructional television, with the aim of teaching children in their classrooms or adults who have a yen to learn foreign languages at six in the morning in their living room.

What the Commission calls Public Television encompasses everything in between: programs intended for the general viewer in his home, and for specialized groups in a variety of settings. Two kinds of audience make up the potential viewers of Public Television. The first is concerned with local interests, the second with special subject matter: new plays, new science, sports, music, and so on. Although it is unlikely that such programs would appeal to millions of viewers simultaneously, limited audiences for them exist in all parts of the country.

To serve the needs of these two kinds of audience, which presumably would include, at one time or another, almost everybody, the Commission proposes a "comprehensive system that will ultimately bring Public Television to all the people of the United States: a system which in its totality will become a new and fundamental institution in our American culture.

"This institution is different from any now in existence. It is not the educational television that we now know; it is not patterned after the commercial system or the British system or the Japanese system. . . . We propose an indigenous American system arising out of our own traditions and responsive to our own needs."

The Commission, which started its work fourteen months ago, was asked by Carnegie Corporation to "conduct a broadly conceived study of noncommercial television," and to "focus its attention principally, although not exclusively, on community-owned stations and their services to the general public."

The Commission was headed by James R. Killian, Jr., chairman of the corporation of Massachusetts Institute of Technology, and the following were its members: James B. Conant, Lee A. DuBridge, Ralph Ellison, John S. Hayes, David D. Henry, Oveta Culp Hobby, J. C. Kellam, Edwin H. Land, Joseph H. McConnell, Franklin Patterson, Terry Sanford, Robert Saudek, Rudolf Serkin, and Leonard Woodcock. A professional staff was directed by Franklin Patterson until June of 1966, when he assumed full-time duties as president of Hampshire College. He was succeeded by Hyman H. Goldin. Stephen White served as assistant to the chairman of the Commission.

Both the Commission and its staff made extensive use of the advice and information provided by consultants. More than 225 individuals and organizations expressed themselves to the Commission, either in person or in writing. Members of the Commission, its staff, or consultants visited 92 educational television stations in 35 states, as well as the television systems of 7 foreign countries. In addition, they conducted extensive statistical surveys.

THE MORE DIVERSITY THE MERRIER

To many, the most interesting—and controversial—of the Commission's recommendations will be its proposal for the establishment of a public corporation through which governmental as well as funds from other sources may be used for developing and supporting a comprehensive system of public television. There is no direct precedent in our history for such a channeling of public funds to a medium so intimately involved in First Amendment considerations. The Commission comes up with an ingenious solution aimed at insulating television programming from the fact or suggestion of political control while at the same time permitting Congress to remain accountable for the expenditure of public funds.

But the public corporation idea is only a means to an end or ends, so let us consider them first. Two major themes and concerns run through the 254 pages of the Commission's report.[1] This is what the gist of them seems to be.

The Commission is more concerned with programming than technology, and it is as interested in diversity as it is in mere "quality" or "level" of program fare. Obviously it wishes the programs offered by public television to be of excellent quality—that goes without saying—but its emphasis is on variety.

By this emphasis, the Commission broadens the terms of discussion as it is customarily carried on about television as we now know it. A current view often runs as follows:

"Commercial television is commercial. Therefore it must appeal to mass audiences. Therefore what it offers, by and large, must by definition be of low level."

There is a flaw in this syllogism, because occasionally programs that are good by anybody's standards do appeal to large audiences. The Commission's point, however, is that any system, commercial or non-commercial, which tries to appeal primarily to large audiences is inadequate not because its program fare must necessarily be "bad" but because it does not cater to the special interests of small audiences, and therefore does not provide a richness and diversity of programming commensurate with the richness and diversity of the American society. If ten million people can be per-

[1] The report, *Public Television: A Program for Action,* is available in paperback (Bantam Books) for $1 and in hard cover (Harper & Row, March 1967) for $4.95.

suaded to watch a good production of *Hamlet*, fine, the Commissioners would say, but they would also say that the few hundreds of thousands who might wish to see some Japanese *noh* plays or a motorcycle race should also be appealed to. Such productions—and one can think of many other kinds of examples—would not only satisfy the interests of various segments of the population (and perhaps stimulate the interests of other segments). They would also deepen the experience and skill of the directors, artists, and actors involved in producing them, and would contribute to the development of television itself as a medium.

The Commissioners' insistent emphasis on the importance of diversity rather than mere "quality" reflects their conviction about the direction non-commercial television should take. Implicit in their report is the assumption that non-commercial television will be greatly expanded, whether through mechanisms such as the ones they propose or other kinds. They suggest that unless a conscious decision is made now to opt for serving specialized as well as regional and local interests, what may result is a non-commercial system which seeks primarily to "elevate" public taste. In this case we would end up with two television systems that are essentially similar: commercial television serving the "vulgar" mass, and non-commercial television offering fare with more intellectual bite and artistic elegance but still seeking the large audience. The Commissioners clearly believe that this kind of development would result in the misuse, or under-use, of the potentialities of television.

Another strong theme of the report concerns the importance of serving local as well as specialized interests. "The heart of the system" should be in the community, the Commissioners declare. "Educational television is to be constructed on the firm foundation of strong and energetic local stations."

The stations should be given the capabilities to produce programs of more than local interest, which could be exchanged with other regions, but they should also have the resources to produce programs of only hometown interest concerning local news, personalities, sports, artistic and theatrical events, and the like. Here again, technical advances in transmission and distribution could by default lead to increased centralization of programming rather than the reverse, unless the decision is made now to emphasize the importance of the local role and local stations.

Naturally each station will use a good deal of material that has been produced nationally, but the Commission believes that real control of programming should remain with the local stations. That means that they must be offered enough programs so that they can choose some and reject others and that in general, except under unusual circumstances, they should be in a position to schedule them in accordance with their own assessment of local needs and interests at the moment. The Commission therefore does not lay stress on the virtues of live simultaneous transmission, though it acknowledges that there are many instances in which it would be appropriate.

THE SYSTEM AS IT EXISTS TODAY

There are now 124 educational television stations on the air. They are all alike in one respect: they are badly underfinanced. The Commission estimates that capital investment and operating expenses, which are now approximately $60 million a year for the entire system, amount to just about half what they should be if the stations were to be able to do properly the jobs they now attempt to do. This is not even to mention the kind of expansion of Public Television which the Commission proposes.

The ETV stations fall into four categories. There are 21 "school stations," licensed to school systems or districts and brought into being primarily to serve elementary and secondary education. There are 23 "state stations," licensed to state boards of education. They too serve elementary and secondary education primarily. There are 35 "university stations," licensed in most cases to public universities, which act as an extension of ordinary university activities, including continuing education. Some of them are also used for instruction in television or communications. Finally, there are 41 "community stations," licensed to non-profit corporations, which predominate in the large metropolitan areas and in general are the largest and best-financed of the stations. They offer more of what the Commission calls Public Television than the others, though all offer some.

There are three major sources of programs for general viewing. Most important is National Educational Television (NET), which produces or procures from other (including foreign) sources five hours of new programs a week, equally divided between public affairs and cultural programs; two-and-a-half hours a week of children's programs; and makes available its large library of programs for rerun.

The second principal source is represented by programs produced for local use.

For the average station, NET provides about half the general programming, and free programs from industrial concerns and local production represent about one-sixth each. Stations also use programs supplied by government agencies and from their own libraries.

State and local governments are the largest source of financial support for educational television, and the federal government provides some support, though it is heavily circumscribed with limitations and can be used only for the construction of certain facilities.

The Ford Foundation has provided enormous moral and financial support to educational television since its beginning in 1953. It provides $6 million of NET's annual budget of about $8 million, and it has made grants on a matching basis to the stations.

The ETV system as it exists is a going—and growing—system, although it also has "grave deficiencies," the Commission states, "not all of which are financial. We have found, as we examined the stations, many which are weak and must be strengthened, many which have not yet dared to elevate their sights."

The Commission nonetheless believes that it is possible to build upon the existing system "a new, vigorous, ambitious system of Public Television." How to do it is the question.

LOCAL STATIONS AND LOCAL INTERESTS

The first task is to enlarge the number and strengthen the capabilities of the local stations. Funds from all levels of government, augmented and stimulated by private initiative, will be required.

Like commercial television, Public Television must have central sources of programming (the Commission suggests two, or one in addition to the already existing NET, plus as many as twenty enlarged stations producing for more than local use). Unlike commercial television, however, it will depend upon a strong component of local and regional programming. To be truly local, says the report, "a station must arise out of a sense of need within a community, must have roots in the community, and must be under community control. It is not the physical location of studio and transmitter that is significant but the degree to which the station identifies itself with the people it seeks to serve."

In addition to federal help, some support should come from state and local sources. But this must go beyond the provision of funds. The stations must establish strong ties with local leadership and institutions. The Commission finds that many stations do not possess ties of this sort, neither with their boards of directors, who sometimes lend little more than their names to the effort, nor with colleges and universities which have so much to offer educational television. Even stations operated by universities are frequently ignored by the parent institution.

FUNDS AND FACILITIES

The federal funds that will be needed in addition to state and local government and private sources can be divided into roughly two kinds: those which could be provided under existing or expanding legislation for the building of facilities and the improvement of instructional television; and those not available under existing legislation for purposes which are politically sensitive, namely program production and all that goes with it, such as research, experimentation, and training.

The Commission recommends that broadened legislation be enacted to enable the Department of Health, Education, and Welfare to provide adequate facilities for stations now in existence, assist in increasing the number of stations (so that roughly 95 per cent of the population will be within range as compared with about 60 per cent now), and to help support the basic operations of all stations. (Many of them now lack the funds to remain on the air during the hours when Public Television would be of the greatest service.)

The Commission recognizes that its attitude toward instructional television is "ambivalent." It concedes that occasionally it has made impressive contributions to formal education, but believes that it could make "massive" contributions, and that so far "the role played in formal education by instructional television has been on the whole a small one." (Conclusions reached in the recent Ford Foundation publication, "Learning by Television," were similar.)

The deficiencies in instructional television "go far beyond matters of staff and equipment," the Commissioners say. "It is of much greater consequence that instructional television has never been truly integrated into the educational process. . . . With minor exceptions, the total disappearance of instructional television would leave the educational system fundamentally unchanged." They therefore recommend that a major program of research and development be launched, "designed to discover, within the full context of education, the manner in which television can best serve education."

THE CORPORATION FOR PUBLIC TELEVISION

Public Television will deal, must deal, with the expression of artistic and cultural values and with the communication of news and commentary. It must therefore be wary of the smallest threat of government control or political interference. On the part of government itself, there is a reciprocal desire to stand clear of even the appearance of control over freedom of expression.

The central recommendation of the Commission's report calls for Congress to establish a non-profit, non-governmental corporation to receive and disburse governmental and other funds. The Corporation for Public Television would stand in that delicate relation toward Congress that enables Congress to exercise stewardship over public funds and yet insulates Congress from the charge that it might be exercising control over freedom of expression. The proposed Corporation would be governed by a board of directors of twelve "distinguished and public-spirited citizens." Six would be appointed by the President with the advice and consent of the Senate, and the remaining six would be initially elected by those previously appointed. The first appointments and elections would be for terms of two, four, and six years; thereafter the President would appoint two members each two years, for six-year terms.

The funds to be made available to the Corporation should be raised, the Commission recommends, by means of a manufacturers' excise tax on television sets, beginning at 2 per cent and rising to a ceiling of 5 per cent. (From 1950 to 1965 an excise tax of 10 per cent was imposed on TV sets.) Ultimately, about $100 million a year would be available under this plan.

This is a mechanism which Congress has used repeatedly in the past: revenues from clearly specified sources are collected by the government and held in the Treasury for carrying out specified purposes. "Such mechanisms maintain intact the ultimate Congressional control over the use of public funds but permit the funds to be transferred outside the usual budgeting and appropriations procedures."

The reason for using the mechanism is that it and the Corporation together insulate Public Television from the dangers of political control which would be present if government scrutinized television's day-to-day operations. Such scrutiny is a natural consequence of annual budgeting and appropriations procedures by which funds are given from general revenues. At the same time, however, since Congress retains the power to terminate the arrangement at any time, its ultimate control over the expenditure of public funds is not impaired.

The principal responsibility of the Corporation will lie in supporting program production. The Commission suggests that it provide support, on the national level, to NET, to another production center to be established, and to twenty local stations which would produce programs for national use and that it also provide funds to local stations to produce programs of only local interest. In short, individual stations would be encouraged to produce on their own, programs that might be used in other parts of the country. The major community stations now do this, though on far too small a scale.

Ultimately, then, each station would have a body of national, special, and local programs from which to make out its weekly schedule. It would have a wide range of choice even at the beginning, and as time goes on a sort of television library would be amassed by the system—great plays or musical productions which would never outlive their usefulness.

In addition to the support of program production, which would consume about $80 million of the $100 million ultimately to be available each year, the Commission recommends that the Corporation provide the educational television system with facilities for interconnection, that it sponsor a center or centers for experimentation and development leading to the improvement of program production, that it support research to improve present television technology, and that it provide means by which technical, artistic, and other specialized personnel may be recruited and trained.

The Corporation for Public Television, if it comes into being, will represent a new kind of institution on the American scene, as did, for example, the National Science Foundation when it was created to meet a special need in a special way. There is no exact precedent for the corporation. But, as one of the Commission's consultants remarked, "we *do* have a precedent in this country for doing something about the things we have neglected, and for figuring out a way to do it."

THE FIFTH FREEDOM—FREEDOM TO VIEW

"We have attempted to design something that corresponds to American traditions and American goals, that can co-exist amicably with commercial television, and that together with commercial television can meet the highest needs of our society," the Commission declares.

"A stimulus to art and technology is brought into being from which the rewards are incalculable. In every major city in the United States there will be a television station operating with funds commensurate with its needs, and likely to be receptive to the writer, the producer, the director, the performer, and the artist who believes he has something to contribute to the culture or the perception of his fellow citizens. There will be room for the young man and woman in a developing stage, for the experimenter, the dissenter, the visionary. The innovator will be able to try his art without being subject to the tyranny of the ratings."

The system will be firmly rooted in the community but will not be parochial.

Public Television should be "historian in addition to the act of being daily journalist." It should "be sensitive to the long groundswells of civilization as well as to its earthquakes." It should be "a mirror of the American style" in all its diversity and humor and tragedy.

"If we were to sum up our proposal with all the brevity at our command," the Commissioners write in conclusion, "we would say that what we recommend is freedom. We seek freedom from the constraints, however necessary in their context, of commercial television. We seek for educational television stations freedom from the pressures of inadequate funds. We seek for the artist, the technician, the journalist, the scholar, and the public servant freedom to create, freedom to innovate, freedom to be heard in this most far-reaching medium. We seek for the citizen freedom to view, to see programs that the present system by its incompleteness denies him."